Watermarking Systems
Engineering

Signal Processing and Communications

Additional Volumes in Preparation

Watermarking Systems Engineering
Enabling Digital Assets Security and Other Applications

MAURO BARNI
University of Siena
Siena, Italy

FRANCO BARTOLINI
University of Florence
Florence, Italy

MARCEL DEKKER, INC.   NEW YORK · BASEL

Library of Congress Cataloging-in-Publication Data
A catalog record for this book is available from the Library of Congress.

ISBN: 0-8247-4806-9

This book is printed on acid-free paper.

Headquarters
Marcel Dekker, Inc., 270 Madison Avenue, New York, NY 10016, U.S.A.
tel: 212-696-9000; fax: 212-685-4540

Distribution and Customer Service
Marcel Dekker, Inc., Cimarron Road, Monticello, New York 12701, U.S.A.
tel: 800-228-1160; fax: 845-796-1772

Eastern Hemisphere Distribution
Marcel Dekker AG, Hutgasse 4, Postfach 812, CH-4001 Basel, Switzerland
tel: 41-61-260-6300; fax: 41-61-260-6333

World Wide Web
http://www.dekker.com

The publisher offers discounts on this book when ordered in bulk quantities. For more information, write to Special Sales/Professional Marketing at the headquarters address above.

Series Introduction

Over the past 50 years, digital signal processing has evolved as a major engineering discipline. The fields of signal processing have grown from the origin of fast Fourier transform and digital filter design to statistical spectral analysis and array processing, image, audio, and multimedia processing, and shaped developments in high-performance VLSI signal processor design. Indeed, there are few fields that enjoy so many applications—signal processing is everywhere in our lives.

When one uses a cellular phone, the voice is compressed, coded, and modulated using signal processing techniques. As a cruise missile winds along hillsides searching for the target, the signal processor is busy processing the images taken along the way. When we are watching a movie in HDTV, millions of audio and video data are being sent to our homes and received with unbelievable fidelity. When scientists compare DNA samples, fast pattern recognition techniques are being used. On and on, one can see the impact of signal processing in almost every engineering and scientific discipline.

Because of the immense importance of signal processing and the fast-growing demands of business and industry, this series on signal processing serves to report up-to-date developments and advances in the field. The topics of interest include but are not limited to the following:

- Signal theory and analysis
- Statistical signal processing
- Speech and audio processing
- Image and video processing
- Multimedia signal processing and technology
- Signal processing for communications
- Signal processing architectures and VLSI design

We hope this series will provide the interested audience with high-quality, state-of-the-art signal processing literature through research monographs, edited books, and rigorously written textbooks by experts in their fields.

Preface

Since the second half of the 1990's, digital data hiding has received increasing attention from the information technology community. To understand the reason for such interest, it may be useful to think about the importance that the ability to hide an object or a piece of information, has in our everyday life. To do so, consider the basic question: Why hide? Without claiming to be exhaustive, the most common answers can be summarized as follows. One may want to hide something:

1. To protect important/valuable objects. It is more difficult to damage, destroy or steal a hidden object than an object in plain sight; suffice it to think of the common habit of hiding valuables in the home to protect them from thieves.

2. To keep information secret. In this case, data hiding simply aims at denying indiscriminate access to a piece of information, either by keeping the very existence of the hidden object secret, or by making the object very difficult to find.

3. To set a trap. Traps are usually hidden for two reasons: not to let the prey be aware of the risk it is running (see the previous point about information secrecy), or to make the prey trigger the trap mechanism as a consequence of one of its actions.

4. For the sake of beauty. However strange it may seem, hiding an object just to keep it out of everyone's sight because its appearance is not a pleasant one, or because it may disturb the correct vision of something else, can be considered the most common motivation to conceal something.

5. A mix of the above. Of course, real life is much more complicated than any simple schematization; thus, many situations may be thought of where a mixture of the motivations discussed above explains the willingness to hide something.

The increasing interest in digital data hiding, i.e., the possibility of hiding a signal or a piece of information within a host digital signal, be it an image, a video, or an audio signal, shares the same basic motivations. Research in digital data hiding was first triggered by its potential use for copyright protection of multimedia data exchanged in digital form. In this kind of application, usually termed digital watermarking, a code conveying some important information about the legal data owner, or the allowed uses of data, is hidden within the data itself, instead of being attached to the data as a header or a separate file. The need to carefully hide the information within the host data is explained by the desire not to degrade the quality of the host signal (i.e., for the sake of beauty), and by the assumption that it is more difficult to remove the information needed for copyright protection without knowing exactly where it is hidden.

Data authentication is another common application of digital data hiding. The authenticity and integrity of protected data are obtained by hiding a fragile signal within them. The fragile signal is such that the hidden data is lost or altered as soon as the host data undergoes any modification: loss or alteration of the hidden data is taken as an evidence that the host signal has been tampered with, whereas the recovery of the information contained within the data is used to demonstrate data authenticity. In this case, the hidden data can be seen as a kind of trap, since a forger is likely to modify it inadvertently, thus leaving a trace of its action (be it malicious or not). Of course, the need to not alter the quality of the host signal is a further motivation behind the willingness to conceal carefully the authenticating information.

In addition to security/protection applications, many other scenarios exist that may take advantage of the capability of effectively hiding a signal within another. They include: image/video indexing, transmission error recovery and concealment, hidden communications, audio in video for automatic language translation, and image captioning. In all of these cases, hiding a piece of data within a host signal is just another convenient - it is hoped - way of attaching the concealed data to the host data. Hiding the data here is necessary because we do not want to degrade the quality of the hosting signal. As a matter of fact, embedding a piece of information within the cover work instead of attaching it to the work as a header or a separate file presents several advantages, including format independence and robustness against analog-to-digital and digital-to-analog conversion.

Having described the most common motivations behind the development of a data hiding system, we are ready to answer a second important question: what is this book about? We mostly deal with digital watermarking systems, i.e. data hiding systems where the hidden information is required to be robust against intentional or non-intentional manipulations of the host signal. However, the material included in the book encompasses many aspects that are common to any data hiding system. In addition, as the title indicates, we look at digital watermarking from a system perspective by describing all the main modules of which a watermarking system consists, and the tools at one's disposal to design and assemble such modules. Apart for some simple examples, the reader will not find in this book any cookbook recipes for the design of his/her own system, since this is impossible without delving into application details. On the contrary, we are confident that after having read this book, readers will know the basic concepts ruling the design of a watermarking (data hiding) system, and a large enough number of solutions to cover most of their needs. Of course, we are aware that watermarking, and data hiding in general, is an immature field, and that more effective solutions will be developed in the years to come. Nevertheless we hope our effort represents a good description of the state of the art in the field, and a good starting point for future research as well as for the development of practical applications.

As to the subtitle of this book, its presence is a clue that our main focus will be on security applications, that is, applications where the motivations for resorting to data hiding technology belong to the first three points of the foregoing motivation list. Nevertheless, the material discussed in the book covers other applications as well, the only limit of applicability being the imagination of researchers and practitioners in assembling the various tools at their disposal and in developing ad hoc solutions to the problems at their hands.

This book is organized as follows. After a brief introductory chapter, chapter 2 describes the main scenarios concerning data hiding technology, including IPR (Intellectually Property Rights) protection, authentication, enhanced multimedia transmission, and annotation. Though the above list of applications is by no means an exhaustive one, it serves the purpose of illustrating the potentialities of data hiding in different contexts, highlighting the different requirements and challenges set by different applications. Chapter 3 deals with information coding, describing how the to-be-hidden information is formatted prior to its insertion within the host signal. The actual embedding of the information is discussed in chapter 4. The problem of the choice of a proper set of features to host the watermark is first addressed. Then the embedding rule used to tie the watermark to them is considered by paying great attention to distinguish between blind and informed embedding schemes. The role played by human perception in the design of an effective data hiding system is discussed in chapter 5. After a brief description of the Human Visual System (HVS) and the Human Auditory System (HAS), the exploitation of the characteristics of such systems to effectively conceal the to-be-hidden information is considered for each type of media.

Having described the embedding part of a watermarking system, chapter 6 describes how to recover the hidden information from a watermarked signal. The recovery problem is cast in the framework of optimum decision/decoding theory for several cases of practical interest, by assuming ideal channel conditions, i.e., in the absence of attacks or in the presence of white Gaussian noise. Though these conditions are rarely satisfied in practice, the detector/decoder structures derived in ideal conditions may be used as a guide to the design of a watermarking system working in a more realistic environment. The set of possible manipulations the marked asset may undergo is expanded in chapter 7, where we consider several other types of attack, including the gain attack, filtering, lossy compression, geometric manipulations, editing, digital-to-analog and analog-to-digital conversion. In the same chapter, the design of a benchmarking system to compare different watermarking systems is introduced and briefly discussed, by the light of the current state of the art. Chapter 7 considers only general attacks, i.e., those attacks that operate in a blind way, without exploiting any knowledge available about the watermarking technique that was used. This is not the case with chapter 8, where watermark security is addressed. In this case, the attacker is assumed to know the details of the watermarking algorithm, and to explicitly exploit such knowledge to fool the watermarking system.

The book ends with a rather theoretical chapter (chapter 9), where the characteristics of a watermarking system are analyzed at a very general level, by framing watermarking in an information-theoretic/game-theory context. Though the assumptions underlying the theoretical analysis deviate, sometimes significantly, from those encountered in practical applications, the analysis given in this last chapter is extremely insightful, since it provides some hints on the ultimate limits reachable by any watermarking system. Additionally, it opens the way to a new important class of algorithms that may significantly outperform classical ones as long as the operating conditions resemble those hypothesized in the theoretical framework.

Each chapter ends with a further reading section, where, along with some historical notes, a number of references to additional sources of information are given, to allow the reader to learn more about the main topics touched upon by this book.

The content of this book is the result of several years of research in digital watermarking. During these years we interacted with several people to whom we are in debt for fruitful discussions and cooperation. Among them a prominent role has been played by Alessandro Piva, Roberto Caldelli and Alessia De Rosa of the Communications and Images Laboratory of the

Department of Electronics and Telecommunications of the University of Florence: no doubt that much of the content of this book derives from the continuous interaction with them. We are also indebted to all the thesis students who during these years stimulated us with their observations, questions and ideas. We are thankful to all the watermarking researchers with whom we came into contact during these years, since the discussions with all of them largely contributed to widen our points of view and to improve our research. Among them special thanks go to Ton Kalker of Philips Research, Fernando Perez-Gonzalez of the University of Vigo, Matthew Miller of NEC Research, Sviatoslav Voloshynovskiy of the University of Geneva, Teddy Furon now with IRISA/INRIA, and Jessica Fridrich of Binghamton University.

From a more general perspective we are in debt to our parents, and to all our teachers, from the primary school through University, for having given us the instruments and the curiosity necessary to any good researcher to carry out and love his work.

Finally, we sincerely thank our respective families, Francesca, Giacomo and Margherita, and Danila, Giovanni, and Tommaso for the encouragement and help they gave us throughout this effort and, more in general, for supporting all our work.

Mauro Barni
Franco Bartolini

Contents

1

Introduction

In this chapter we introduce the main elements of a digital watermarking system, by starting from data embedding until data recovery. We give a description which is as general as possible, avoiding to focus on copyright and data protection scenarios, so to encompass as many as possible data hiding applications. In spite of this, readers must me aware that some data hiding scenarios like steganography for covert communications are not properly covered by our models.

We also give some fundamental definitions regarding the various actors involved in the watermarking problem, or to better say, the watermarking game, and some fundamental properties of the watermarking algorithms which have a fundamental impact on the applicability of such algorithms in practical application scenarios. For example, we pay great attention to distinguish between different approaches to watermark recovery, since it has been proven that, in many cases, it is the way the hidden information is extracted from the host signal that determines whether a given algorithm is suitable for a particular application or not.

Even if this book is mainly concerned with the signal processing level of digital watermarking, in this first chapter (and part of chapter 2) we briefly touch the protocol level of the system, i.e. we consider how digital watermarking may be conveniently used, together with other complementary technologies, such as cryptography, to solve some practical problems, e.g. copyright protection, ownership verification, and data authentication.

1.1 Elements of a watermarking system

According to a widespread point of view, a watermarking system is much like a communication system consisting of three main elements: a trans-

1

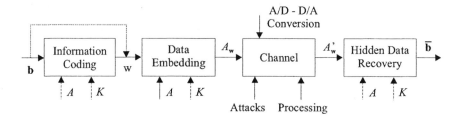

Figure 1.1: Overall picture of a data hiding system. The watermark code b represents the very input of the chain. Then, b is transformed in a watermark signal **w** (optionally $\mathbf{b} = \mathbf{w}$), which is embedded into the host asset A, thus producing the watermarked asset $A_{\mathbf{w}}$. Due to possible attacks, $A_{\mathbf{w}}$ is transformed into $A'_{\mathbf{w}}$. Finally the decoder/detector recovers the hidden information from $A'_{\mathbf{w}}$. Note that embedding and watermark recovery may require the knowledge of a secret key K, and that recovery may benefit from the knowledge of the original, non-marked asset A.

mitter, a communication channel, and a receiver. To be more specific, the embedding of the to-be-hidden information within the host signal plays the role of data transmission; any processing applied to the host data after information concealment, along with the interaction between the concealed data and the host data itself, represents the transmission through a communication channel; the recovery of the hidden information from the host data acts the part of the receiver. By following the communication analogy, any watermarking system assumes the form given in figure 1.1.

The information to be hidden within the host data represents the very input of the system. Without loosing generality, we will assume that such an information is given the form of a binary string

$$\mathbf{b} = (b_1, b_2 \ldots b_k), \tag{1.1}$$

with b_i taking values in $\{0, 1\}$. We will refer to the string **b** as the watermark code[1] (not to be confused with the watermark signal which will be introduced later on).

At the transmitter side, a data embedding module inserts the string **b**

---

[1]Some authors tend to distinguish between watermarking, fingerprinting and data hiding in general, depending on the content and the role of the hidden information within the application scenario. Thus, for example, the term watermarking is usually reserved for copyright protection applications where the robustness of the hidden data plays a central role. Apart from some examples, in this book we will not deal explicitly with applications, thus we prefer to always use the term watermark code, regardless of the semantic content of **b**. In the same way we will use the terms watermarking and data hiding interchangeably, by paying attention to distinguish between them only when we will enter the application level.

within a piece of data called host data or host signal[2]. The host signal may be of any media type: an audio file, a still image, a piece of video or a combination of the above[3]. To account for the varying nature of the host signal we will refer to it as the host digital asset, or simply the host asset, denoted by the symbol A. When the exact nature of A can not be neglected, we will use a different symbol, namely I for still images and video, and S for audio. The embedding module may accept a secret key K as an additional input. Such a key, whose main goal is to introduce some secrecy within the embedding step, is usually used to parameterize the embedding process and make the recovery of the watermark impossible for non-authorized users which do not have access to K.

The functionality's of the data embedding module can be further split into three main tasks: (i) information coding; (ii) watermark embedding; (iii) watermark concealment.

1.1.1 Information coding

In many watermarking systems, the information message **b** is not embedded directly within the host signal. On the contrary, before insertion vector **b** is transformed into a watermark signal $\mathbf{w} = \{w_1, w_2 \ldots w_n\}$ which is more suitable for embedding. In a way that closely resembles a digital communication system, the watermark code **b** may be used to modulate a much longer spread-spectrum sequence, it may be transformed into a bipolar signal where zero's are mapped in $+1$ and one's in -1, or it may be mapped into the relative position of two or more pseudo-random signals in the case of position-encoded-watermarking. Eventually, **b** may be left as it is, thus leading to a scheme in which the watermark code is directly inserted within A. In this case the watermark signal **w** coincides with the watermark code **b**.

Before transforming the watermark code into the watermark signal, **b** may be channel-coded to increase robustness against possible attacks. As a matter of fact, it turns out that channel coding greatly improves the performance of any watermarking system.

1.1.2 Embedding

In watermark embedding, or watermark casting, an embedding function $\mathcal{E}$ takes the host asset A, the watermark signal **w**, and, possibly, a key K,

---

[2]Sometimes the host signal is referred to as the cover signal.

[3]Though many of the concepts described in this book can be extended to systems in which the host signal is a piece of text, we will not deal with such a case explicitly

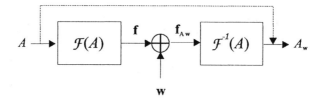

Figure 1.2: Watermark embedding via invertible feature extraction.

and generates the watermarked asset $A_\mathbf{w}$:

$$\mathcal{E}(A, \mathbf{w}, K) = A_\mathbf{w}. \tag{1.2}$$

Note that the above equation still holds when the watermark code is embedded directly within A, since in this case we simply have $\mathbf{w} = \mathbf{b}$. The definition of $\mathcal{E}$ usually goes through the selection of a set of asset features, called host features, that are modified according to the watermark signal. By letting the host features be denoted by $\mathcal{F}(A) = \mathbf{f}_A = \{f_1, f_2 \ldots f_m\} \in \mathbb{F}^{m\,4}$, watermark embedding amounts to the definition of an insertion operator $\oplus$ which transforms $\mathcal{F}(A)$ into the set of watermarked features $\mathcal{F}(A_\mathbf{w})$, i.e.:

$$\mathcal{F}(A_\mathbf{w}) = \mathcal{F}(\mathcal{E}(A, \mathbf{w}, K)) = \mathcal{F}(A) \oplus \mathbf{w}. \tag{1.3}$$

In general $m \neq n$, that is the cardinality of the host feature set needs not be equal to the watermark signal length.

Though equations (1.2) and (1.3) basically describe the same process, namely watermark casting within A, they tend to view the embedding problem from two different perspectives. According to (1.2), embedding is more naturally achieved by operating on the host asset, i.e. $\mathcal{E}$ modifies A so that when the feature extraction function $\mathcal{F}$ is applied to $A_\mathbf{w}$, the desired set of features $\mathbf{f}_{A_\mathbf{w}} = \{f_{w,1}, f_{w,2} \ldots f_{w,m}\}$ is obtained.

Equation (1.3) tends to describe the watermarking process as a direct modification of $\mathbf{f}_A$ through the embedding operator $\oplus$. According to this formulation, the watermark embedding process assumes the form shown in figure 1.2. First the host feature set is extracted from A, then the $\oplus$ operator is applied producing $\mathbf{f}_{A_\mathbf{w}}$, finally the extraction procedure is inverted to obtain $A_\mathbf{w}$:

$$A_\mathbf{w} = \mathcal{F}^{-1}(\mathbf{f}_{A_\mathbf{w}}). \tag{1.4}$$

The necessity of ensuring the invertibility of $\mathcal{F}^{-1}$ may be relaxed by allowing $\mathcal{F}^{-1}$ to exploit the knowledge of A to obtain $A_\mathbf{w}$, that is (weak

---

[4]We will use the symbology $\mathcal{F}(A)$ and $\mathbf{f}_A$ interchangeably depending on whether we intend to focus on the extraction of host features from A or on the host features themselves.

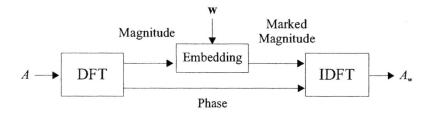

Figure 1.3: Watermark embedding in the magnitude of DFT. After embedding, the original phase information is used to go back in the asset domain.

invertibility):

$$A_{\mathbf{w}} = \mathcal{F}^{-1}(\mathbf{f}_{A_{\mathbf{w}}}, A). \tag{1.5}$$

As an example, let us consider a system in which the watermark is embedded into the magnitude of the DFT coefficients of the host asset. The feature extraction procedure is not strictly invertible, since it discards phase information. Phase information, though, can be easily retrieved from the original asset A, a possibility which is admitted by formulation (1.5) (see figure 1.3 for a schematic description of the whole process).

It is worth noting, though, that neither strict, nor weak invertibility of $\mathcal{F}$ is requested in general, since $\mathcal{E}$ may always be defined as a function operating directly in the asset domain (equation (1.2)).

A detailed discussion of the possible choices of $\mathcal{E}$, $\mathcal{F}(A)$ and $\oplus$ will be given in chapter 4.

1.1.3 Concealment

The main concern of the embedding part of any data hiding system is to make the hidden data imperceptible. This task can be achieved either implicitly, by properly choosing the set of host features and the embedding rule, or explicitly, by introducing a concealment step after watermark embedding. To this aim, the properties of the human senses must be carefully studied, since imperceptibility ultimately relies on the imperfections of such senses. Thereby, still image and video watermarking will rely on the characteristics of the Human Visual System (HVS), whereas audio watermarking will exploit the properties of the Human Auditory System (HAS).

A detailed description of the main phenomena underlying the HVS and the HAS, is given in chapter 5.

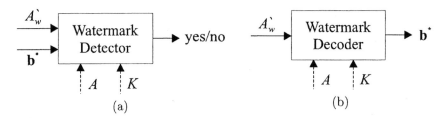

Figure 1.4: With detectable watermarking (a) the detector just verifies the presence of a given watermark within the host asset. With readable watermarking (b) the prior knowledge of $\mathbf{b}^*$ is not necessary.

1.1.4 Watermark impairments

After embedding, the marked asset $A_{\mathbf{w}}$ enters the channel, i.e. it undergoes a series of manipulations. Manipulations may explicitly aim at removing the watermark from $A_{\mathbf{w}}$, or may pursue a completely different goal, such as data compression, asset enhancement or editing. We will denote the output of the channel by the symbol $A'_{\mathbf{w}}$.

1.1.5 Recovery of the hidden information

The receiver part of the watermarking system may assume two different forms. According to the scheme reported in figure 1.4a, the watermark detector reads $A'_{\mathbf{w}}$ and a watermark code $\mathbf{b}^*$, and decides whether $A'_{\mathbf{w}}$ contains $\mathbf{b}^*$ or not. The detector may require that the secret key K used to embed the watermark is known. In addition, the detector may perform its task by comparing the watermarked asset $A'_{\mathbf{w}}$ with the original, non-marked, asset A, or it may not need to know A to take its decision. In the latter case we say that the detector is *blind*[5], whereas in the former case the detector is said to be *non-blind*.

Alternatively, the receiver may work as in figure 1.4b. In this case the watermark code $\mathbf{b}^*$ is not known in advance, the aim of the receiver just being that of extracting $\mathbf{b}^*$ from $A'_{\mathbf{w}}$. As before, the extraction may require that the original asset A and the secret key K are known.

The two different schemes given in figure 1.4 lead to a distinction between algorithms embedding a mark that can be *read* and those inserting a code that can only be *detected*. In the former case, the bits contained in the watermark can be read without knowing them in advance (figure 1.4b). In the latter case, one can only verify if a given code is present in the document, i.e. the watermark can only be revealed if its content is known

---

[5]Early works on watermarking used the term *oblivious* instead than blind.

in advance (figure 1.4a). We will refer to the extraction of a readable watermark with the term watermark decoding, whereas the term watermark detection will be used for the recovery of a detectable watermark.

The distinction between readable and detectable watermarking can be further highlighted by considering the different form assumed by the decoding/detection function $\mathcal{D}$ characterizing the system. In blind, detectable watermarking, the detector $\mathcal{D}$ is a three-argument function accepting as input a digital asset A, a watermark code $\mathbf{b}$, and a secret key K (the secret key is an optional argument which may be present or not). As an output $\mathcal{D}$ decides whether A contains $\mathbf{b}$ or not, that is

$$\mathcal{D}(A, \mathbf{b}, K) = yes/no. \tag{1.6}$$

In the non-blind case, the original asset A_{or} is a further argument of $\mathcal{D}$:

$$\mathcal{D}(A, A_{or}, \mathbf{b}, K) = yes/no. \tag{1.7}$$

In blind, readable watermarking, the decoder function takes as inputs a digital asset A and, possibly, a keyword K, and gives as output the string of bits $\mathbf{b}$ it reads from A:

$$\mathcal{D}(A, K) = \mathbf{b}, \tag{1.8}$$

which obviously assumes the form

$$\mathcal{D}(A, A_{or}, K) = \mathbf{b}, \tag{1.9}$$

for non-blind watermarking. Note that in readable watermarking, the decoding process always results in a decoded bit stream, however, if the asset is not marked, decoded bits are meaningless. Even with readable watermarking, then, it may be advisable to investigate the possibility of assessing whether an asset is watermarked or not.

Detectable watermarking is also known as 1-bit watermarking (or 0-bit watermarking), since, given a watermark, the output of the detector is just *yes* or *no*. As the 1-bit designation says, a drawback with detectable watermarking is that the embedded code can convey only one bit of information. Actually, this is not the case, since if one could look for all, say N, possible watermarks, then the detection of one of such watermarks would convey $log_2 N$ information bits. Unfortunately, such an approach is not computationally feasible, since the number of possible watermarks is usually tremendously high.

1.2 Protocol considerations

Even if this book aims mainly at describing how to hide a piece of information within a host asset and how to retrieve it reliably, it is interesting to take a look at some protocol-level issues. In other words, once we know how to hide a certain amount of data within a host signal, we still need to investigate how the hidden data can be used in real applications such as, for example, copyright protection or data authentication. Moreover, it is instructive to analyze the requirements that protocol issues set on data hiding technology and, viceversa, how technological limitations impact protocol design.

The use of digital watermarking for copyright protection is a good example to clarify the close interaction between data hiding and protocol-level analysis. Suppose, for example, that watermarking has to be used to unambiguously identify the owner of a multimedia document. One may simply insert within the document a watermark code with the identity of the document owner. Of course, the watermark must be as robust as possible, otherwise an attacker could remove the watermark from the document and replace it with a new watermark containing his/her name. However, more subtle attacks can be thought of, thus calling for a more clever use of watermarking. Suppose, for example, that instead of attempting to remove the watermark with the true data owner, the attacker simply adds his/her own watermark to the watermarked document. Even by assuming that the new watermark does not erase the first one, the presence within the document of two different watermarks makes it impossible to determine the true document owner by simply reading the watermark(s) contained in it.

To be specific, let us assume that to protect a work of her (the asset A), Alice adds to it a watermark with her identification code $\mathbf{w}_A$[6], thus producing a watermarked asset $A_{\mathbf{w}_A} = A + \mathbf{w}_A$[7], then she makes $A_{\mathbf{w}_A}$ publicly available. To confuse the ownership evidence provided by the watermark, Bob takes the watermarked image and adds to it his own watermark $\mathbf{w}_B$, producing the asset $A_{\mathbf{w}_A\mathbf{w}_B} = A + \mathbf{w}_A + \mathbf{w}_B$. It is now impossible to decide whether $A_{\mathbf{w}_A\mathbf{w}_B}$ belongs to Bob or Alice since it contains both Alice's and Bob's watermarks. To solve the ambiguity, Alice and Bob could be asked to show if they are able to exhibit a copy of the asset that contains their watermark but does not contain the watermark of the other contender. Alice can easily satisfy the request, since she owns the original asset without Bob's identification code, whereas this should not be possible for Bob, given that the asset in his hands is a copy of the asset with Alice's watermark. However, further precautions must be taken, not to be susceptible to a more subtle attack known as the SWICO attack (Single-Watermarked-Image-Counterfeit-Original)[8]. Suppose, in fact, that

---

[6]We assume, for simplicity, that $\mathbf{w}_A = \mathbf{b}_A$

[7]The symbol $+$ is used to indicate watermark casting since we assume, for simplicity, that the watermark is simply added to the host image

[8]The attack described here is a simplified version of the true SWICO attack which

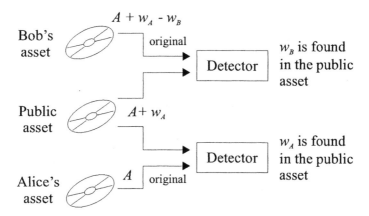

Figure 1.5: The SWICO attack, part (a). Bob subtracts his watermark $\mathbf{w}_B$ from the asset in his hands, maintaining that this is the true original asset. In this way the public asset seems to contain Bob's watermark.

the watermarking technique used by Alice is not blind, i.e. to reveal the presence of the watermark the detector needs to compare the watermarked asset with the original one. For instance, we can assume that the watermark is detected by subtracting the original asset from the watermarked one. Alice can use the true original asset to show that Bob's asset contains her watermark and that she possesses an asset copy, $A_{\mathbf{w}_A}$ containing $\mathbf{w}_A$ but not $\mathbf{w}_B$, in fact:

$$A_{\mathbf{w}_A \mathbf{w}_B} - A = A + \mathbf{w}_A + \mathbf{w}_B - A = \mathbf{w}_A + \mathbf{w}_B, \qquad (1.10)$$

$$A_{\mathbf{w}_A} - A = A + \mathbf{w}_A - A = \mathbf{w}_A, \qquad (1.11)$$

which proves that $A_{\mathbf{w}_A \mathbf{w}_B}$ contains $\mathbf{w}_A$ (as well as $\mathbf{w}_B$), and that $A_{\mathbf{w}_A}$ contains $\mathbf{w}_A$ but does not contain $\mathbf{w}_B$.

The problem is that Bob can do the same thing by building a fake original asset A_f to be used during the ownership verification procedure. By referring to figures 1.5 and 1.6, it is sufficient that Bob subtracts his watermark from $A_{\mathbf{w}_A}$, maintaining that the true original asset is $A_f = A_{\mathbf{w}_A} - \mathbf{w}_B = A + \mathbf{w}_A - \mathbf{w}_B$. In this way Bob can prove that he possesses an asset, namely the public asset $A_{\mathbf{w}_A}$, that contains $\mathbf{w}_B$ but does not contain $\mathbf{w}_A$:

$$A_{\mathbf{w}_A} - A_f = A + \mathbf{w}_A - (A + \mathbf{w}_A - \mathbf{w}_B) = \mathbf{w}_B. \qquad (1.12)$$

will be described in more detail in 1.2.7

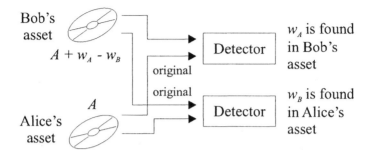

Figure 1.6: The SWICO attack, part (b). Bob subtracts his watermark $\mathbf{w}_B$ from the asset in his hands, maintaining that this is the true original asset. In this way the original asset in Alice's hands seems to contain Bob's watermark.

As it can be seen, the plain addition of a non blind watermark to a piece of work is not sufficient to prove ownership, even if the watermark can not be removed without destroying the host work.

More details about the characteristics that a watermark must have in order to be immune to the SWICO attack will be given below (section 1.2.7), here we only want to stress out that watermarking by itself is not sufficient to prevent abuses unless a proper protection protocol is established. In the same way, the exact properties a watermarking algorithm must satisfy can not be defined exactly without considering the particular application scenario the algorithm has to be used in.

Having said that an exact list of requirements of data hiding algorithms can not be given without delving into application details, we now discuss the main properties of data hiding algorithms from a protocol perspective. In most cases, a brief analysis of such properties permits to decide whether a given algorithm is suitable for a certain application or not, and can guide the system designer in the choice of an algorithm rather than another.

1.2.1 Capacity of watermarking techniques

Although in general the watermarking capacity does not depend on the particular algorithm used, but it is rather related to the characteristics of the host signal, of the embedding distortion and of the attack strength (this will be more evident in chapter 9), it makes also sense to speak about the capacity of a given technique, as the amount of information bits that it is able to, more or less reliably, convey. As it can be readily understood, capacity is a fundamental property of any watermarking algorithm, which very often determines whether a technique can be profitably used in a given context or not. Once again, no requirements can be set without consid-

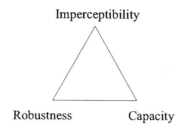

Figure 1.7: The watermarking trade-off.

ering the application the technique has to serve in. Possible requirements range from some hundreds of bits in security-oriented applications, where robustness is a major concern, through several thousands of bits in applications like captioning or labeling, where the possibility of embedding a large number of bits is a primary need.

Generally speaking, capacity requirements always struggle against two other important requirements, that is watermark imperceptibility and watermark robustness (figure 1.7). As it will be clear from subsequent chapters, a higher capacity is always obtained at the expense of either robustness or imperceptibility (or both), it is thereby mandatory that a good trade-off is found depending on the application at hand.

1.2.2 Multiple embedding

In some cases the possibility of inserting more than one watermark is requested. Let us consider, for example, a copyright protection scheme, where each protected piece of data contains two watermarks: one with the identity of the author of the work and one indicating the name of the authorized consumer. Of course, algorithms enabling multiple watermark embedding must grant that all the watermarks are correctly read by the decoder. In addition, the insertion of several watermarks should not deteriorate the quality of the host data. In applications where watermark robustness is required, the necessity of allowing the insertion of several watermarks also derives from the observation that the insertion of a watermark should not prevent the possibility of reading a preexisting watermark. If this was the case, in fact, watermark insertion would represent an effective mean at everyone's disposal to make a preexisting watermark unreadable without perceptible distortion of the host signal, thus nullifying any attempt to make the watermark robust.

Though necessary in many cases, the possibility of inserting more than one watermark must be carefully considered by system designers, since it may produce some ambiguities in the interpretation of the information hidden within the protected piece of work (see the SWICO attack described previously).

1.2.3 Robustness

Watermark robustness accounts for the capability of the hidden data to survive host signal manipulation, including both non-malicious manipulations, which do not explicitly aim at removing the watermark or at making it unreadable, and malicious manipulations, which precisely aims at damaging the hidden information.

Even if the exact level of robustness the hidden data must possess can not be specified without considering a particular application, we can consider four qualitative robustness levels encompassing most of the situations encountered in practice:

- *Secure watermarking*: in this case, mainly dealing with copyright protection, ownership verification or other security-oriented applications, the watermark must survive both non-malicious and malicious manipulations. In secure watermarking, the loss of the hidden data should be obtainable only at the expense of a significant degradation of the quality of the host signal. When considering malicious manipulations it has to be assumed that attackers know the watermarking algorithm and thereby they can conceive ad-hoc watermark removal strategies. As to non-malicious manipulations, they include a huge variety of digital and analog processing tools, including lossy compression, linear and non-linear filtering, cropping, editing, scaling, D/A and A/D conversion, analog duplication, noise addition, and many others that apply only to a particular type of media. Thus, in the image case, we must consider zooming and shrinking, rotation, contrast enhancement, histogram manipulations, row/column removal or exchange; in the case of video we must take into account frame removal, frame exchange, temporal filtering, temporal resampling; finally, robustness of an audio watermark may imply robustness against echo addition, multirate processing, reverb, wow-and-flutter, time and pitch scaling. It is, though, important to point out that even the most secure system does not need to be perfect, on the contrary, it is only needed that a high enough degree of security is reached. In other words, watermark breaking does not need to be impossible (which probably will never be the case), but only difficult enough.

- *Robust watermarking*: in this case it is required that the watermark be resistant only against non-malicious manipulations. Of course, robust watermarking is less demanding than secure watermarking. Application fields of robust watermarking include all the situations in which it is unlikely that someone purposely manipulates the host data with the intention to remove the watermark. At the same time, the application scenario is such that the, so to say, normal use of data comprises several kinds of manipulations which must not damage the hidden data. Even in copyright protection applications, the adoption of robust watermarking instead than secure watermarking may be allowed due to the use of a copyright protection protocol in which all the involved actors are not interested in removing the watermark[9].

- *Semi-fragile watermarking*: in some applications robustness is not a major requirement, mainly because the host signal is not intended to undergo any manipulations, but a very limited number of minor modifications such as moderate lossy compression, or quality enhancement. This is the case, for example, of data labelling for improved archival retrieval, in which the hidden data is only needed to retrieve the host data from an archive, and thereby it can be discarded once the data has been correctly accessed. It is likely, though, that data is archived in compressed format, and that the watermark is embedded prior to compression. In this case, the watermark needs to be robust against lossy coding. In general, we say that a watermark is semi-fragile if it survives only a limited, well-specified, set of manipulations leaving the quality of the host document virtually intact.

- *Fragile watermarking*: a watermark is said to be fragile, if the information hidden within the host data is lost or irremediably altered as soon as any modification is applied to the host signal. Such a loss of information may be global, i.e. no part of the watermark can be recovered, or local, i.e. only part of the watermark is damaged. The main application of fragile watermarking is data authentication, where watermark loss or alteration is taken as an evidence that data has been tampered with, whereas the recovery of the information contained within the data is used to demonstrate data origin[10].

---

[9]Just to give an example, consider a situation in which the ownership of a digital document is demonstrated by verifying that the owner name is hidden within the document by means of a given watermarking technique. Of course, the owner is not interested in removing his/her name from the document. Here, the main concern of system designer is not robustness, but to make it impossible that a fake watermark is built and inserted within the document. At the same time, the hidden information must survive all the kinds of non-malicious manipulations the rightful owner may want to apply to the host document.

[10]Interesting variations of the previous paradigm, include the capability to localize tampering, or to discriminate between malicious and innocuous manipulations, e.g. mod-

Even without going into much details (which will be the goal of next chapters), we can say that robustness against signal distortion is better achieved if the watermark is placed in perceptually significant parts of the signal. This is particularly evident if we consider the case of lossy compression algorithms, which operate by discarding perceptually insignificant data not to affect the quality of the compressed image, audio or video. Consequently, watermarks hidden within perceptually insignificant data are likely not to survive compression.

Achieving watermark robustness, and, to a major extent, watermark security, is one of the main challenges watermarking researchers are facing with, nevertheless its importance has sometimes been overestimated at the expense of other very important issues such as watermark capacity and protocol-level analysis.

1.2.4 Blind vs. non-blind recovery

A watermarking algorithm is said *blind* if it does not resort to the comparison between the original non-marked asset and the marked one to recover the watermark. Conversely, a watermarking algorithm is said *non-blind* if it needs the original data to extract the information contained in the watermark. Sometimes *blind* techniques are referred to as *oblivious*, or *private* techniques. However, we prefer to use the term *blind* (or *oblivious*) for algorithms that do not need the original data for detection and leave the term *private* watermarking to express a different concept (see next subsection).

Early works in digital watermarking insisted that blind algorithms are intrinsically less robust than non-blind ones, since the true data in which the watermark is hidden is not known and must be treated as disturbing noise. However, this is not completely true, since the host asset is known by the encoder and thus it should not be treated as ordinary noise, which is not known either by the encoder or by the decoder. Indeed, it can be demonstrated (see chapter 9) that, at least in principle, and under some particular hypotheses, blindness does not cause any loss of performance, neither in terms of capacity nor robustness. At a more practical level, blind algorithms are certainly less robust than non-blind ones, even if the loss of performance is not as high as one may expect. For example, by knowing the original, non-marked, non-corrupted asset some preprocessing can be carried out to make watermark extraction easier, e.g. in the case of image watermarking, rotation and magnification factors can be easily estimated and compensated for if the non-marked image is known.

Very often, in real-world scenarios the availability of the original host asset can not be warranted, thus making *non-blind* algorithms unsuitable

---

erate lossy compression, through semi-fragile watermarking.

for many practical applications. Besides, as it is summarized below, this kind of algorithms can not be used to prove rightful ownership, unless additional constraints regarding the *non-quasi-invertibility* of the watermark are satisfied.

In the rest of this book we will focus only on *blind* watermarking, being confident that the extension of most of the concepts we will expose to the *non-blind* case is trivial.

1.2.5 Private vs. public watermarking

A watermark is said *private* if only authorized users can recover it. In other words, in private watermarking a mechanism is envisaged that makes it impossible for unauthorized people to extract the information hidden within the host signal. Sometimes by private watermarking, non-blind algorithms are meant. Indeed, non-blind techniques are by themselves private, since only authorized users (e.g. the document owner) can access the original data needed to read the watermark. Here, we extend the concept of privateness to techniques using any mechanism to deny the extraction of the watermark to unauthorized personnel. For instance, privateness may be achieved by assigning to each user a different secret key, whose knowledge is necessary to extract the watermark from the host document. In contrast to private watermarking, techniques allowing anyone to read the watermark are referred to as *public*.

Due to Kerkhoff's principle that security can not be based on algorithm ignorance, but rather on the choice of a secret key, it can be concluded that private watermarking is likely to be significantly more robust than public watermarking, in that, once the embedded code is known, it is much easier for an attacker to remove it or to make it unreadable, e.g. by inverting the encoding process or by encoding an *inverse* watermark. Note that the use of cryptography does not help here, since once the embedded bits have been read, they can be removed even if their meaning is not known because they have been previously encrypted.

1.2.6 Readable vs. detectable watermarks

As stated in section 1.1 (see figure 1.4), an important distinction can be made between data hiding schemes where the embedded code can be *read*, and those in which the embedded information can only be *detected*. In the former case (*readable* watermarking), the bits contained in the watermark can be *read* without knowing them in advance, whereas in the latter case (*detectable* watermarking), one can only verify if a given code is present in the document. In other words, with detectable watermarking, the watermark presence can only be revealed if the watermark content is known

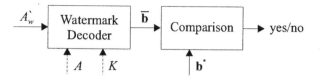

Figure 1.8: Construction of a readable watermark by starting from a detectable one.

in advance. Of course, detectable watermarking techniques are intrinsically private, since it is impossible for an attacker to guess the content of the watermark without knowing anything about it, this being especially true if the information to be embedded in the data is encrypted prior to watermark insertion.

The readable/detectable nature of the hidden data heavily affects the way such data can be used in practical applications. Indeed readable watermarking is by far more flexible than detectable watermarking, since the *a priori* knowledge of the watermark content can not always be granted from an application point of view, thus making the usage of this kind of algorithms in practical scenarios more cumbersome. On the contrary, a detectable watermark is intrinsically more robust than a readable one, both because it conveys a smaller payload and because of its inherently private nature. As an example, let us consider a situation in which one wants to know the owner of a piece of work downloaded somewhere in Internet. Suppose that the owner identification code has been hidden within the work itself. If a detectable scheme was used, there would be no mean to read the owner name, since the user does not know in advance which watermark he has to look for. On the contrary, this would be possible if readable watermarking was used.

Note that given a readable watermarking scheme, the construction of detectable scheme is straightforward; it only needs to add a module that compares the retrieved information $\overline{b}$ against the to-be-searched code b^* (figure 1.8). As it will be shown in chapter 3, several methods also exist to build a readable watermarking scheme by starting from a detectable one.

1.2.7 Invertibility and quasi-invertibility

The concept of watermark *invertibility* arises when analyzing at a deeper level the SWICO attack described previously. At the heart of the attack there is the possibility of reverse engineering the watermarking process, i.e. the possibility of building a fake original asset and a fake watermark such that the insertion of the fake watermark within the fake original asset produces a watermarked asset which is equal to the initial one. To be more specific, let A be a digital asset, and let assume that a non-blind, detectable watermarking scheme is used to claim ownership of A. Let the watermarking scheme be characterized by an embedding function $\mathcal{E}$ and a detector function $\mathcal{D}$. We say that:

Definition: *the watermarking scheme is invertible if for any asset A it exists an inverse mapping $\mathcal{E}^{-1}$ such that $\mathcal{E}^{-1}(A) = \{A_f, \mathbf{w}_f\}$ and $\mathcal{E}(A_f, \mathbf{w}_f) = A$, where $\mathcal{E}^{-1}$ is a computationally feasible mapping, and the assets A and A_f are perceptually similar. Otherwise the watermarking scheme is said to be* non-invertible.

We call A_f and $\mathbf{w}_f$ respectively fake original asset and fake watermark. In the simplified version of the SWICO attack described at the beginning of this section, it simply was:

$$\mathcal{E}^{-1}(A) = \{A - \mathbf{w}_f, \mathbf{w}_f\}, \tag{1.13}$$

with $\mathbf{w}_f = \mathbf{w}_B$. Note that, unlike in our simplified example, in general the design of the inverse mapping $\mathcal{E}^{-1}$ involves two degrees of freedom, since both the fake original asset and the fake watermark can be adjusted to reverse engineer the watermarking process.

A more sophisticated version of the SWICO attack (TWICO attack, from the acronym of Twin-Watermarked-Images-Counterfeit-Original) leads to the extension of the invertibility concept to the concept of *quasi-invertibility*. The extension of the SWICO attack relies on the observation that, in order to be effective, such an attack does not need that the insertion of the fake watermark within the fake original asset produces an asset which is identical to the initial one, i.e. A. On the contrary, it is only needed that when the watermark detector is applied to A by using the fake original asset as original non-marked document, the presence of the fake watermark is revealed. Stated in another way, we need that:

$$\mathcal{D}(A, A_f, \mathbf{w}_f) = yes, \tag{1.14}$$

thus yielding the following:

Definition: *a non-blind watermarking scheme, characterized by an embedding function $\mathcal{E}$ and a detector function $\mathcal{D}$, is* quasi invertible *if for any asset A it exists an inverse mapping $\mathcal{E}^{-1}$ such that $\mathcal{E}^{-1}(A) = \{A_f, \mathbf{w}_f\}$ and $\mathcal{D}(A, A_f, \mathbf{w}_f) = yes$, where $\mathcal{E}^{-1}$ is a computationally feasible mapping, and the assets A and A_f are perceptually similar. Otherwise the watermarking scheme is said to be* non-quasi-invertible.

The analysis carried out so far applies to detectable, non-blind techniques, however, the concept of watermark invertibility can be easily extended to readable watermarking as well. As to blind schemes, given an asset A, the main difference with respect to non-blind watermarking, is that inversion reduces to finding a fake watermark $\mathbf{w}_f$ such that its presence is revealed in A:

$$\mathcal{D}(A, \mathbf{w}_f) = yes. \tag{1.15}$$

As it can be seen, inversion of a blind watermark has only one degree of freedom, thus making it easier to prevent it by acting at a protocol level, e.g. by requiring that watermarks are assigned by a trusted third party, thus avoiding the use of ad-hoc fake watermarks. A similar strategy could be conceived in the non-blind case, however more attention is needed, since the two degrees of freedom implicit in the inversion of a non-blind watermarking scheme, could make it possible to handle a situation in which $\mathbf{w}_f$ is fixed a priori and pirates only act on A_f.

1.2.8 Reversibility

We say that a watermark is strict-sense reversible (SSR) if once it has been decoded/detected it can also be removed from the host asset, thus making it possible the exact recovery of the original asset. Additionally, we say that a watermark is wide-sense reversible (WSR) if once it has been decoded/detected it can be made undecodable/undetectable without producing any perceptible distortion of the host asset. It is obvious by the above definitions that strict-sense reversibility implies wide-sense reversibility, whereas the converse is not true. Watermark reversibility must be carefully considered when robustness/security of the hidden information is a major concern, since it implies that only trusted users should be allowed to read/detect the watermark, thus complicating considerably the design of suitable application protocols.

1.2.9 Asymmetric watermarking

Watermark reversibility is a serious threat especially because most of the watermarking schemes developed so far are *symmetric*, where by symmetric watermarking we mean that the decoding/detection process makes use of the same set of parameters used in the the embedding phase. These parameters include the possible usage of a secret key purposely introduced to bring in some secrecy in watermark embedding, and all the parameters defining the embedding process, e.g. the number and position of host features. All of them are generally included in the secret key K appearing in equation (1.2) and figure 1.1. Indeed, the general watermarking scheme depicted in figure 1.1 implicitly assumes that the secret key K used in the decoding process, if any, is the same used for embedding. This may lead to security problems, especially if the detector is implemented in consumer

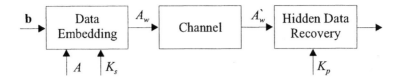

Figure 1.9: In asymmetric watermarking two distinct keys, K_s and K_p, are used for embedding and retrieving the watermark.

devices that are spread all over the world. The knowledge of this set of parameters, in fact, is likely to give pirates enough information to remove the watermark from the host document, hence such an information should be stored safely. The above care is only necessary with wide-sense reversible watermarking; however, this is likely to be always the case, since the effective possibility of developing a symmetric watermarking algorithm which is not WSR has not been demonstrated yet.

In order to overcome the security problems associated with symmetric watermarking, increasing attention has been given to the development of *asymmetric* schemes. In such schemes two keys are present (figure 1.9), a private key, K_s, used to embed the hidden information within the host data, and a public key, K_p, used to detect/decode the watermark (often, K_p is just a subset of K_s). Knowing the public key, it should be neither possible to deduce the private key nor to remove the watermark[11]. In this way, an asymmetric watermarking scheme is not WSR by definition.

A thorough discussion of asymmetric watermarking is given in chapter 8.

1.3 Audio vs image vs video assets

In the attempt to be as general as possible, we discuss how to hide a piece of information within all the most common kinds of media: still images, image sequences, video signals[12] or audio signals. We will not consider text or graphic files, since hiding a piece of data within a text or a graphic raises completely different problems which fall outside the scope of this introductory book. We will not consider data hiding within 3D objects as well, since research in this field is still in its infancy and neither an

---

[11]Unlike in asymmetric cryptography, knowledge of K_s may be sufficient to derive K_p; additionally, the roles of the private key and the public key can not be exchanged.

[12]To be more precise, we will use the term image sequence or moving pictures, to indicate a sequence of frames without audio, and the term video signal to denote the multimedia signal obtained by considering an image sequence and its corresponding audio signal together.

established theory nor efficient algorithms have been developed yet.

Though hiding a watermark signal within a still image is a different piece of work than hiding the same information within an image sequence or an audio signal, most of the concepts employed are the same. Media-independent issues include: coding of the to-be-hidden information, definition of the embedding rule, informed embedding, decoding/detection theory, information theoretic analysis. For this reason, we decided not to discuss the watermarking of each different media separately, on the contrary we tried to be as general as possible, thus presenting the main concepts without explicitly referring to the watermarking of a particular type of signal. Of course, when needed we will distinguish between still images, image sequences and audio, by paying attention to highlight the peculiarities of each type of media. This will be the case with host feature selection, practical concealment strategies, description of possible attacks and benchmarking, description of practical algorithms.

From a general point of view, it has to be said that, so far, most of the research in digital data hiding has been focused on image watermarking, up to a point that many of the techniques proposed for moving pictures and audio watermarking closely resemble the algorithms developed in the still image case. This is particularly evident in the case of image sequences, where some of the most powerful techniques proposed so far simply treat video frames as a sequence of still images, and watermark each of them accordingly. No need to say, though, that data hiding techniques which fully exploit the peculiarities of image sequences (and to a major extent of audio signals) are more promising, thus justifying, here and there in the book, a separate treatment of moving pictures and audio.

In some applications, the need for universal data hiding techniques that can be applied to all kinds of media has been raised, nevertheless doubts exist that such kind of techniques can be developed, and in fact, no practical universal watermarking algorithm has been proposed so far.

Finally, it is often called for that true multimedia watermarking techniques, exploiting the cross-properties of different media, are developed, to be applied in all the cases where still images, image sequences and audio signals are just assets of a more complex multimedia signal or document, e.g. a video signal. Even in this case, though, research is still at a very early stage, thus we will always assume that different media assets are marked separately.

1.4   Further reading

Though digital watermarking is a young discipline, steganography, i.e. the art of secretely hiding a piece of information into an apparently innocuous message, is as old as the human kind. For a brief, easy-to-read, history of steganography, readers may refer to the paper by F. A. P. Petitcolas, R. J. Anderson and M. G. Kuhn [177]. An interesting overview of watermarking covering the second half of the twentieth century is also given in [55].

As data hiding becomes a mature field, terminology and symbolism tend to get more and more uniform; this was not the case in the early days of research. Even now, after ten years have passed since digital watermarking first came to the attention of researchers, a complete agreement on a common terminology has not been reached. A first attempt to define data-hiding terminology and symbolism can be found in [178]. A noticeable effort to define a non-ambiguous terminology, which somewhat differs from that used in this book, is also done in [56].

Protocol issues were brought to the attention of watermarking researchers by S. Craver, N. Memon, B. L. Yeo and M. M. Yeung [59, 60]. More specifically, they first introduced the SWICO and TWICO attacks, thus demonstrating the problems deriving from the adoption of a non-blind watermark detection strategy.

The interest in asymmetric watermarking was triggered by the works by R. G. van Schyndel, A. Z. Tirkel and I.D. Svalbe [220], J. J. Eggers, J. K. Su and B. Girod [73], T. Furon and P. Duhamel [82]. Since then researchers have investigated the potentiality offered by asymmetric schemes [58, 154], however a ultimate answer on whether asymmetric watermarking will permit to overcome some of the limitations of conventional methods has not been given yet.

In this book we do not cover explicitly steganography applications, where the ultimate goal of the embedder is to create a so called stego-channel whereby information can be transmitted without letting anyone be aware that a communication is taking place. For a good survey of steganographic techniques, the reader is referred to the introductory paper by N. F. Johnson and S. Katzenbeisser [113].

2

Applications

Though steganography, i.e. the art of keeping information secret by hiding it within innocuous messages, is as old as the human kind, a renewed interest in digital data hiding was recently triggered by its potential use as an effective mean for copyright protection of multimedia data exchanged in digital form.

Early research, in mid nineties, was mainly focused on robust watermarking, i.e. the insertion within the to-be-protected data of an imperceptible code bearing some information about data itself, e.g. data owner or its allowed uses. In addition to be imperceptible, the code should be robust, in that it should survive any possible manipulations applied to the data, at least until the degradation introduced in the attempt to remove it does not reduce significantly the commercial value of data.

As research has gone on, it has become evident that several other application scenarios exist where digital data hiding could be used successfully. First, the authentication of digital documents was considered, that is the possibility of using the embedded information to prove data integrity or to discover possible, malicious or non-malicious, modifications applied to it. Then a number of other applications has emerged including, just to mention some, data indexing, transmission error recovery and concealment, hidden communications, audio in video for automatic language translation, image captioning. In some cases, data hiding represents a new solution to unsolved problems raised by the wider and wider diffusion of digital technologies. In other cases, data hiding is just another way to tackle a problem that could be faced with by resorting to other technologies as well. In any case, it is not wise, if at all possible, to ignore the possibilities offered by digital data hiding, since it may provide elegant problem solutions virtually at no, or at a very low, cost.

In this chapter, we run through a number of possible applications of data hiding. Our aim is just to give some examples of the possibilities data hiding makes available, and of the issues usually encountered when trying to apply general data hiding concepts to practical situations. We first examine protection of property rights, since this may still be considered the main application field of digital watermarking. As a second topic, we discuss data authentication, an application that can be more easily served by data hiding due to its less demanding requirements, especially in terms of robustness. Then, we briefly discuss other emerging applications including transmission error recovery, annotation, captioning, compression, and covert communications.

2.1 IPR protection

The protection of the rights possessed by the creator, or the legitimate owner, of a multimedia piece of work encompasses many different aspects including copyright protection and moral rights protection, e.g. the insurance that the integrity of the work is respected not to violate the moral beliefs of the owner/creator. In the sequel we will refer globally to such rights as Intellectually Property Rights (IPR), even if, rigorously speaking, IPR protection should consider topics such as patents and trademarks as well. Due to the wide variety of situations encountered in practical applications, to the large number of objectives possibly pursued by an IPR protection system, and to the different legislations holding in different countries, it is impossible (and beyond the scope of this book) to give a unified treatment of watermarking-based IPR protection. We then present here only the major tasks watermarking may be used for and the corresponding watermarking paradigms, by keeping in mind that any practical IPR protection system will need to address these tasks, and a considerable number of other security and economic issues all together[1].

2.1.1 Demonstration of rightful ownership

This is the most classical scenario served by watermarking: the author of a work wishes to prove that he/she is the only legitimate owner of the work. To do so, as soon as he/she *creates* the work, he/she also embeds within it a watermark identifying him/her unambiguously. Unfortunately, this simple scheme can not provide a valid proof in front of a court of law, unless the

---

[1]In general, the design of an IPR protection system goes through the definition of a Business Model (BM) describing the way electronic transactions are performed, an Electronic Copyright Management System defining how IPRs are handled within the BM, and the specification of how the BM and the ECMS are implemented in practice, e.g. through digital watermarking or cryptography.

non-invertibility (non-quasi-invertibility) of the watermarking algorithm is demonstrated (see section 1.2.7). Nevertheless, the watermark may still be used by the rightful owner for his/her own purposes. For example, the author may wish to detect suspicious products existing in the distribution network. Such products could be individuated by an automated search engine looking for the watermark presence within all the works accessible through the network. Then, the author may rely on more secure mechanisms to prove that he/she was the victim of a fraud, e.g. by depositing any new creation to a registration authority.

A common way to confer the watermark verification procedure a legal value, is to introduce the presence of a Trusted Third Party (TTP) in the watermarking protocol. For example, the watermark identifying the author may be assigned to him/her by a trusted registration authority, thus preventing the possibility to use the SWICO attack to fool the ownership verification procedure. In this way, in fact, it would be by far more difficult to invert the watermarking operation, especially when blind watermarking is used, since pirates can not rely on the design of an ad hoc fake original work.

As to the requirements a watermarking algorithm to be used for rightful ownership verification must satisfy, it is obvious that for any scheme to work, the watermark must be a secure one, given that pirates are obviously interested in removing the watermark, possibly by means of computationally intensive procedures. In addition, private watermarking is preferable, due to its inherently superior security. Finally, capacity requirements depend on the number of different author identification codes the system must accommodate for.

2.1.2 Fingerprinting

A second classical application of digital watermarking is copy protection. Two scenarios are possible here; according to the first one, a mechanism is envisaged to make it impossible, or at least very difficult, to make illegal copies of a protected work (see section 2.1.3 for a discussion on copy control mechanisms). In the second scenario, a so called *copy deterrence* mechanism is adopted to discourage unauthorized duplication and distribution. Copy deterrence is usually achieved by providing a mechanism to trace unauthorized copies to the original owner of the work. In the most common case, distribution tracing is made possible by letting the seller (owner) inserting a distinct watermark, which in this case is called a *fingerprint*, identifying the buyer, or any other addressee of the work, within any copy of data which is distributed. If, later on, an unauthorized copy of the protected work is found, then its origin can be recovered by retrieving

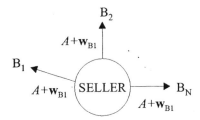

Figure 2.1: To take into account buyer's right, it is necessary that the situation depicted in the figure, where several copies of the host asset containing the identification code of client B_1 are distributed to other purchasers, is avoided.

the unique watermark contained in it.

Of course, the watermark must be secure, to prevent any attempt to remove it, and readable, to make its extraction easier. Note that the readability requirement may be relaxed if the owner has the possibility to guess in advance the watermark content.

A problem with the plain fingerprinting protocol described above, is that it does not take into account buyer's rights, since the watermark is inserted solely by the seller. Thus, a buyer whose watermark is found in an unauthorized copy can not be inculpated since he/she can claim that the unauthorized copy was created and distributed by the seller. The possibility exists, in fact, that the seller is interested in fooling the buyer. Let us consider, for example, the situation depicted in figure 2.1, where the seller is not the original owner of the work, but an authorized reselling agent. The seller may distribute many copies of a work containing the fingerprint of buyer B_1 without paying the due royalties to the author, and claim that such copies were illegally distributed or sold by B_1.

As in the case of rightful ownership demonstration, a possible solution consists in resorting to a trusted third party. The simplest way to exploit the presence of a TTP to confer a legal value to the fingerprint protocol, is to let the TTP insert the watermark within the to-be-protected work, and retrieve it in case a dispute resolution protocol has to be run. Despite its simplicity, such an approach is not feasible in practical applications, mainly because the TTP must do too much work, then it may easily become the bottleneck of the whole system. In addition, the protected work must be transmitted from the seller to the TTP and from the TTP to the customer, or, in an even worse case, from the TTP to the seller and from the seller to the customer, thus generating a very heavy traffic on the communication channel.

An ingenious way to avoid the above difficulties and still ensure that buyer's rights are respected, relies on the joint exploitation of watermarking and cryptography, as suggested by the Interactive Buyer-Seller (IBS) protocol. Even in this case, the the presence of a TTP is envisaged, however TTP's role is minimized, thus making the IBS protocol more suited to practical applications. Data exchange is kept to a minimum as well, resulting in a very low communication overhead. The basic idea the IBS protocol relies on, is that attention is paid not to let the seller get to know the exact watermarked copy received by the buyer, hence he/she can not distribute or sell copies of the original work containing the buyer's identification watermark. In spite of this, the seller can identify the buyer from whom unauthorized copies originated, and prove it by using a dispute resolution protocol. The same protocol can be used by the buyer to demonstrate his/her innocence. In order to exemplify the IBS protocol, let Alice be the author of the work and Bob the buyer. We assume that Alice and Bob possess a pair of public/private keys denoted by K_A, K_B (public keys) and K'_A, K'_B (private keys). Let the encryption of a message with a key K be indicated by E_K. After sending an identification of his identity, Bob requests the TTP to send him a valid watermark $\mathbf{w}$ (once again we assume that $\mathbf{w}$ coincides with $\mathbf{b}$). The TTP checks Bob's credentials and generates the watermark $\mathbf{w}$. It then sends back to Bob $\mathbf{w}$ encrypted with Bob's public key:

$$E_{K_B}(\mathbf{w}) = \{E_{K_B}(w_1), E_{K_B}(w_2) \ldots E_{K_B}(w_n)\}, \qquad (2.1)$$

along with a signature of $E_{K_B}(\mathbf{w})$, $S_{TTP}(E_{K_B}(\mathbf{w}))$. For example,

$$S_{TTP}(E_{K_B}(\mathbf{w})) = E_{K'_{TTP}}(H(E_{K_B}(\mathbf{w}))), \qquad (2.2)$$

where H is a proper hash function. Note that we assumed that watermark components w_i's are watermarked independently by using the same encryption key.

As a second step, Bob sends Alice $E_{K_B}(\mathbf{w})$ and $S_{TTP}(E_{K_B}(\mathbf{w}))$, so that Alice can verify that $E_{K_B}(\mathbf{w})$ is a valid encrypted watermark. Let A be the digital asset Bob wants to buy. Before sending A to Bob, Alice inserts within it two distinct watermarks. For the first watermark $\mathbf{v}$, which conveys a distinct ID univocally identifying the buyer, Alice can use the watermarking scheme she prefers, since such a watermark is used by Alice only to identify potentially deceitful customers through plain fingerprinting. The second watermark is built by relying on $E_{K_B}(\mathbf{w})$. As for E_K, we require that the watermarking scheme acts on each host feature independently, that is we require that:

$$\mathbf{f}_{A_\mathbf{w}} = \{f_1 \oplus w_1, f_2 \oplus w_2 \ldots f_n \oplus w_n\}, \qquad (2.3)$$

where $\mathbf{f}_A = \{f_1, f_2 \ldots f_n\}$ represents the set of non-marked host features. As a second requirement, we ask that the cryptosystem used by the IBS protocol is a *privacy homomorphism* with respect to $\oplus$, that is:

$$E_K(x \oplus y) = E_K(x) \oplus E_K(y), \qquad (2.4)$$

where x and y are any two messages. Strange as it may seem, the privacy homomorphism requirement is not difficult to satisfy. For instance, it is known that the popular RSA cryptosystem is a privacy homomorphism with respect to multiplication.

To insert the second watermark within A, Alice performs the following steps. First she permutes the watermark components through a secret permutation σ:

$$\sigma(E_{K_B}(\mathbf{w})) = E_{K_B}(\sigma(\mathbf{w})), \qquad (2.5)$$

where the equality immediately follows from equation (2.1). Then she inserts $E_{K_B}(\sigma(\mathbf{w}))$ within A directly in the encrypted domain. This is possible due to (2.4) and because Alice knows Bob's public key. Stated in another way, Alice sends to Bob an encrypted version of A containing $\sigma(\mathbf{w})$:

$$E_{K_B}(A_{\mathbf{v}, \sigma(\mathbf{w})}) = E_{K_B}(A_{\mathbf{v}}) \oplus E_{K_B}(\sigma(\mathbf{w})). \qquad (2.6)$$

It is worth stressing again that in order to produce $E_{K_B}(A_{\mathbf{v}, \sigma(\mathbf{w})})$, Alice does need to access the plain watermark $\mathbf{w}$, since watermark casting is performed in the encrypted domain.

When Bob receives $E_{K_B}(A_{\mathbf{v}, \sigma(\mathbf{w})})$, he decrypts it by using his private key K_B', thus obtaining $A_{\mathbf{v}, \sigma(\mathbf{w})}$. Note that Bob can not read the watermark $\sigma(\mathbf{w})$, thus it is not necessary to ensure the non reversibility of the watermarking scheme.

In order to recover the identity of potential copyright violators, Alice first looks for the presence of $\mathbf{v}$. Upon detection of an illegal copy of A, say A', she can use the second watermark to effectively prove that such a copy originated from Bob. To do so, Alice must reveal to a judge the permutation σ, the encrypted watermark $E_{K_B}(\mathbf{w})$, and $S_{TTP}(E_{K_B}(\mathbf{w}))$. After verifying $S_{TTP}(E_{K_B}(\mathbf{w}))$, the judge asks Bob to reveal its private key K_B' to calculate $\mathbf{w}$ (actually it is not necessary that Bob reveals K_B', it is only necessary that he reveals $\mathbf{w}$ whose validity can be verified by applying K_B to it and checking whether it equals $E_{K_B}(\mathbf{w})$). Now it is possible to check A' for the presence of $\sigma(\mathbf{w})$: if such a presence is verified, then Bob is judged guilty, otherwise Bob's innocence is been proven. Note that if $\sigma(\mathbf{w})$ is found in A', Bob can not maintain that A' originated from Alice, since to do so Alice should have known either $\mathbf{w}$ to insert it within the plain asset A, or K_B' to decrypt $E_{K_B}(A_{\mathbf{v}, \sigma(\mathbf{w})})$ after having inserted the watermark in the encrypted domain.

The protocols described in this and in previous section, are just two examples of how illegal copy deterrence can be achieved by relying on watermarking technology. With regard to the effective value of such mechanisms as proofs in front of a judge, it must be said that the current state-of-the-art allows this possibility only if a watermarking/certification authority acting as a TTP is included within the copyright protection protocol. Nevertheless, it is important to stress out that, even if plain fingerprinting may not be considered a proof from a legislative point of view, it may useful in several situations. For instance, the seller may use it to identify potentially deceitful customers and break off any further business with them.

2.1.3 Copy control

When copy deterrence is not sufficient to effectively protect legitimate rightholders, a true copy protection mechanism must be envisaged. Having said that a comprehensive solution of copy protection mechanisms goes well beyond watermarking technology, we describe a mechanism which has been considered for protection of DVD video. This scenario, in fact, represents a good example of how watermarking can be integrated in a complex copy protection system and effectively contribute to its efficacy.

The DVD copy protection system outlined below, is the result of the efforts of many important companies, including IBM, NEC, Sony, Hitachi, Pioneer, Signafy, Philips, Macrovision and Digimarc. Though the systems proposed by various companies differ with respect to many important issues such as, for example, the choice of the underlying watermarking technology, the overall protection scheme and the role of watermarking within it are very similar, thus allowing us to briefly describe them without delving into implementation details.

The mechanism employed to make illegal duplication and distribution difficult enough to keep losses caused by missed revenues sustainable, relies on the distinction between copyright compliant devices (CC-devices) and non compliant devices (NC-devices). In particular, the DVD copy protection system is designed in such a way that the CC world and the NC world are kept as distinct as possible, for example, by allowing NC devices to play only illegal disks and CC devices to play only legal disks. In this way, users willing to draw from both the worlds must buy two series of devices, one for legal and one for illegal disks, in the hope that this will prevent massive, unauthorized, copying, as it happened in the case of audio.

A first important feature of a protected DVD is that its content is scrambled through a Content Scrambling System (CSS). Descrambling requires a pair of keys, one of which is unique to the video file, while the other is unique to the DVD. Keys are stored on the lead-in area of the DVD, an area

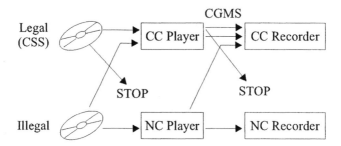

Figure 2.2: The use of CSS only prevents that legal data are not passed to the non-compliant world, whereas the converse is still possible (dashed lines).

that is only read by CC devices. The use of CSS results in the situation depicted in figure 2.2: a protected DVD can only be played and recorded in the CC world. It is not possible, in fact, that the output of a CC player is connected to a NC recorder, since CC devices are not allowed to dialog with NC-devices. On the other side, recording through CC devices is governed by a Copy Generation Management System (CGMS) which allows copying only if this is permitted for that particular disk. Simply speaking, CGMS relies on two bits stored in the header of an MPEG stream, encoding one of the following three indications: *copy-freely*, *copy-never* and *copy-once*, where the result of the *copy-once* indication is that the video can be copied but after copying, the CGMS bits are changed to *copy-never*.

CSS and CGMS prevent the flow from the legal world toward the NC world, nevertheless, in order to discourage illegal copying the reverse must also be true, i.e., it should not be possible to use a CC device to play or record an illegal disk. Otherwise the whole protection mechanism would only succeed in stimulating the diffusion of CC devices. To this aim, the sole CSS is not sufficient. Consider, for example, the case of a pirate using the analog RGB output of a compliant to make an unencrypted copy of the video by means of a NC recorder. Such a copy can be played, and recorded, on CC devices as well, since they would mistake the illegal video for a free video without protection. This is because both scrambling and CGMS bits are no longer present. Data hiding can help solving this problem, it suffices that CGMS bits are embedded within the video in the form of a secure watermark. It is obvious that the presence of CGMS bits prevents video recording on a CC recorder, since, upon reading the CGMS bits, the CC devices refuse to copy the video if CGMS bits indications do not allow it. At the same time, CC players can be designed so to recognize as illegal a DVD copy without CSS, yet containing the CGMS watermark, and refuse playing it. A summary of the effect of embedding CGMS bits within DVD

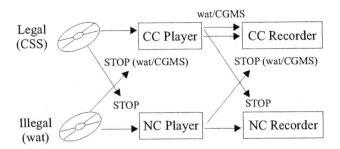

Figure 2.3: The addition of a robust CGMS watermark surviving digital to analog conversion permits to avoid the flow of data from the non-compliant to the compliant world.

video by means of digital watermarking is given in figure 2.3. As desired, the worlds of CC- and NC-devices are kept separate, since illegal disks can only be managed by NC devices and legal disks by CC devices.

2.2 Authentication

One of the (undesired) effects of the availability of more and more effective signal processing tools, and of their possible use to modify the visual or audio content of digital documents without leaving any perceptible traces of the modification, is the loss of credibility of digital data, since doubts always exist that they have been tampered with, in a way that substantially changes the initial data content[2]. To overcome such a problem, it is necessary that proper countermeasures are taken to authenticate signals recorded in digital form, i.e. to ensure that signals have not been tampered with (data integrity) and to prove their true origin. As it is explained below, data authentication through digital watermarking is a promising solution to both the above problems.

2.2.1 Cryptography vs watermarking

A straightforward way to authenticate a digital signal, be it a still image, an image sequence or an audio signal, is by means of cryptography, namely through the joint use of asymmetric-key encryption and a digital hash function. Let us assume that the device used to produce the digital signal, e.g. a scanner or a video camera, is assigned a public/private key pair, and that

---

[2]Such a loss of credibility is dramatic if digital data has to be used legally, for instance as a proof in front of a court of law; however, it may have an important impact in everyday's life as well.

the private key is hardwired within the acquisition device (which, of course, should be as tamper-proof as possible). Before recording the digital signal, the acquisition device calculates a digital summary (digest) of the signal by means of a proper hash function. Then, it encrypts the digest with the private key, thus obtaining a signed digest which is stored together with the digital signal. Later on, the digest can be used to prove data integrity or to trace back to its origin: one only needs to read the signed digest by using the public key of the electronic device which produced the signal and check if it corresponds to the actual signal content. For long signals, e.g. audio or video signals, the digest should be computed on suitable signal sub-parts, e.g. a video frame, rather than on the whole signal.

Though cryptography may provide a valuable mean for digital signal authentication, the development of alternative approaches is desirable in order to deal with some potential weaknesses of the cryptographic approach. Let us consider, for example, the digest-based approach outlined previously. This approach requires that the signal digest is tied to the signal itself, e.g. by defining a proper format allowing the usage of authentication tools (see for example the MPEG21 effort of ISO). In this way, however, the possibility of authenticating the signal is constrained to the use of a particular format, thus making impossible to use a different format, or to authenticate the signal after digital-to-analog conversion. This is not the case if authentication is achieved through digital data hiding, since the authenticating information is embedded within the signal itself. Another drawback with digest-based authentication is that the digest changes dramatically as soon as any modification, be it a small or a large one, is applied to the signal, thus making impossible to distinguish between malicious and innocuous modifications. Moreover, if the basic scheme outlined above is used, cryptographic authentication does not allow a precise localization of tampering.

Data-hiding-based authentication represents a feasible and very elegant solution to the above problems. It must be remembered, though, that despite all the reasons usually produced to justify the resort to data hiding authentication with respect to conventional cryptography, the main difference between the two approaches is the way the authenticating information is tied to the to-be-authenticated signal. More specifically, if the data hiding approach is adopted, no header or separate file has to be used to ensure data integrity, in addition digital-to-analog and analog-to-digital conversion is allowed. Conversely, the main drawbacks of data-hiding-authentication derive from the relative immaturity of watermarking technology with respect to cryptography.

In the following section, we describe a general authentication framework whereby providing authentication through data hiding. Such a framework is a very general one since it encompasses both schemes using (semi-) fragile and robust watermarking.

2.2.2 A general authentication framework

Generally speaking, authentication of the host signal may accomplished either by means of (semi-)fragile or robust watermarking.

As we stated in section 1.2.3, with fragile watermarking the hidden information is lost or altered as soon as the host signal undergoes any modification: watermark loss or alteration is taken as an evidence that data has been tampered with, whereas the recovery of the information contained within the data is used to demonstrate data integrity and, if needed, to trace back to data origin. Interesting variations of the previous paradigm, include the capability to localize tampering, or to discriminate between malicious and innocuous manipulations (e.g. moderate image compression). In the latter case, a semi-fragile watermarking scheme has to be used, since it is necessary that the hidden information survives only a certain kind of allowed manipulations.

The use of robust watermarking for data authentication relies on a different mechanism: a summary of the host signal is computed and inserted within the signal itself by means of a robust watermark. Information about the data origin is embedded together with the summary. To prove data integrity, the information conveyed by the watermark is recovered and compared with the actual content of the sequence: their mismatch is taken as an evidence of data tampering. The capability to localize manipulations will depend on the accuracy of the embedded summary. If tampering is so heavy that the watermark is lost, watermark absence is simply taken as an evidence that some manipulations occurred and the output of the authentication procedure is a negative one. Note that in this case watermark security is not a pressing requirement, since it is unlikely that someone is interested in intentionally removing the watermark. On the contrary, pirates would be interested in modifying the host data without leaving any trace of the modification.

Though the approaches to data authentication relying on (semi-) fragile and robust watermarking may seem rather different, it is possible to describe both of them by means of the same mathematical framework. Let us start by assuming that the watermark authentication relies on is a blind[3] and readable one.

During the embedding phase, the watermark signal is generated by a suitable watermark generation function $\mathcal{G}$, taking as input a secret key K_g

---

[3]Data authentication through non-blind techniques does not make sense. If the original signal is available, in fact, data authentication is a trivial task.

and, possibly, the to-be-authenticated asset A.

$$\mathbf{w} = \mathcal{G}(A, K_g). \tag{2.7}$$

The watermarking signal $\mathbf{w}$ is then hidden within A, thus producing a watermarked asset $A_\mathbf{w}$ (for sake of simplicity we assume that $\mathbf{w}$ coincides with $\mathbf{b}$):

$$A_\mathbf{w} = \mathcal{E}(A, \mathbf{w}, K), \tag{2.8}$$

where the secret key K used for watermark embedding must not be confused with the secret key K_g used to generate the watermark.

To describe the verification procedure, let us indicate by $A'_\mathbf{w}$ a possibly corrupted copy of $A_\mathbf{w}$. In order to verify the integrity of $A'_\mathbf{w}$, a watermark signal $\mathbf{w}'$ is computed by means of the generation function $\mathcal{G}$.

$$\mathbf{w}' = \mathcal{G}(A'_\mathbf{w}, K_g). \tag{2.9}$$

Then the watermark embedded within $A'_\mathbf{w}$ is extracted, producing the watermark signal $\mathbf{w}''$. Finally, the signals $\mathbf{w}'$ and $\mathbf{w}''$ are compared: if they are equal the integrity verification procedure succeeds, otherwise it fails[4]:

$$\mathbf{w}'' = \mathcal{D}(A'_\mathbf{w}, K), \tag{2.10}$$

If $\mathbf{w}' = \mathbf{w}''$ Then

the Asset is authentic

Else

the Asset has been tampered with.

$$\tag{2.11}$$

Authentication algorithms allowing tampering localization, infer the position of tampering by giving $\mathbf{w}$ a suitable form and by looking at the positions where $\mathbf{w}'$ and $\mathbf{w}''$ differ.

The above framework is valid both for fragile and robust watermarking. The difference between the two approaches resides in the mechanism at the basis of manipulation detection: while fragile techniques assume that any manipulations modify the embedded watermark, robust techniques assumes that the watermark is not affected by any manipulations; on the contrary, it is the watermark generation function that, in this case, produces a watermark that does not correspond to the embedded one. More formally, we can say that, when a manipulation occurs, for fragile techniques we expect that:

$$\begin{cases} \mathbf{w}' = \mathbf{w} \\ \mathbf{w}'' \neq \mathbf{w}; \end{cases} \Rightarrow \quad \mathbf{w}' \neq \mathbf{w}'' \tag{2.12}$$

---

[4]Here we are mainly concerned with integrity verification, nevertheless the recovery of signal origin is rather easy, e.g. by including such an information within $\mathbf{w}$.

that is, the generation function is not affected by manipulations, whereas the decoding function is. Conversely, in the robust watermarking case we expect that:

$$\begin{cases} \mathbf{w}' \neq \mathbf{w} \\ \mathbf{w}'' = \mathbf{w}, \end{cases} \Rightarrow \quad \mathbf{w}' \neq \mathbf{w}'' \qquad (2.13)$$

i.e. manipulations only affect the output of the generation function $\mathcal{G}$.

To introduce a certain degree of tolerance in the integrity verification phase, e.g. to discriminate between allowed and non-allowed manipulations, the dependence of $\mathcal{G}$ (in the robust watermarking case) or $\mathcal{D}$ (in the fragile watermarking case) upon asset manipulations has to be relaxed. In the fragile scheme, this leads to the use of semi-fragile watermarking, whereas in the robust approach, this implies the design of a function $\mathcal{G}$ that depends only on certain asset features[5]. Alternatively, the possibility of distinguishing between different types of manipulations can rely on a clever comparison between $\mathbf{w}'$ and $\mathbf{w}''$. For instance, if $\mathbf{w}$ coincides with a low resolution version of A, the comparison between $\mathbf{w}'$ and $\mathbf{w}''$ can be performed manually, thus letting a human operator decide whether revealed modifications are admissible or not.

As to authentication through fragile watermarking, the easiest way to achieve the conditions expressed in equation (2.12) is to let $\mathcal{G}$ depend only on K_g. In this way, in fact, the watermark signal $\mathbf{w}$ does not depend on the host asset, hence it does not depend on asset manipulations as well. In the case of robust watermarking, the most common choice for the generation function $\mathcal{G}$, is to let its output correspond to a summary of the to-be-authenticated asset. More specifically, to focus the authentication procedure on meaningful modifications only, it is rather common to design $\mathcal{G}$ so that it grasps the semantic content of A, e.g. by letting $\mathcal{G}(A, K_g)$ coincide with a low resolution version of A.

The authentication framework described above applies to readable watermarking, however its extension to detectable watermarking is straightforward. It only needs to replace equations (2.10) and (2.11), with the following authenticity check:

If $\mathcal{D}(A'_{\mathbf{w}}, \mathbf{w}', K) = yes$ Then

the Asset is authentic

Else

the Asset has been tampered with.
$$(2.14)$$

---

[5]It is worth noting that, in any case, $\mathcal{G}$ must be insensible at least to watermark addition, otherwise it would always be $\mathbf{w}' \neq \mathbf{w}''$, since even in the absence of manipulations $A \neq A_{\mathbf{w}}$

where $\mathbf{w}'$ is still computed as in equation (2.9). As for readable watermarking, the possibility of distinguishing between allowed and non allowed manipulations resides in the sensibility of $\mathcal{G}$ or $\mathcal{D}$ on asset manipulations.

2.2.3 Requirements of data-hiding-based authentication

As we already noted, it is impossible to define an exact list of requirements a data hiding algorithm must fulfill without taking into account the application scenario; nevertheless, when restricting the analysis to data authentication, the following general considerations hold:

- *Blindness*: of course, if the original asset A is available, checking the integrity of a copy of A is a trivial task, since it only needs to compare the copy with A. As to data origin, when disentangled from integrity verification, it can be treated by the same standard as annotation watermarks (see section 2.4).

- *Readability/detectabily*: by following the discussion carried out so far, it can be concluded that no particular preference can be given to readable or detectable watermarking with respect to integrity verification.

- *Robustness*: data authentication can be achieved both by means of fragile and robust watermarking. Moreover, both the approaches permit, at least in principle, to discriminate between different classes of manipulations. Trying to summarize the pro's and con's of the two methods, we can say that with (semi-)fragile techniques it is more difficult to distinguish between malicious and innocuous modifications, whereas the robust watermarking approach seems more promising, since the final judgement on tampering usually relies on a visual comparison between the asset summary conveyed by the watermark and the to-be-authenticated copy. Conversely, the need of ensuring a high watermark capacity without loosing robustness is the Achille's heel of robust techniques; the need for a high capacity deriving from the large number of bits needed to produce a meaningful asset summary.

- *Imperceptibility*: due to the particular nature of the authentication task, it is usually necessary that watermark imperceptibility is guaranteed. Nevertheless, some applications may exist in which a slightly perceptible watermark is allowed. This is the case, for example, of Video Surveillance (VS) data authentication, where authentication is needed to keep the legal value of VS data intact: in most cases, it is only necessary that the hidden information does not disturb the correct behavior of the automatic visual inspection process the VS system relies on (see section 5.5.1 for more details).

2.3 Data hiding for multimedia transmission

Among the possible applications of data hiding, the exploitation of a hidden communication channel for improved transmission, particularly video transmission, is gaining more and more consensus. Data hiding can be helpful for video transmission in several ways. From the source coding point of view, it can help to design more powerful compression schemes where part of the information is transmitted by hiding it in the coded bit stream. For instance, chrominance data could be hidden within the bit stream conveying luminance information. Alternatively, the audio data could be transmitted by hiding it within the video frame sequence (audio in video). From a channel coding perspective, data hiding can be exploited to improve the resilience of the coded bit stream with respect to channel errors (self-correcting or self-healing images/video). As a matter of fact, redundant information about the transmitted video could be hidden within the coded bit-stream and used for video reconstruction in case channel errors impaired the bit-stream.

2.3.1 Data compression

Traditionally, data hiding and data compression are considered contradictory operations, in that each of them seems to obstruct the goal of the other. As a consequence, a great deal of research has been devoted to finding an appropriate compromise between the two goals, e.g. by designing a watermarking algorithm which survives data compression. Despite this apparent contradiction, some authors started investigating the possibility of improving the effectiveness of existing data compression schemes by encoding only part of the information and hiding the remaining information within the coded bit-stream itself. For instance, the possibility of hiding the chrominance part of an image within luminance information has been investigated with rather good results, in that for a given compression rate, a better fidelity to the original image is obtained. Another possibility consists in hiding the audio signal of a video within the visual part of the video stream.

From a theoretical point of view, one may wonder whether, in the presence of ideal, perceptually lossless compression, data hiding is still possible or not. The answer to this question is not easy at all. First of all the notion of ideal perceptually lossless compression must be clarified. For example, we may say that a perceptually lossless compression algorithm is ideal if it removes all the information contained in the digital asset which can not be perceived by a human observer. It is readily seen, that the above definition precludes the possibility that data hiding and ideal compression coexist together, given that the ultimate goal of any data hiding scheme is to modify the host asset in such a way that it contains a piece of information which, though present, can not be perceived by the human senses. Unfortunately (or fortunately, depending on the point of view), the above definition is by far too heuristic to lead to a mathematical formulation of the coding problem. When a more practical definition of perceptually lossless ideal coding is given, the possibility of designing a data hiding scheme which survives ideal compression remains an open issue, mainly due to the particular nature of perceptual equality between assets. As a matter of fact, perceptual equality is not an equality in a strict mathematical sense since the transitive property is not satisfied, nor the perceptual distance between assets is a true distance in a mathematical sense, since the triangular inequality does not hold, thus making the theoretical analysis of data hiding in the presence of ideal, perceptually lossless compression extremely difficult.

At a more practical level, data hiding can represent a new way to overcome the imperfections of current compression algorithms, which, far from being ideal as they are, do not remove all the perceptual redundancy[6] contained within the to-be-compressed asset.

From an application perspective, the most stringent requirement is watermark capacity, since the larger the capacity the higher the effectiveness of the source coding algorithm. On the contrary, robustness is not an issue at all, provided that data hiding is performed in the compressed domain, or simultaneously to data compression. This is not the case, if data hiding precedes compression, since in this case the hidden data must survive compression. Protocol level requirements are rather obvious: blind watermark detection is required, as well as the adoption of a readable watermarking scheme.

2.3.2 Error recovery

A problem with the transmission of data in compressed form is the vulnerability of the coded bit stream to transmission errors. This is the case with most of the compression standards, including JPEG for still images, MPEG and H.263 for digital video or MP3 for audio. For example, in MPEG-2 video, a single bit error can cause a loss of synchronization that will be visible over an entire group of pictures (GOP). To cope with the fragility of compressed data, channel coding is usually adopted to enable error detection or correction. This always corresponds to the introduction of a controlled amount of redundancy. Redundancy can either be introduced at the transmission level, by relying on error correcting codes, or at

---

[6]Perceptually redundant information may be defined as the part information contained within the asset which is not perceived by a human observer.

the application level, i.e. by modifying the syntax of the coded bit stream, in the attempt to make it more resilient against errors. Though the above solutions considerably improve the quality of the reconstructed data in the presence of errors, all of them share two basic drawbacks: i) usually they are not standard-compliant (even if new standards make provision for error resilient compression, backward compatibility with previous standard is often lost); ii) the net available bit-rate decreases to make room for the redundancy.

A possible alternative consists in performing error detection and concealment at the decoder side. For instance, in video transmission, temporal concealment may be applied in the attempt to reconstruct the missed information from past frames, or the data in the present frame may be used to reconstruct lost information (spatial concealment). Nevertheless, it is obviously impossible to exactly recover the original content of a video frame, e.g. in the presence of occlusions, once the corresponding part of the bit stream has been lost.

Data hiding represents an alternative approach to the problem: the redundant information is hidden within the compressed stream and, possibly, used by the decoder to recover from errors. For instance, a low quality version of the compressed asset may be transmitted through the hidden channel to enable the reconstruction of the information that was lost because of channel errors. In some cases, it is only important to detect errors, e.g. to ask the retransmission of data, then the hidden data can be used as in authentication applications, with tampering being replaced by transmission errors. Note that with data hiding, backward standard compliance is automatically achieved, since the hidden data is simply ignored by a decoder which is not designed to exploit it. As to the preservation of the net bit-rate available for payload transmission, it has to be noted that, though unperceivable, the watermark always introduces a certain amount of distortion which decreases the PSNR of the encoded data. Such a loss in PSNR should be compared to the PSNR loss caused by the reduction of the net bit-rate consequent to the use of conventional forward error correction techniques, or to transmission of the redundant information at the application level.

As for joint source coding and data hiding, even in this case, the actual possibility of replacing error correcting codes with data hiding (or improving the capability of error correcting codes via data hiding methodologies) is not easy to asses from a theoretical point of view. As a matter of fact, results from rate distortion theory and Shannon's theorem on channel coding seem to indicate that no improvement has to be expected by using data hiding for error correction. Nevertheless, real data transmission conditions are far from the ideal conditions assumed in information theory: the channel is not AWGN, source and channel coding are not ideal, asymptotic analysis does not always hold, PSNR is not a correct measure of perceptual degradation. At a more practical level, then, data hiding is likely to bring some advantages with respect to conventional error handling techniques.

Even in this case, applications requirements are less demanding than in copyright protection applications. The most stringent requirement regards capacity, in that the higher the capacity the larger amount of redundancy can be transmitted, thus increasing robustness against errors. For example, the transmission of a low resolution version of a 512×512 gray level image may require the transmission of 4096 pixel values, for a total required capacity of about 10 Kbit (we assumed that each pixel requires at least 2.5 bits to be coded), which is a rather high value. Conversely, robustness is not a major concern, even if it is obviously required that the hidden information survives transmission errors. It is also obvious that the use of a blind watermarking algorithm is required. The adoption of readable watermarking is also mandatory, unless the hidden information is only used to detect errors, without attempting to correct them.

2.4 Annotation watermarks

Despite digital watermarking is usually looked at as a mean to increase data security (be it related to copyright protection, authentication or reliable data transmission), the ultimate nature of any data hiding scheme can be simply regarded as the creation of a side transmission channel, associated to a piece of work. Interestingly, the capability of the watermark to survive digital to analog and analog to digital conversion leads to the possibility of associating the side channel to the work itself, rather than to a particular digital instantiation of the work. This interpretation of digital watermarking paves the way for many potential applications, in which the watermark is simply seen as annotation data, inserted within the host work to enhance its value. The range of possible applications of annotation watermarks is a very large one, we will just describe a couple of examples to give the reader a rough idea of the potentiality of digital watermarking when this wider perspective is adopted. Note that the requirements annotation watermarks must satisfy, can not be given without carefully considering application details. In many cases, watermark capacity is the most important requirement, however system performance such as speed or complexity may play a predominant role. As to robustness, the requirements for annotation watermarks are usually much less stringent that those raised by security or copyright protection applications.

2.4.1 Labelling for data retrieval

Content-based access to digital archives is receiving more and more attention, due to the difficulties in accessing the information stored in very large, possibly distributed, archives. By letting the user specify the work he is looking for, by roughly describing its content at a semantic level, many of the difficulties usually encountered during the retrieval process can be overcome. Unfortunately, it is very difficult for a fully automated retrieval engine to analyze the data at a semantic level, thus virtually all content-based retrieval systems developed so far fail to provide a true access to the content of the database. A possibility to get around this problem consists in attaching to each work a description of its semantic content. Of course, producing a label describing the semantic content of each piece of work is a very time consuming operation, thus it is essential that such a label is indissolubly tied to the object it refers to, regardless of the object format, and its analog or digital nature. In this context, digital watermarking may provide a way whereby the labelling information is indissolubly tied to the host work, regardless of the format used to record it. When the work moves from an archive to a new one, possibly passing from the analog domain, the information describing the content of the work travels with the work itself, thus avoiding information loss due to format modification. To exemplify the advantages of data hiding with respect to conventional data labelling, let us consider the archival of video sequences in MPEG-4 format. An annotation watermark could be hidden within each video object forming the MPEG-4 stream. For instance, the name of an actor could be hidden within the corresponding video object. If the marked object is copy-edited to create a different video sequence, the hidden label is automatically copied with the object thus avoiding the necessity of labelling it again. Similarly, if the object is pasted to a new video after going in the analog and back to the digital domain, the annotation watermark is not lost, thus making the semantic labelling of the new video easier.

2.4.2 Bridging the gap between analog and digital objects

A clever way to exploit the side communication channel made available by digital watermarking, consists in linking any analog piece of work to the digital world. The smart image concept, derived by the MediaBridge system developed by Digimarc Corporation, is an example of such a vision of digital watermarking. According to the smart image paradigm, the value of any image is augmented by embedding within it a piece of information that can be used to link the image to additional information stored on the Internet. For example, such an information can be used to link a picture on a newspaper to a web page further exploring the subject of the article the image appears in. The actual link to the Internet is activated by showing the printed picture to a video camera connected to a PC; upon watermark extraction the URL of the web site with the pertinent information is retrieved and the connection established. More generally, the information hidden within the piece of work is dormant until a suitable software reads it, then it may be used to control the software which retrieved the watermark, to link the object to additional information, to indicate the user how to get additional services, or to provide the user with a secret information to be used only upon the payment of a fee. Watermark retrieval itself, may be conditioned to the payment of a fee, thus providing a conditional access mechanism that can be exploited in commercial applications, e.g. bonus programme applications, where the gathering of a certain number of watermarks is the access key to a discount programme.

2.5   Covert communications

Covert communication is the most ancient application of data hiding, since it traces backs at least to the ancient Greeks, when the art of keeping a message secret was used for military applications. Indeed, it is often invoked that the first example of covert communication is narrated by Herodotus, who tells the story of a message tattooed on the shaved head of a slave: the slave was sent through the enemy's lines after his hair was grown again, thus fooling the enemy. Even if it is likely that the history of covert communication started well before Herodotus' time, the art of keeping a message secret is called *steganography*, from the Greek words $\sigma\tau\varepsilon\gamma\alpha\nu o\varsigma$ (covered) and $\gamma\rho\alpha\varphi\varepsilon\iota\nu$ (writing). As opposed to cryptography, the ultimate goal of a covert communication scheme is to hide the very existence of the hidden message. In this case, the most important requirement is the imperceptibility requirement, where imperceptibility assumes a wider sense, in that it is essential that the presence of the message can not be revealed by any means, e.g. through statistical analysis. In steganography, the most important requirement after security (undetectability) is capacity, even if it is obvious that the less information is embedded into the carrier signal, the lower the probability of introducing detectable artifacts during the embedding process.

A covert communication scheme is often modelled by considering the case of a prisoner who wants to communicate with a party outside the prison. To avoid any illegal communication, the warden inspects all the messages sent by the prisoner and punishes him every time he discovers that a secret message was hidden within the cover message (even if he is not able to understand the meaning of the hidden message). Once casted in a statistical framework, the prisoner problem can be analyzed by using tools derived from information theory, and the possibility of always establishing a secure covert channel demonstrated. The capacity of the covert channel can also be calculated. It is important to point out that according to the *prisoner and the warden* model, the prisoner is free to design the host message so to facilitate the transmission of the hidden message, a condition which does not hold in many practical applications where the sender is not allowed to choose the host message.

Despite its ancient origin, and although a great deal of research has been carried out aiming at designing robust watermarking techniques, very little attention has been paid to analyzing or evaluating the effectiveness of such techniques for steganographic applications. Instead, most of the work developed so far has focused on analyzing watermarking algorithms with respect to their robustness against various kinds of attacks attempting to remove or destroy the watermark. However, if digital watermarks are to be used in steganography applications, the detectability of watermark presence must be investigated carefully, since detection by an unauthorized agent would defeat the ultimate purpose of the covert communication channel.

2.6 Further reading

The necessity of considering buyer's rights in addition to those of the seller in fingerprinting-based copy-protection systems was first pointed out by [186]. Such a problem was lately analyzed by N. Memon and P. W. Wong in [149], where they first introduced the IBS copy protection protocol.

The DVD copy protection protocol we briefly discussed in section 2.1.3, is part of a complex system devised by an international pool of consumer electronics companies, to protect the digital distribution of copyrighted video. More details about such a system may be found in [32, 141].

An early formalization of data authentication relying on (semi-)fragile watermarking may be found in [126], whereas for a detailed list of requirements fragile authentication-oriented watermarking must satisfy the reader is referred to [79].

Authentication of video surveillance data through digital watermarking is thoroughly discussed in [23]. In the same paper, the general mathematical framework for data authentication discussed in section 2.2.2 was first introduced.

The notion of compressive data hiding, i.e. the possibility of exploiting data hiding technology to improve coding efficiency, was formalized by P. Campisi, D. Kundur, D. Hatzinakos and A. Neri in [35], where the advantages obtained by hiding the chrominance components of an image within the luminance bit stream are shown. Such a concept, though, was already present in earlier works in which the possibility of hiding the audio component of a video within its visual component was advanced [209].

The possibility of exploiting data hiding to improve the reliability of multimedia transmission in the presence of errors has been explored by several researchers. For some practical examples illustrating the potentiality of such a strategy, readers may refer to [17, 80, 188, 203].

The potentialities of annotation watermarking are still largely unexplored, partly because research was mainly focused on security-oriented applications, partly because for this kind of application watermarking just represents an additional way of solving problems which could be addressed through different technologies as well. Readers interested in this particular kind of application may refer to [68] where a survey of possible applications of annotation watermarks is given, and [65, 198], where the smart image concept is illustrated.

An insightful mathematical formalization of the covert communication problem may be found in the seminal work by C. E. Shannon [197]. For a good survey of covert communication through digital watermarking, the reader may refer to [113].

3

Information coding

According to the paradigm described in chapter 1, the aim of any data hiding scheme consists in the imperceptible embedding of a string of bits, namely the watermark code **b**, within a host asset A. Embedding is achieved by modifying a set of host features $\mathcal{F}(A)$ according to the content of **b**.

In some cases, it is convenient to transform the watermark code **b** in a signal **w**, called the watermark signal, which can be more easily hidden within A. In this case, watermark embedding amounts to the insertion of **w** within $\mathcal{F}(A)$. In detectable watermarking the detection process consists in assessing whereas **w** is contained in A or not. With readable watermarking, a further step is needed, since the bit string **b** must be inferred from **w**. By following the digital communication analogy, the transformation of the information string **b** before its injection within the host asset, can be paralleled to line coding or digital modulation, where the bits to be transmitted are associated to a set of suitable waveform signals which are more easily transmitted through the channel. Unlike in digital communication, though, we will intend information coding in a wider sense, thus letting the transformation from **b** to **w** encompass operations such as channel coding or message repetition as well.

The interpretation of the watermarking process as the insertion of a watermark signal **w** within the host asset, stems from the analogy with digital communication, where the to-be-transmitted signal is first injected within the channel, then it is recovered by the receiver. Though widely used in the early years of watermarking research, such a viewpoint is not the only possible. One could first specify how the detector works, defining exactly the detection region in the asset space, then look at data embedding as the mapping of the host digital asset into a point inside the detection

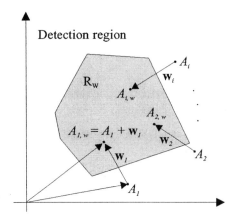

Figure 3.1: Direct embedding paradigm. Given a watermark detection region R_w (grey region) in the feature space, watermark embedding is simply seen as the mapping of the host asset within a point inside R_w. The same action can be seen as the addition of an asset dependent signal w_i (informed embedding).

region. According to this perspective, watermark embedding can be seen as a direct modification of the host features, rather than their mixing with a watermark signal.

Figure 3.1 summarizes the above concepts. Instead of being defined by a signal **w**, the watermark is seen as a detection region R_w in the feature space. Marking a host asset A_i, then reduces to mapping it into a point within R_w (direct embedding). Note that the same action can be seen as the insertion (here exemplified by a vector addition) of an asset-dependent signal w_i within the host asset A_i.

From the point of view of information coding, the procedure described in figure 3.1 can be interpreted in two equivalent ways. According to the former, information coding reduces to the definition of the detection region R_w and to channel coding[1]. Alternatively, the role of the watermark signal **w** may be kept, and information coding seen as the mapping of **b** into an asset-dependent signal **w**. In this book we will use the first viewpoint, in that the choice of the asset-dependent signal w_i will be seen as part of the watermark embedding process, rather than as part of information coding. Accordingly, we will discuss channel coding in this chapter (section 3.4) and leave the mapping of the host asset into a point in R_w to the next chapter. As to the definition of detection regions, the discussion of such a topic is postponed to chapters 4 and 6.

---

[1] In readable watermarking a detection region is defined for each possible watermark code.

In the rest of the book, we will refer to schemes for which watermarking passes through the definition of an asset-independent signal $\mathbf{w}$ as blind-embedding or waveform-based schemes, in contrast to informed embedding schemes which allows the use of an asset-dependent signal $\mathbf{w}_i$. Informed embedding schemes will also be referred to as direct embedding schemes, according to the interpretation that sees the watermarking process as the mere mapping of the host asset A within the proper detection region[2].

With regard to this chapter, information coding in detectable watermarking is discussed in section 3.1, where particular attention is given to waveform-based coding. Readable watermarking is treated in sections 3.2 and 3.3, where waveform-based and direct embedding schemes are considered respectively. Finally, the use of channel coding as a mean to improve the reliability of readable watermarking algorithms is described in section 3.4.

3.1 Information coding in detectable watermarking

In waveform-based detectable watermarking, information coding is straightforward. Let $\mathbf{B} = \{\mathbf{b}_1, \mathbf{b}_2, \ldots \mathbf{b}_{2^k}\}$ be the set of possible watermark codes. The information coding process corresponds to the definition of a set of digital waveforms $\mathbf{W} = \{\mathbf{w}_1, \mathbf{w}_2, \ldots \mathbf{w}_M\}$ ($M \geq 2^k$), and a coding rule Φ that maps each $\mathbf{b} \in \mathbf{B}$ into a distinct element of $\mathbf{W}$. At the same time, watermark detection can be seen as a classical problem of signal detection within noise. For each element of $\mathbf{W}$, the detector defines a detection region and a non-detection region, if the analyzed asset lies within the detection region, the detector decides for the watermark presence, otherwise the watermark absence is established. In the following, we will review a set of different ways of defining $\mathbf{W}$ and Φ. More specifically, we will consider information coding through PN-sequences and orthogonal sequences (sections 3.1.1 through 3.1.3). We will also discuss the use of colored (section 3.1.5) pseudo-random sequences, of periodic self-synchronizing sequences (section 3.1.4), and information coding by means of chaotic sequences (section 3.1.6). The section ends with a brief discussion of direct embedding watermarking, leaving a more detailed description on this kind of techniques to the next chapter.

3.1.1 Spread spectrum watermarking

By relying on the observation that digital communications through very noisy channels, possibly affected by intentional disturbs such as jamming

---

[2]Direct embedding schemes are also referred to as substitutive schemes, since host features are completely replaced by watermarked ones.

or interferences, are usually based on spread spectrum technology, many of the watermarking algorithms developed so far use a similar technique to code the to-be-hidden information. To be more specific, each message $\mathbf{b} \in \mathbf{B}$ is transformed into a pseudo-random sequence $\mathbf{w}$ of proper length n (in agreement with the spread spectrum paradigm, n is usually much larger than k). In this case, w_i's are random variables drawn from a given probability density function (pdf) $p_{w_i}(w)$[3]. In most of the cases w_i's are identically distributed variables, that is $p_{w_i}(w) = p_w(w) = p(w)$, where we omitted the subscript w for sake of simplicity. As to the choice of $p(w)$, possible solutions include a normal pdf, $N(0, \sigma^2)$, for which:

$$p(w) = \frac{1}{\sqrt{2\pi\sigma^2}} \exp\left(-\frac{w^2}{2\sigma^2}\right) ; \tag{3.1}$$

a uniform pdf:

$$p(w) = \begin{cases} \frac{1}{2\alpha} & w \in [-\alpha, \alpha] \\ 0 & w \notin [-\alpha, \alpha] \end{cases} \tag{3.2}$$

and a bipolar pdf, for which w_i's take value $+\alpha$ or $-\alpha$ with equal probability. In some applications, the bipolar distribution is conveniently replaced by a three-valued distribution:

$$w = \begin{cases} -\alpha & \text{with probability} \quad p \\ 0 & \text{with probability} \quad 1 - 2p \\ +\alpha & \text{with probability} \quad p \end{cases} \tag{3.3}$$

Each of the above solutions presents a number of advantages and drawbacks, however no definitive arguments exist demonstrating the superiority of one pdf over the others. Sometimes, it is necessary that the watermark signal be limited, e.g. to ensure invisibility, to exactly control the maximum modification of host features, or to avoid that the host features change sign thus loosing their meaning[4]. In this case, the normal pdf can not be used, thus limiting the choice to the uniform and the bipolar distributions. A drawback with the bipolar distribution is that, for a given peak distortion, it results in a higher average distortion than the uniform distribution, due to its larger variance. On the other side, it also ensures a higher transmission rate, in that the bipolar pdf is the capacity-achieving distribution under a peak distortion constraint.

A problem with discrete valued watermarks is their weakness against the collusion attack. In the collusion attack, it is assumed that t copies of the same asset, each marked with a different watermark, are available[5].

---

[3]We use the same symbol w_i both to indicate the random variable and the values it takes, the exact meaning being easily recoverable from the specific context the symbol is used in.

[4]This is the case, for example, of multiplicative, DFT-domain, watermarking.

[5]A situation easily occurring in fingerprinting applications.

The attacker tries to remove the watermark by averaging the t assets in his/her hands. One possible defense against the collusion attack consists of making it unfeasible by trying to increase the minimum number of copies necessary to remove the watermark as much as possible. This is not the case with a binary valued watermark. Suppose, for example, that watermark insertion is achieved by either adding α or $-\alpha$ to f_i. Then, an attacker only needs to find out two documents in which f_i takes different values, and averaging them, since in this way the watermark is completely erased. It is clear that the use of continuous valued watermarks can give greater robustness to this kind of attack. More specifically, it is found that the normal distribution greatly outperforms the uniform one, since a much larger number of copies are needed to effectively remove the watermark. Such a behavior can be explained by noting that the collusion attack may be thought of as a problem of signal estimation in noise, where the host coefficient is the constant-valued signal and the various instances of w_i represent noise. It is known that, at least for the additive case, the gaussian noise leads to the worst estimation accuracy.

¿From a general point of view, it is essential that $p(w)$ has a zero mean, since it is known from digital communication theory that in this way the transmission power can be minimized without increasing the error probability. In a data hiding scheme this results in a lower distortion, for a given level of robustness. Moreover, some detection schemes explicitly exploit the knowledge that the expected value of w_i, $\mu_{w_i} = E[w_i]$, is equal to zero.

In addition to being identically distributed, watermark coefficients are usually designed so to be independent of each other, i.e. w_i's are independent random variables. This leads to a white watermark, in that the power spectrum of the watermark signal is a flat one. Though very popular, the adoption of a white watermark signal is not always the best choice. In some cases, in fact, it is preferable to accurately shape the watermark spectrum in order to make it more robust and less perceptible at the same time. Power shaping of $\mathbf{w}$ is treated in more detail in section 3.1.5.

Continuous pseudo-random sequence generation

According to the above ideal formulation of spread spectrum watermarking, it is necessary that a sequence $\{w_1, w_2 \ldots w_n\}$ of random numbers following a given pdf is generated. This is known to be a very hard problem, that has received considerable attention due to its importance in many different fields including computer simulations, Monte Carlo methods, and CDMA[6] digital communication.

---

[6]Code Division Multiple Access.

Let us consider first the continuous case. Our aim is to generate a sequence of random numbers (RN sequence) with a given, continuous, distribution. In the present state of the art, however, this can not be done directly. On the contrary, an RN sequence $\mathbf{z}$ uniformly distributed in an interval $(0, m)$ is first generated, then, a new RN sequence having the desired pdf is obtained by properly operating on $\mathbf{z}$. Note that $\mathbf{z}$ is not truly continuous, since it only takes integer values (all the integers in $(0, m)$), nevertheless if m is sufficiently large, an RN sequence which closely approximates an RN sequence uniformly distributed in $(0, 1)$ can be obtained by dividing $\mathbf{z}$ by m:

$$u_i = \frac{z_i}{m}. \tag{3.4}$$

The most general algorithm for generating an RN sequence has the following form:

$$z_i = g(z_{i-1} \ldots z_{i-r}) \mod m, \tag{3.5}$$

where $g(z_{i-1} \ldots z_{i-r})$ is a function depending on the last r values of the RN sequence. Note that, due to the presence of the mod operator, z_i is the remainder of the division of $g(z_{i-1} \ldots z_{i-r})$ by m. The simplest and one of the most effective RN generators is the Lehmer's algorithm based on the following recursion formula:

$$\begin{cases} z_i = a z_{i-1} \mod m, \quad i \geq 1 \\ z_0 = 1 \end{cases} \tag{3.6}$$

where m is a large prime number and a is an integer. Alternatively, equation (3.6) can be written as

$$z_i = a^i \mod m. \tag{3.7}$$

Due to the mod operator, the sequence z_i takes values between 1 and $m - 1$; hence after m steps at least two equal values are found, thus permitting us to conclude that z_i is a periodic sequence with period $m_0 \leq m - 1$. Of course, the periodic nature of z_i does not agree with the randomness requirement, nevertheless, if the required number of samples is lower than m_0, periodicity is not a problem. To take into account the imperfect randomness of z_i, RN sequences generated through mathematical formulas such as those expressed in equations (3.5) through (3.7) are usually referred to as pseudo-random or pseudo-noise sequences (PN sequences for short). A possible choice of m, suggested by Lehmer in 1951, is $2^{31} - 1$, leading to a period m_0 which is large enough for most practical applications. To complete the specification of the Lehmer generator, the value of the multiplier a must be chosen. A first consideration leads to searching for a number a which maximizes the period of z_i, i.e. $m_0 = m - 1$. To do so, let us introduce the notion of primitive root of m.

Definition: *Given a number m, an integer a is said the* primitive root *of m, if the smallest n such that*

$$a^n = 1 \mod m, \tag{3.8}$$

is $n = m - 1$.

It can be shown that each prime number has at least one primitive root[7]. It is obvious from the above definition, that $m_0 = m - 1$ if and only if a is a primitive root of m. Unfortunately, letting a be a primitive root of m only ensures that the period of z_i is $m - 1$, but it does not guarantee that z_i is a good PN sequence. Without going into much details about the definition of what a *good* PN sequence is, we can say that a good PN sequence should pass a number of randomness tests aiming at assessing whether the characteristics of the sequence agree with those of an ideal RN sequence. For a selection of the specific randomness tests z_i has to be subjected to, the particular application served by z_i must be considered. An example of a good choice of a and m, which has been used effectively in a variety of applications, is:

$$a = 16807, \quad m = 2^{31} - 1 = 2147483647. \tag{3.9}$$

In the framework of the data hiding scenario, the most important properties of z_i are the adherence to the uniform distribution, and the lack of correlation between subsequent samples. In many applications it is also necessary that a sufficiently large number of sequences can be generated. This follows directly from the condition that the number K of watermark signals must be larger than 2^k. The generation of all the sequences $\mathbf{w}_i \in \mathbf{W}$ by starting from the Lehmer algorithm goes through the observation that when a is chosen in such a way that $m_0 = m - 1$, the initial state of the generator z_0 can assume any integer value in $[1, m - 1]$, with different choices resulting in sequences that are cyclically shifted version of the same sequence. If n is much smaller than m, then the sequences $\mathbf{w}_i$'s can be obtained by running Lehmer's algorithm several times, each time by varying z_0. For example, a common way to vary z_0 consists in generating it randomly, e.g. by using the computer clock. Of course, the larger the number of sequences, the higher the probability that two sequences are generated exhibiting large cross-correlation values, possibly invalidating the effectiveness, and security, of the whole data hiding system. Alternatively, z_0 could be varied systematically. Suppose, for example, that the length n of the watermark signal is 16,000, and that the Lehmer generator is used with

---

[7]Actually it can be shown that if a is a primitive root of m and b is prime with respect to $m - 1$, then a^b is a primitive root of m, hence demonstrating that each prime number has an infinite number of primitive roots

$m = 2^{31} - 1$. The generator can be run $m/n \approx 128,000$ times, each time producing a completely different sequence, thus accommodating approximately 2^{17} bit-strings.

By relying on the pseudo-random sequence u_i, having a uniform distribution in $(0, 1)$, a random sequence w_i following any desired pdf can be built. For example, to generate a uniform sequence taking values in any finite interval (a, b) it only needs to compute $w_i = u_i(b - a) + a$. The construction of a normally distributed sequence is more tricky. Due to the importance that the normal distribution has in many applications, several methods have been developed to build a normal PN sequence by starting from a uniform one. We will review the most popular ones, namely the central-limit method and the Box-Muller method.

Given l independent identically distributed random variables $x_1, x_2 \ldots x_l$, having mean μ_x and variance σ_x^2, it is known from the central limit theorem[8] that, for $l \to \infty$, the random variable:

$$y_l = \frac{1}{\sqrt{l}} \sum_{i=1}^{l} \left(\frac{x_i - \mu}{\sigma} \right) \tag{3.10}$$

tends to the standardized normal distribution, i.e.:

$$\lim_{l \to \infty} p(y_l) = N(0, 1). \tag{3.11}$$

The above equations suggest that if we take l pseudo-random numbers uniformly distributed in $(0, 1)$ and we consider

$$y_l = \sqrt{\frac{12}{l}} \sum_{i=1}^{l} \left(x_i - \frac{1}{2} \right), \tag{3.12}$$

then, for l large enough (commonly $l = 12$ is used), y_l can be considered to be normally distributed with zero mean, and unitary variance. A drawback with the above method is that computing time is rather high, since only one normal sample is obtained from 12 uniformly distributed numbers. Additionally, the normal sequence produced by equation (3.12) takes values in $(-l, l)$ thus failing to approximate the normal distribution for large values of y.

The Box-Muller algorithm represents a valuable alternative to the usage of the central limit theorem. This method, originally introduced by G.E.P.Box and M.E.Muller in 1958, relies on the following observation.

---

[8]Actually the central limit theorem does not require that x_i's are identically distributed, however in practical applications such a condition is always satisfied.

Given two random variables u and v uniformly distributed in $(0,1)$, the random variables

$$x = (-2\ln u)^{\frac{1}{2}} cos 2\pi v$$
$$y = (-2\ln u)^{\frac{1}{2}} sen 2\pi v$$

(3.13)

follow a standardized normal distribution $N(0,1)$[9]. Then, to generate a normally distributed sequence, it only needs to generate two uniformly distributed sequences, and use equations (3.13), to build the normal random variables x and y. The Box-Muller method is usually preferred to methods based on the central limit theorem, because of its lower complexity and, most of all, because it produces a sequence which better approximates the normal distribution.

Binary PN sequences

A binary PN sequence can be easily obtained by starting from a continuous pseudo-random number generator. For example, if a PN sequence u_i which is uniformly distributed in $(0,1)$ is available, one can build a new sequence w_i by letting:

$$w_i = \begin{cases} +1 & if \quad u_i \geq 0.5 \\ -1 & if \quad u_i < 0.5 \end{cases}$$

(3.14)

The sequence w_i is a bipolar i.i.d. sequence taking value -1 or $+1$ with equal probability.

The use of pseudo-random generators ensures that the desired characteristics of the watermark signal are satisfied on a probabilistic basis.

A possible alternative to the use of a continuous pseudo-random number generator consists in the usage of maximum length sequences (m-sequences for short).

Maximum length sequences are generated by means of an m-stage shift register with feedback, as exemplified in figure 3.2. Due to the finite number of states of the shift register, the output sequence is a periodic one, with maximum period equal to $2^m - 1$, where m is the number of stages of the shift register. The actual period of the output sequence depends on feedback connections. When the connections are chosen in such a way that the output period is maximum, the sequence of bits produced by the shift register is called a maximum-length sequence. The configurations of feedback connections ensuring a maximum period can be derived by relying on cyclic coding theory and Galois fields and have been extensively studied for digital communication applications. In Table 3.1, possible shift register

---

[9]A proof can be found in [166]

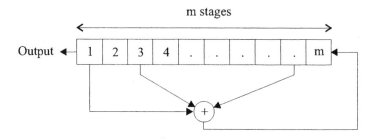

Figure 3.2: M-sequences generation through m-stage shift register with linear feedback.

connections for generating m-sequences with m ranging from 2 through 34 are shown. Of course, the output of an m-stage shift register such as that depicted in figure 3.2, takes value in $\{0, 1\}$. However, it is trivial to map this sequence into a bipolar sequence with elements in $\{-1, 1\}$. From now on, when speaking about m-sequences we will always mean the bipolar sequence associated to the $\{0, 1\}$ sequence at the output of the shift register.

Table 3.1: Possible shift register connections for generating m-sequences of different lengths.

m	Connected stages	m	Connected stages	m	Connected stages
2	1,2	13	1,10,11,13	24	1,18,23,24
3	1,3	14	1,5,9,14	25	1,23
4	1,4	15	1,15	26	1,21,25,26
5	1,4	16	1,5,14,16	27	1,23,26,27
6	1,6	17	1,15	28	1,26
7	1,7,	18	1,12	29	1,28
8	1,5,6,7	19	1,15,18,19	30	1,8,29,30
9	1,6	20	1,18	31	1,29
10	1,8	21	1,20	32	1,11,31,32
11	10	22	1,22	33	1,21
12	1,7,9,12	23	1,19	34	1,8,33,34

Maximum length sequences possess a number of interesting properties resembling those of ideal RN sequences. First, each period has exactly 2^{m-1} positive and $2^{m-1} - 1$ negative coefficients, thus ensuring that the sequence average is very close to 0 (the average value of a bipolar m-

sequence exactly equals $1/(2^m - 1)$. Ideally a PN sequence should also have correlation properties that are similar to those of white noise. That is the autocorrelation $R_w(l)$ should be L for $l = 0$ and 0 for $1 \leq l \leq L - 1$, with $L = 2^m - 1$ denoting the sequence period. In the case of m sequences, we have

$$R_w(l) = \begin{cases} L & if \ l = 0 \\ -1 & if \ 1 \leq l \leq L - 1 \end{cases} \tag{3.15}$$

It is evident that for large values of m m-sequences are very close to ideal RN sequences from the autocorrelation point of view. A set of different m-sequences having length m can be generated by using different feedback connections. However, the number of possible m-sequences for a given m is not unlimited, as it is shown in table 3.2. System designers must take carefully into account the limits expressed in table 3.2, since they directly impact the maximum number of different watermarks that can be accommodated by the system.

Table 3.2: Number of different m-sequences for several values of m.

m	number of sequences	m	number of sequences
3	2	8	16
4	2	9	48
5	6	10	60
6	6	11	176
7	18	12	144

In most cases, the cross-correlation properties of different watermark signals are as important as the autocorrelation properties. For example, when two or more watermarks have to be inserted within the same asset, it is desirable that the two watermarks do not interfere each other, a con-

Table 3.3: Peak cross-correlation of m-sequences.

m	$R_{max}/R(0)$	m	$R_{max}/R(0)$
3	0.71	8	0.37
4	0.60	9	0.22
5	0.35	10	0.37
6	0.36	11	0.14
7	0.32	12	0.34

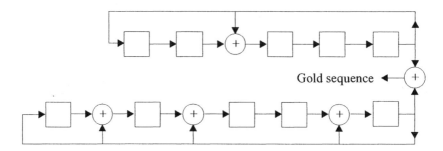

Figure 3.3: Shift register arrangement for the generation of Gold sequences of length 31.

dition that is usually met if the PN sequences corresponding to different watermarks are mutually uncorrelated. Unfortunately, m-sequences do not meet this constraint, since they may exhibit a rather high cross-correlation. As it can be seen by observing the values reported in table 3.3, the max value of $R(l)$ for $l \neq 0$, say R_{max}, of the cross-correlation function may be rather high (30% of the peak value, i.e. $R(0)$, on the average).

A possible solution consists in selecting a sub-set of all possible m-sequences that have a smaller cross-correlation, however in this way the number of admissible watermarks may become too small. Gold sequences are a valid alternative to m-sequences. Gold sequences are generated by selecting a pair of m-sequences, called *preferred m-sequences*, and summing them modulo-2, as depicted in figure 3.3.

If m is odd, the maximum value of the cross-correlation between any two pairs of sequences is $R_{max} = \sqrt{2L}$, whereas for m even, we have $R_{max} = \sqrt{L}$. Given a set of M binary sequences of period L, it is known that a lower bound on their maximum cross-correlation is given by

$$R_{max} \leq L\sqrt{\frac{M-1}{LM-1}}, \tag{3.16}$$

then, for large values of L and M, Gold sequences are almost optimal.

Another advantage of Gold sequences with respect to m-sequences, is their abundance. For a fixed length L, in fact, up to $L + 2$ different Gold sequences can be generated (see table 3.4), a number that greatly outperforms the number of possible m-sequences of the same length.

3.1.2   Orthogonal waveforms watermarking

Though the use of Gold sequences improves the cross-correlation properties of binary PN-sequences, better results can be obtained by using watermarking sequences which are expressly designed to be orthogonal of each other.

Table 3.4: Number of different Gold sequences for different values of m.

m	number of sequences	m	number of sequences
3	9	8	257
4	17	9	513
5	33	10	1025
6	65	11	2049
7	129	12	4097

In this way, the cross correlation between different signal is equal to zero and the interference between two watermarks which are simultaneously present in the same asset minimized.

A solution that is commonly adopted consists in letting **W** coincide with the set of Walsh-Hadamard sequences of length n.

Walsh sequences may be generated in many different ways, however, the easiest passes through the definition of Hadamard matrices (hence the name Hadamard-Walsh sequences). Hadamard matrices, are squared matrix, whose possible orders are limited to the powers of two, hence Walsh sequences will be limited to lengths of $n = 2^m$. Hadamard matrices are recursively obtained by starting from the lowest order matrix, defined by:

$$H_2 = \begin{bmatrix} 1 & 1 \\ 1 & -1 \end{bmatrix} \qquad (3.17)$$

Higher order matrices are obtained through the recursive relationship

$$H_n = H_{n/2} \otimes H_2; \qquad (3.18)$$

where $\otimes$ indicates the Kronecker matrix product, whereby each element of the matrix $H_{n/2}$ is multiplied by the matrix H_2. For instance, for $n = 4$ and $n = 8$, we have:

$$H_4 = \begin{bmatrix} H_2 & H_2 \\ H_2 & -H_2 \end{bmatrix} \cdot = \begin{bmatrix} 1 & 1 & 1 & 1 \\ 1 & -1 & 1 & -1 \\ 1 & 1 & -1 & -1 \\ 1 & -1 & -1 & 1 \end{bmatrix}; \qquad H_8 = \begin{bmatrix} H_4 & H_4 \\ H_4 & -H_4 \end{bmatrix}.$$

$$(3.19)$$

The difference between Walsh sequences and Hadamard matrices only consists in the order in which the codes appear. The rows in the Hadamard matrices are first re-ordered according to the number of zero-crossings in each row: Walsh sequences correspond to the rows of these re-ordered matrices. For instance, for $n = 4$, Walsh sequences correspond to the rows of the matrix:

$$W_4 = \begin{bmatrix} 1 & 1 & 1 & 1 \\ 1 & 1 & -1 & -1 \\ 1 & -1 & -1 & 1 \\ 1 & -1 & 1 & -1 \end{bmatrix}. \tag{3.20}$$

Due to orthogonality, in a perfectly synchronized environment, interference between Walsh sequences is zero[10].

To increase the number of sequences (and hence the number of admissible users) and the average distance between signals in $\mathbf{W}$, bi-orthogonal sequences may be used, where each orthogonal signal in $\mathbf{W}$ is further modulated by a binary antipodal symbol ± 1. More specifically, given a set of n orthogonal sequences $\mathbf{W}$, a set $\mathbf{W}'$ consisting of $2n$ bi-orthogonal sequences is built as follows:

$$\begin{cases} \mathbf{w}'_i = \mathbf{w}_i & i = 1 \dots n \\ \mathbf{w}'_i = -\mathbf{w}_{i-n} & i = n+1 \dots 2n \end{cases} \tag{3.21}$$

The use of bi-orthogonal sequences permits to double the number of watermarking signals with performances which are very close to those of orthogonal sequences.

3.1.3 Orthogonal vs PN watermarking

In the attempt to summarize the pro's and con's of PN and orthogonal watermarking, we will take into account several points of view, referring to different practical scenarios.

Single user watermarking

It is well known from digital communications theory that for single user scenarios, no advantage is gained from the use of spread spectrum modulation, unless the possible presence of a jammer is considered (see below in the text). On the contrary, it is advisable to deterministically design the watermarking signals in such a way that they are as far apart as possible, thus minimizing the possibility that the presence of the wrong watermark is erroneously detected. From this perspective, the use of bi-orthogonal sequences is the most advisable solution.

---

[10]Among the drawbacks of Walsh sequences it has to be mentioned that for these sequences autocorrelation and cross-correlation sidelobes have considerably larger magnitudes than those of Gold and PN sequences.

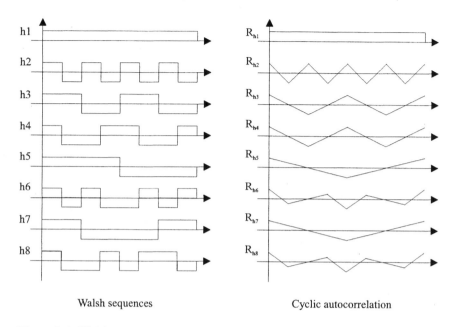

Figure 3.4: Walsh sequences of order 8 with corresponding cyclic autocorrelations.

Non-synchronized watermarking

The above conclusion does not hold when the possibility that the watermark embedder and the watermark detector are not synchronized. As we will see later on in the book, such a situation is very common, when the watermarked asset is processed, either by a malevolent or a non-malevolent user.

Due to their peaked autocorrelation function, PN sequences provide a simple mean to recover the synchronism between embedder and detector. In most cases, it only needs to compute the correlation between the searched watermark and the watermarked asset. A pronounced peak in the correlation, in fact, indicates that the watermark contained within the asset and the one the detector is looking for are synchronized.

In general, this is not possible with orthogonal watermarking sequences, since in this case the autocorrelation function is not necessarily peaked. Let us consider, for example, the Walsh sequences depicted in figure 3.4 together with their autocorrelation. As it can be seen, in this case the autocorrelation functions do not present a unique, well pronounced, peak, thus making synchronization more cumbersome. Actually, this result was expected, since the good synchronization properties of PN sequences, ultimately, relies on the spreading property (wide spectrum) of these sequences, a property which does not hold for orthogonal sequences, such as the Walsh-Hadamard ones.

Multiple watermarking

When the possibility that more than one watermark is embedded within the same asset is taken into account, the choice of the watermarking sequence is more critical[11]. Such a situation resembles a classical multiple access scenario, where the same communication medium is shared between different users. By relying on digital communication theory, we can divide multiple access techniques into two main categories: orthogonal waveform multiple access (OWMA), including Frequency Division Multiple Access (FDMA), Time Division Multiple Access (TDMA) and Orthogonal-Code Division Multiple Access (OCDMA), and Pseudo-Noise Code Division Multiple Access (PN-CDMA), including conventional direct sequence and frequency hopping CDMA. In the watermarking framework, it is readily seen that the use of orthogonal watermarking sequences may be paralleled to OWMA (OCDMA in particular), whereas spread spectrum watermarking relies on the operating principle of PN-CDMA.

When the number of users that can be accommodated on a given channel is taken into account, the two different approaches show a different behavior. In OWMA no interference is present, up to a number of n users, which is the hard limit on the maximum number of users admitted by the system. In PN-CDMA, interference appears as soon as two active users are present, however no hard limit on the maximum number of users exists, such a limit depending on the desired performance and the allowed Signal to Interference Ratio (SIR). As a matter of fact, this is the reason why PN-CDMA is often used in overloaded environments, but it is only rarely adopted in situations where the number of simultaneous users is lower than n. Apart from the above general principles, it must be pointed out that in some cases it is not easy to generate a large number of spread sequences having a *reasonably* low cross-correlation, thus limiting the number of users that can be practically accommodated by the system. For example, if Gold sequences are used, the maximum number of users that is allowed is roughly the same as for orthogonal sequences ($n - 1$ vs $n + 1$). This is not the case for continuous-valued signals, where the number of spread sequences is much larger than n, which is the maximum number of possible, continuous-valued, orthogonal sequences of length n.

---

[11]Multiple watermarking may be due to the explicit need that two or more watermarks are embedded within the same asset, or to the action of a pirate embedding a false watermark to fool the watermarking system either at a signal or at a protocol level.

Watermarking in hostile environments

A (heuristic) rationale for using a PN-CDMA comes from the original applications such a technology was devised for. Actually, PN-CDMA derived directly from direct sequence spread spectrum systems originally developed for military communications system. More specifically, three properties of spread spectrum communication, make it suitable for the hostile environments typical of military applications: discreteness, low intercept probability and robustness against jamming. When considering the peculiarities of security-oriented applications of watermarking, it immediately turns out that the same properties make PN-CDMA an ideal technology for a watermarking system to rely on. Of course, many important differences exist between watermarking and secure digital communications, nevertheless, the intrinsic security of PN-CDMA technology suggests that some advantages may be got from its use in terms of robustness and overall system security.

A summary of the main properties of PN- and orthogonal-sequence watermarking is given in table 3.5.

Table 3.5: Comparison between PN and orthogonal sequences. In both cases n indicates the sequence length and m the number of simultaneous watermarks possibly embedded within the host asset.

	Spread spectrum watermarking	Orthogonal sequence watermarking
Multiple access $(m < n)$	Interference grows linearly with m	No interference
Multiple access $(m > n)$	Interference grows linearly with m	Not possible
Synchronization	Easy due to wide spectrum	Cumbersome
Robustness/security	Good security and robustness (not theoretically proved)	Lack of security

In order to overcome some of the limitations of orthogonal sequence watermarking, this approach may be combined with spread spectrum. To be more specific, let $\mathbf{w}$ be a binary antipodal spread spectrum sequence and $\{\mathbf{h}_i\}_{i=1}^{n}$ be a set of orthogonal sequences, e.g. the Walsh-Hadamard sequences. By starting from $\mathbf{w}$ and $\mathbf{h}_i$ a new set of sequences, $\{\mathbf{q}_i\}_{i=1}^{n}$, is built by multiplying each sequence $\mathbf{h}_i$ by $\mathbf{w}$, i.e.:

$$\mathbf{q}_i = \mathbf{h}_i \cdot \mathbf{w}. \tag{3.22}$$

Note the the sequences $\{\mathbf{q}_i\}$ are still orthogonal since

$$\langle \mathbf{q}_i, \mathbf{q}_j \rangle = \sum_{k=1}^{n} h_{i,k} h_{j,k} w_k w_k = \sum_{k=1}^{n} h_{i,k} h_{j,k} = E_h \delta(i-j), \tag{3.23}$$

where by $\delta(\cdot)$ the Kronecker impulsive function is meant[12], and E_h is the energy of the sequence $\mathbf{h}$. The new set of sequences $\mathbf{q}_i$ has now a more peaked autocorrelation function and is more secure than $\mathbf{h}_i$ since its knowledge is constrained to the knowledge of the spreading sequence $\mathbf{w}$. It still remains valid, though, the the maximum number of users that can be accommodated through orthogonal signalling is upper bounded by the sequence length n.

3.1.4   Self-synchronizing PN sequences

As it will be discussed later on in the book (chapter 7), loss of synchronization in one of the most serious threat to watermark retrieval. Though the watermark is virtually intact, the detector is not able to recover it, since it does not know its exact geometric configuration, e.g. its position in the host asset or its scale factor. A possible way to recover the synchronism between the encoder and the detector, consists in the use of auto-synchronizing watermark signals. In the most common case, an auto-synchronizing watermark is nothing but the periodic repetition of the same PN signal[13]. Note that the exact nature of the periodic repetition depends on the host asset media, thus in the audio case we have a 1D periodic repetition, whereas in the image case, repetition has to be intended in a 2D sense. The periodic nature of the watermark may be exploited to infer important geometric properties of the watermark such as orientation[14] or scale. To exemplify how watermark periodicity can be exploited to estimate the watermark scale, let us consider the simple case of a zero-mean, mono-dimensional watermark which is added to a host audio asset S. Also assume that watermark embedding is performed in the time domain, i.e. the set of host features corresponds to the audio sample s_i's. We have:

$$S_\mathbf{w} = S + \mathbf{w}, \tag{3.24}$$

---

[12]$\delta(k) = 0$ for $k \neq 0$, and $\delta(k) = 1$ for $k = 0$.

[13]Orthogonal Walsh-Hadamard sequences can not be used here, since their autocorrelation is not sufficiently peaked.

[14]Of course, watermark orientation only makes sense in the image case.

or, alternatively:

$$s_{w,i} = s_i + w_i, \tag{3.25}$$

where $s_{w,i}$ denotes the i-th watermarked audio sample. If S and $\mathbf{w}$ are assumed to be uncorrelated, the autocorrelation function of $S_\mathbf{w}$ is:

$$R_{S_\mathbf{w}}(l) = R_S(l) + R_w(l). \tag{3.26}$$

By assuming that S is not periodic and that the period L of the watermark signal is much lower than the cardinality of S, $R_{S_\mathbf{w}}(l)$ will exhibit a set of periodic peaks corresponding to the periodic peaks of $R_w(l)$ (see figure 3.5). The distance between such peaks is a clear indication of the period of $\mathbf{w}$. By comparing the original period of $\mathbf{w}$ and that estimated by looking at the autocorrelation function, an estimate of the current watermark scale with respect to the original one can be obtained, thus making watermark detection easier.

3.1.5 Power spectrum shaping

The power spectrum of the watermark signal, affects the performance of a data hiding system in a twofold way: first it determines the perceptibility of the watermark, second it impacts watermark robustness. Unfortunately, a direct dependency between the watermark and the signal power spectrum can not be easily established without taking into account the exact embedding rule. Consider, for example, the case of an additive watermark embedded in the time domain of an audio asset S. In this case, the watermark power spectrum directly modifies the power spectrum of S. If we assume that $\mathbf{w}$ and S are independent of each other, in fact, the power spectrum of the watermark and that of S sum together to form the power spectrum of the watermarked asset. It is rather easy, then, to analyze the influence of the power spectrum shape on watermark audibility and robustness. This is not the case if a multiplicative embedding rule is adopted, since the power spectrum of the resulting watermarked asset is a complex combination of the spectra of S and $\mathbf{w}$ (also involving spectrum phase). Similar considerations hold if the set of host features does not coincide with plain asset samples. If watermark insertion is performed in the frequency domain, for example, the shape of the spectrum of the watermarked asset depends on the frequency location of host features, regardless of the power spectrum of $\mathbf{w}$.

In spite of the above difficulties, properly shaping the power spectrum of $\mathbf{w}$ has received considerable attention in the technical literature, both because of its impact on watermark robustness, and because of the importance of time/spatial domain watermarking. If early research focused on

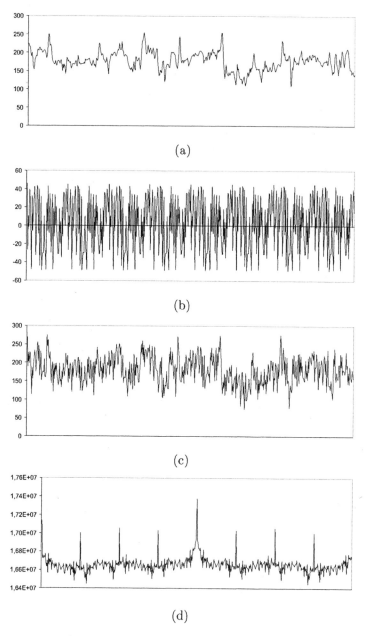

Figure 3.5: Watermark self-synchronization. A periodic watermark (b) is added to the original signal (a), thus producing a watermarked signal with a periodic component (c). Such a periodic component produces a set of equally spaced peaks in the autocorrelation function (d).

white watermarks, as a direct extension of spread spectrum communication, it soon became evident that a superior robustness (and a lower perceptibility) could be achieved by adopting colored watermark signals. Thus some authors used high-frequency watermarks, since they can be more easily separated from the host signal. Others proposed to embed watermarks which are *perceptually similar to the host asset*, pointing out that in this way watermark masking is more easily achieved and that an attacker can not modify the watermark without severely degrading the host asset as well. Eventually, by relying on the observation that low-pass watermarks may result too perceptible and that high-pass watermarks are too susceptible to attacks, it has been proposed to use band-pass watermarking signals. As a matter of fact, when a rigorous analysis of the trade-off between perceptibility, robustness and capacity is carried out, the necessity of carefully shaping the watermark power spectrum is confirmed. The optimal shape of $\mathbf{w}$, however, heavily depends on the possible attacks, the embedding rule, and the distortion metric used to measure asset degradation as a consequence of attacks and watermark embedding. Possible choices range from a white watermark to a PSC-compliant[15] watermark, where the shape of the watermark power spectrum is derived through complex optimization procedures aiming at maximizing a given measure of the performance of the data hiding system (e.g. watermark capacity or probability of correct detection). In most of the cases, though, the heuristic rule leading to a band-pass watermark is confirmed.

Direct generation of PN sequences having a desired power spectrum, or equivalently a desired autocorrelation function, is not easy to achieve. To get around the problem, a white PN sequence is usually generated, then the white sequence is linearly filtered to obtain the desired autocorrelation characteristics. Note, however, that due to filtering, the pdf of the PN sequence is changed, unless the PN sequence is normally distributed, thus leading to possible difficulties in applications where both the power spectrum and the watermark pdf are exactly specified.

3.1.6   Chaotic sequences

Though the use of pseudo-random sequences is by far the most common way to generate the watermarking signal $\mathbf{w}$, some alternative possibilities have been proposed. Among them, chaotic sequences have received a considerable attention for their good properties in terms of unpredictability, power spectrum design, and ease of use. More specifically, watermarking signals generated through n-way Bernoulli shift maps, seem to present some

---

[15]Power Spectrum Condition compliant

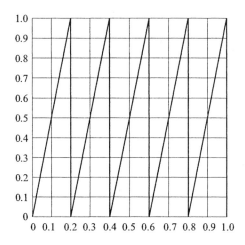

Figure 3.6: Bernoulli shift map with $n = 5$.

advantages with respect to conventional PN sequences[16].

An n-way Bernoulli shift $B_n(x)$ is defined by the following expression:

$$x' = B_n(x) = nx \quad \mod 1 \tag{3.27}$$

where n is an integer. Bernoulli shifts generate a piecewise affine chaotic map defined in the $[0,1]$ interval (see figure 3.6, for an example of a Bernoulli shift map with $n = 5$).

A Bernoulli chaotic watermark is easily generated though the iterative applications of $B_n(x)$, as described by the following recursive equation:

$$w_{i+1} = B_n(w_i) = nw_i \quad \mod 1. \tag{3.28}$$

Note that despite the similarity with PN sequences, chaotic sequences possess rather different properties due to the continuous nature of equations (3.27) and (3.28), whereas the generators at the basis of PN sequences deal with integer numbers[17]. The watermark sequence generated through equation (3.28) heavily depends on the sequence starting point w_0. More specifically, if w_0 is an irrational number the sequence exhibits a chaotic non-periodic behavior. If w_0 is chosen randomly, the sequence itself can

---

[16]Any claim about the superiority of chaotic sequences with respect to PN sequences, though, must be checked with particular care, since the current state of the art in watermarking is not mature enough to draw any ultimate conclusions on the matter.

[17]Of course, the practical implementation of Bernoulli shifts must rely on finite precision arithmetic, nevertheless if double precision, floating point, numbers are used, the continuous assumption may be assumed to hold

be regarded as a random sequence, whose characteristics depend on the particular choice of n. From a watermarking perspective, w_0 corresponds to the watermark key K, thus playing a role similar to that played by the seed used to initialize the PN generator at the basis of PN watermarking. It can be shown that the uniform distribution is an invariant pdf for the n-way Bernoulli shift maps, in that if w_0 is uniformly distributed, then w_i is uniformly distributed too. As a consequence $\mu_{w_i} = 0.5$ regardless of i. Additionally, it can be shown that the autocorrelation function of $\mathbf{w}$, $R_{ww}(i+k, i)$ does not depend on i, since we have:

$$R_{ww}(i+k, i) = E[w_{i+k} w_i] = R_{ww}(k) = \frac{1}{12n^k}, \quad k \geq 0. \tag{3.29}$$

We can conclude that the watermark signal generated by n-way Bernoulli shift maps are wide-sense stationary.

Since it is usually required that the watermarking signal is zero mean, the mean value 0.5 of the sequence generated through equation (3.28) must be subtracted, thus leading to a map defined in the $[-0.5, 0.5]$ interval. The possibility of controlling the autocorrelation function of a chaotic watermark by simply varying n, provides a useful way to shape the watermark power spectrum without changing the watermark pdf. This property may be used, for example, to generate a PSC-compliant watermark or to trade-off between the properties of white and colored watermarks. To be more specific, let us consider equation (3.29) in more details. It is readily seen that for small values of n and k, w_{i+k} and w_i exhibit a high correlation, however. As n increases, $R_{ww}(k)$ approximates a Dirac delta function, that is, the watermark approximates a white random watermark. Given $R_{ww}(k)$, the power spectral density $S_{\mathbf{w}}(f)$ of Bernoulli chaotic watermarks can be easily evaluated. More specifically, by properly exploiting the symmetry of $R_{ww}(k)$ around $k = 0$, we have:

$$S_{\mathbf{w}}(f) = \sum_{k=-\infty}^{\infty} R_{ww}(k) e^{-j2\pi fk} = R_{ww}(0) + \sum_{k=1}^{\infty} R_{ww}(k) \left(e^{j2\pi fk} + e^{-j2\pi fk} \right)$$

$$= \frac{n^2 - 1}{12(n^2 - 2n \cos(2\pi f) + 1)}, \tag{3.30}$$

with $f \in [-0.5, 0.5]$. Note again that for small values of n, a lowpass spectrum is obtained, whereas for large values of n, $S_{\mathbf{w}}(f)$ becomes approximately flat. By varying n one can control the power spectrum of the watermark, trying to achieve the best performance for the application at hand.

With regard to the correlation between two different watermark signals $\mathbf{w}_a$ and $\mathbf{w}_b$, three cases are possible: i) $\mathbf{w}_a$ and $\mathbf{w}_b$ belong to the same chaotic orbit; ii) $\mathbf{w}_a$ and $\mathbf{w}_b$ belong to different chaotic orbits; iii) $\mathbf{w}_a$ and $\mathbf{w}_b$ belong to the same chaotic orbit after an initial number of iterations, a situation occurring when the sequences are initialized with two irrational numbers not belonging to the same chaotic orbit, but map iteration eventually leads to the same orbit. When considering the actual implementation of a chaotic map, though, the finite precision of calculators must be taken into account. Chaotic orbits become periodic (even if the period may be extremely long), and only two possibilities have to be distinguished: i) the watermark length is smaller that the separation l between the starting points of the two sequences; ii) the watermark length is larger than l. In both cases, cross-correlation becomes an autocorrelation and equation (3.29) may be used, by paying attention to shift $R_{ww}(k)$ by l.

3.1.7 Direct embedding

So far we have only considered waveform-based schemes, where watermarking of the host asset A goes through the definition of a watermarking signal $\mathbf{w}$ and its insertion within A. Though such an approach is a very popular one, in many cases watermark insertion is performed directly, without having to introduce the signal $\mathbf{w}$. According to this strategy, usually referred to under the umbrella of informed embedding, substitutive or direct embedding, watermarking simply corresponds to moving the host asset into a point within the proper detection region. When the direct embedding approach is used, information coding loses most of its importance since it reduces to the definition of the detection regions associated to each $\mathbf{b} \in \mathbf{B}$.

Informed coding

In the discussion carried out so far we have always assumed that a unique watermarking signal is associated to each watermark code. When looked at from the point of view of detection regions, such an approach corresponds to associating a unique, usually connected, detection region to each watermark code $\mathbf{b}$. In some cases, though, it may be convenient to associate a pool of signals (or, equivalently, a pool of disjoint detection regions), to each watermark code. The rationale for this choice will be clear in subsequent chapters, after that informed coding/embedding is discussed. Here it only needs to observe that if the overall detection region is spread all over the asset space, it is easier for the embedder to map the host asset within it, since the distance between the host asset and the detection region is, on the average, lower. The simplest way to achieve such a goal is to let more than one signal be available to transmit the same codeword, so that the embedder may choose the signal which better fits the characteristics of the host asset, e.g. the signal which results in a lower asset distortion. This is exactly what the informed embedding/coding principle says: adapt the watermarking signal to the asset at hand, so that either distortion is minimized or robustness is maximized. For a more detailed discussion of the informed embedding/coding approach the reader is referred to chapters 4 and 9.

3.2 Waveform-based readable watermarking

Having discussed information coding in detectable watermarking, we are now ready to pass to the readable case. Even in this case, watermark insertion may either require that a proper watermarking signal is defined, or may be performed directly by mapping the feature sequence into a proper subregion of the asset space. We will treat the first case in this section and leave direct watermark embedding to the next section.

Generally speaking, two opposite approaches can be adopted: sequence coding, where the watermark code $\mathbf{b} = \{b_1 \ldots b_k\}$ is coded/embedded all at once (sections 3.2.1 and 3.2.2), and bit-wise coding, where each bit is embedded separately (section 3.2.3). These approaches have their counterpart in digital modulation where bit transmission may be achieved either through M-ary or binary signalling. Of course, intermediate schemes exist too in which the bit sequences is split into blocks and each block is coded separately.

3.2.1 Information coding through M-ary signaling

We have seen that in detectable watermarking information coding amounts to associate a waveform signal $\mathbf{w} = \{w_1 \ldots w_n\}$ to each possible watermark code $\mathbf{b} = \{b_1 \ldots b_k\}$. A similar approach can be used for readable watermarking. Even in this case, we have to define a set of digital waveforms $\mathbf{W} = \{\mathbf{w}_1, \mathbf{w}_2, \ldots \mathbf{w}_M\}$ ($M \geq 2^k$), and a coding rule Φ that maps each $\mathbf{b}$ into a distinct element of $\mathbf{W}$. The difference between the two classes of watermarking schemes comes into play at the detector/decoder side. In detectable watermarking the detector only has to decide whether the host asset contains a given watermark signal $\mathbf{w}^*$ or not. On the contrary, in readable watermarking, the decoder does not know which watermark signal to look for, thus it has to look for all possible $\mathbf{w}_i \in \mathbf{W}$, and pick out the most probable one. By assuming that M different signals can be generated, the watermark conveys $\log_2 M$ information bits. In order to increase the payload, one may insert more than one watermark signals, and associate each watermark code $\mathbf{b}$ to a different subset of signals. In this way

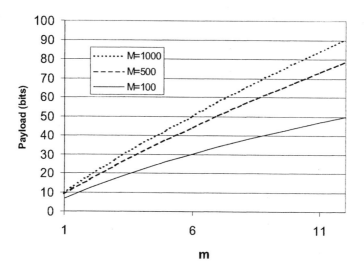

Figure 3.7: Watermark payload for an M-ary signalling scheme embedding m watermarks out of M simultaneously.

the payload is increased up to $\log_2 \binom{M}{m}$, where m is the number of watermarks embedded within A (figure 3.7). Of course, a tradeoff must also be reached between payload and watermark perceptibility, since the higher the number of embedded signals, the higher the distortion introduced by the watermark.

To formalize the above concepts, let us assume that a detectable watermarking scheme $\mathcal{W}_d(\mathcal{E}_d, \mathcal{D}_d)$ is available characterized by an embedding function $\mathcal{E}_d$ and a detection function $\mathcal{D}_d$. A readable watermarking scheme $\mathcal{W}_r(\mathcal{E}_r, \mathcal{D}_r)$ can be built by starting from $\mathcal{W}_d$ in the following way. Define a mapping rule Φ that univocally associates each codeword $\mathbf{b} \in \mathbf{B}$ to a watermark signal $\mathbf{w} \in \mathbf{W}$:

$$\Phi(\mathbf{b}) = \mathbf{w}, \tag{3.31}$$

then let

$$\mathcal{E}_r(A, \mathbf{w}, K) = \mathcal{E}_d(A, \mathbf{w}, K), \tag{3.32}$$

where A is the host asset and K is a secret key used both to embed and retrieve the watermark, and

$$\mathcal{D}_r(A, K) = \{\mathbf{b} \in \mathbf{B} | \mathcal{D}_d(A, \Phi(\mathbf{b}), K) = yes\}. \tag{3.33}$$

where, for simplicity, we have assumed that $\mathcal{W}_d$ is a blind watermarking scheme. A possible problem with the above definition is that $\mathcal{D}_d$ may output a positive answer for more than one watermark, thus making it difficult to retrieve the true watermark inserted within A. Additionally, especially in presence of attacks, the answer of the detector may always be negative, thus failing to retrieve the information bits hidden within A. Such a problem may be alleviated if $\mathcal{D}_d$ works by comparing an observation variable, giving an indication of watermark presence, against a threshold. In this case, the readable decoder may output the information code $\mathbf{b}$ for which the answer of the detector is maximum:

$$\mathcal{D}_r(A, K) = \arg\max_{\mathbf{b} \in \mathbf{B}}(\mathcal{D}_d(A, \Phi(\mathbf{b}), K)). \tag{3.34}$$

Of course, a drawback with this second approach is that the decoder always outputs a decoded bit string, even when the host asset is not watermarked. Equations (3.31) through (3.33) can be easily extended to the case in which the payload is augmented by inserting more than one watermark.

As to the particular choice of the signal set $\mathbf{W}$, the same approaches used for detectable watermarking can be adopted, with the choice of pseudo-random and orthogonal sequences being the most commonly adopted solutions. In particular, the drawbacks and advantages listed in table 3.5 still hold, and may be used as a guide to the choice of the set $\mathbf{W}$ for the application at hand.

3.2.2 Position encoding

A problem with M-ary signalling is that the number of watermarks the decoder has to look for increases exponentially with the payload. Position encoding represents a feasible alternative to orthogonal signalling. According to the position encoding approach, information bits are encoded in the position within the host feature space of M known watermarks $\{\mathbf{w}_1 \ldots \mathbf{w}_M\}$. To be specific, let $\mathbf{f} = \{f_1 \ldots f_n\}$ be the set of host features, and $\mathbf{w} = \{w_1 \ldots w_n\}$ be a watermark sequence. As for M-ary signalling, we assume that a detectable watermarking algorithm $\mathcal{W}_d(\mathcal{E}_d, \mathcal{D}_d)$ is available to embed $\mathbf{w}$ within $\mathbf{f}$ and to decide whether a given set of features contains $\mathbf{w}$ or not.

By relying on $\mathcal{W}_d$, we can construct a readable watermarking algorithm $\mathcal{W}_r$, which conveys $\log_2 n$ information bits. To do so, we first convert the information code $\mathbf{b}$ into an integer number l ranging from 0 to $2^k - 1$ (note that the constraint $2^k - 1 \leq n$ must hold):

$$l = \Phi(\mathbf{b}). \tag{3.35}$$

Then we embed a cyclically shifted version of $\mathbf{w}$ within A:

$$A_{\mathbf{w}(l)} = \mathcal{E}(A, \mathbf{w}(l), K), \tag{3.36}$$

$$\mathbf{f}_{A_\mathbf{w}} = \mathbf{f} \oplus \mathbf{w}(l), \qquad (3.37)$$

where $\mathbf{w}(l) = \{w_l, w_{l+1}, \ldots w_n, w_1, \ldots w_{l-1}\}$ denotes an l-step, cyclically shifted version of the watermark. The second step to transform $\mathcal{W}_d$ into a readable watermarking algorithm consists in changing the decoder structure so that the watermark $\mathbf{w}$ is searched for by considering all possible shifts l. Ideally, the detector will reveal the watermark presence only for one rotation l^*, thus making it possible to transmit up to $\log_2 n$ bits. In order to deal with cases in which a positive answer is obtained for more than one l it is necessary to assume that the detector operates by comparing an intermediate detector response against a threshold. In such a case, in fact, the decoder can just consider the value l resulting in the highest detector response. To further increase the payload, two or more watermarks can be embedded at the same time. More specifically, if M different watermarks are embedded, and each watermark $\mathbf{w}_i$ is cyclically shifted by a different step l_i, then the watermark payload increases up to $M \log_2 n$ bits.

A problem with position-encoded bit embedding is computational complexity, since each of the M watermarks must be looked for at all possible locations within the feature vector. To alleviate the computational burden, the same watermark can be inserted M times, every time by using a different shift. In this way, the decoder complexity is reduced by a factor M, at the expense of payload reduction. Due to the impossibility of distinguishing between multiple copies of the same watermark, in fact, the number of shifts combinations is $\binom{n}{M}$ instead of n^M, thus leading to a payload of $\log_2 \binom{n}{M}$.

Note that in many cases a fast algorithm exists to speed up the exhaustive search of all the possible shifted version of $\mathbf{w}$. This is the case, for example, of correlation-based detection, in which the exhaustive search of the watermark at all possible locations in the feature space can be conveniently carried out by FFT-transforming both the watermark and the host feature sequence.

3.2.3 Binary signaling

As opposed to the previous methods, one may decide to embed each bit of $\mathbf{b}$ independently. As a matter of fact, most of the readable watermarking algorithms proposed in the literature operate by embedding and decoding one bit at a time. In this way, both embedding and decoding are greatly simplified, even if sometimes this may lead to a certain degradation of performance. In addition, it is no longer possible to rely on a pre-existing detectable watermarking scheme to build a readable algorithm. An advantage of independent bit encoding is that classical channel coding techniques, possibly coupled with soft-decision decoding, can be applied to improve watermark robustness or to increase the payload. Transmission diversity techniques or bit repetition may be exploited as well, especially when very high robustness is required. Another advantage of independent bit encoding with respect to sequence coding, is that the decoder performance degrade gracefully as the noise, or attack strength, increases, since errors affect each bit independently. On the contrary, the performance of systems based on M-ary modulation or position encoding degrade more abruptly, since a decoding error usually results in a completely erroneous decoded sequence.

A possible drawback of independent bit encoding, is that the assessment of whether a given asset is watermarked or not is quite problematic, especially in the presence of channel coding. On the contrary, M-ary schemes are often derived from detectable watermarking systems (in many cases the same detector is used), thus the switching between decoding and detection is usually straightforward.

In the following, we will describe the main approach to independent bit encoding in a waveform-based context, namely direct sequence spread spectrum watermarking, whereas we will leave the discussion on direct embedding schemes to the next section.

Direct sequence spread spectrum

Independent bit embedding through direct sequence spread spectrum (DS-spread spectrum) works by associating to the bit string $\mathbf{b}$ a spread spectrum watermark signal which is amplitude modulated by $\mathbf{t}$, the antipodal version of $\mathbf{b}$:

$$\begin{cases} b_i = 0 & \rightarrow & t_i = -1 \\ b_i = 1 & \rightarrow & t_i = +1 \end{cases} \tag{3.38}$$

A pseudo-random sequence is first generated and then split into k chunks, then each chunk is amplitude-modulated by multiplying it by t_i. To improve robustness against manipulations affecting a block of consecutive host features, the spread sequence is split into random non-overlapping subsets $\{S_i\}_{i=1}^{k}$, then each bit modulates the samples of a different subset. This is equivalent to reordering the sequence by means of a random permutation [ ] and then splitting it into k consecutive chunks, i.e. $S_j = \{w_{[i]}\}_{i=(j-1)r+1}^{jr}$, where by r we indicated the number of host feature samples associated to each bit. Note that this is also equivalent to permuting the host feature set as indicated in equation (3.40). To be more precise, by letting $\mathbf{w}'$ denote the watermark pseudo random sequence prior to modulation, the DS-modulated watermark signal is defined as:

$$w_{[i]} = w'_{[i]} t_j, \qquad i = (j-1)r+1, \ldots jr. \tag{3.39}$$

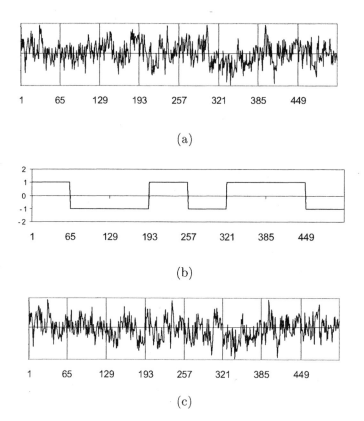

Figure 3.8: Direct sequence spread spectrum modulation. The original pseudo-random sequence (a) is multiplied by the bit antipodal sequence (b) to obtain the final watermarking signal (c).

Note that the same bit t_j is used for all the coefficients in the same water-mark subset. Alternatively, the same subsequence could be used for all the bits, however this would lead to a correlated watermark, thus weakening security.

Information coding through DS spread spectrum modulation is exem-plified in figure 3.8, where a 512 sample long sequence is used to convey 8 bits. The original pseudorandom sequence $\mathbf{w}'$ (a) is multiplied by the antipodal sequence $\mathbf{t}$ (b) to produce the final watermarking signal $\mathbf{w}$ (c).

It is interesting to observe that spread spectrum watermarking is noth-ing but a particular form of bit repetition, and hence similar performance are to be expected in both cases[18].

---

[18]An advantage of spread spectrum watermarking with respect to plain bit repetition

3.3 Direct embedding readable watermarking

So far, we have only considered the case in which the embedding of the information sequence $\mathbf{b}$ within the host asset goes through the definition and the embedding of a watermark signal $\mathbf{w}$. This is not necessarily the case, since in many cases $\mathbf{b}$ is directly hidden within A. In order not to introduce a new symbolism, we still adopt the symbolism introduced in section 1.1.2, equation (1.2), by only noting that now we have $\mathbf{w} = \mathbf{b}$, i.e. there is no more need for the watermark signal $\mathbf{w}$. Of course, in this case information coding loses part of its meaning since it is no more necessary to define the mapping rule between $\mathbf{B}$ and $\mathbf{W}$.

In order to introduce a general framework for direct embedding schemes, let $\mathbb{F}^m$ be the space with the host features. Direct embedding can be formulated as a problem of associating to each different information sequence $\mathbf{b}_i$ a region R_i of $\mathbb{F}^m$. The region R_i is usually referred to as the detection region associated to the i-th information sequence. Given an information string $\mathbf{b}_i$, then, retrieval of the hidden information corresponds to determine (the only) i such that $\mathcal{F}(A) \in R_i$. Direct embedding watermarking simply consists in mapping the host asset A into a point inside the target region R_i. It is readily seen that in this case information coding amounts to the definition of the regions R_i, nevertheless such an operation is closely entangled with the definition of the host feature set, the embedding rule and the algorithm used to decode the watermark, thus we prefer to postpone the discussion of the possible choices of R_i to the next chapters (specifically to chapters 4 and 6). A considerable simplification of the above framework is obtained when each bit of $\mathbf{b}$ is embedded separately, i.e. when a binary signalling strategy is adopted.

3.3.1 Direct embedding binary signalling with bit repetition

The simplest way to inject the watermark code within the host asset A is to embed each bit b_i within a single host feature f_i. In this way a very high payload would be obtained, since the number of host features is generally very high. The embedding rule used to tie b_i to the host feature f_i depends on the specific watermarking algorithm. However, regardless of the embedding rule, hiding each bit within a single feature always results in a very weak watermark, since host features can not be modified too much not to infringe the imperceptibility constraint. Such an approach, then, is only advisable in applications demanding for high capacity and for which

---

concerns security. With spread spectrum watermarking, in fact, it is impossible to read the watermark without knowing the particular spreading sequence used in equation (3.39).

robustness is not an issue at all, e.g fragile watermarking applications.

The simplest way to augment the robustness of direct bitwise embedding schemes, is through bit repetition, whereby the same bit is repeatedly inserted within several host features.

In order for bit repetition to be effective, the same bit must be embedded in independent host features, i.e. features that are likely to undergo different modifications. For this reason, it is rather common to let feature interleaving precede bit insertion. More formally, if each bit is inserted in $r = n/k$ features, we have:

$$f_{w,[i]} = f_{[i]} \oplus b_j, \qquad i = (j-1)r + 1, \ldots jr, \qquad (3.40)$$

where $\mathbf{f}_w = \{f_{w,1} \ldots f_{w,n}\}$ denotes the set of watermarked features and $[\,]$ indicates a permutation of the host feature set.

A more sophisticated way to improve robustness consists in binding each bit to a pool of features, e.g. by varying a collective characteristic of the host features. Once again the specific rule used to bind single bits to host features is highly algorithm-dependent and will be treated in detail in chapter 4.

According to the definition we gave in the introduction of this chapter, we consider channel coding for readable watermarking as part of the information coding process. In direct watermark embedding, channel coding may be simply viewed as a scheme in which not all the information sequences in $\mathbf{B}$ are valid. When considering detection regions, two possible approaches are possible: i) a detection region is associated to all the sequences in $\mathbf{B}$, regardless of whether they are valid sequences or not; ii) detection regions are associated to valid codewords only. The former approach leads to hard channel decoding, whereas the latter amounts to soft decoding. More details about channel coding are given in the next section.

3.4 Channel coding

As for digital data transmission, the use of channel coding significantly improves the reliability of any data hiding system. By referring to figure 3.9, we can see that channel coding acts at a very early stage in the data hiding chain, prior to the generation of the watermark signal, if any, and prior to watermark embedding. As it is customarily in digital communication systems, we will regard to all the blocks intervening after channel coding as an overall binary channel, whose performance depends on the particular algorithm used to embed the coded data within the host asset, to the presence of attacks and to the scheme used to decode the embedded information. Here it is only important to point out that the repetition of the same information bit at several feature positions will be considered as

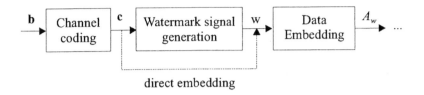

Figure 3.9: Channel coding acts at the earliest stage in the data embedding chain.

part of the binary channel, and will not be treated as part of the channel coding process. Such an assumption will be removed in section 3.4.3 where coding at the feature level will be described. A completely different point of view will be introduced in section 3.4.5 where the informed coding approach will be introduced and its potential benefits over conventional blind coding analyzed.

As described in this section, channel coding only makes sense with watermark decoding (readable watermarking), even if its use as a tool for the assessment of watermark presence in readable watermarking has been recently proposed. We will come back on the usage of error detection codes for watermark presence assessment in chapter 6.

3.4.1 Block codes

With block coding, the k bits of the information sequence are divided into k_c-bit long blocks, each of which is coded into a block of length n_c ($n_c > k_c$). We will indicate the set of all admissible codewords by $\mathcal{C}$. Note that $\mathcal{C}$ is subset of all n_c-bit long binary sequences. A block code having the above characteristics is referred to as a $C(n_c, k_c)$ block code[19]. Block codes are characterized by the code rate R:

$$R = \frac{k_c}{n_c}, \tag{3.41}$$

defining the fraction of information conveyed by each bit in the coded sequence, and by the correcting capability t, defined as the maximum number of errors the code is capable to recover from. The correcting capability of the code t, in turn, is closely related to the minimum distance of the code, d_m, defined as the minimum Hamming distance between any two codewords in $\mathcal{C}$, where the Hamming distance between two sequence of bits is defined as the number of positions for which the two sequences differ.

---

[19]We used the symbols n_c and k_c to distinguish the code block length by the overall length of the watermark sequence.

The performance improvement obtained by applying block channel coding prior to watermark embedding, depends heavily on the particular watermarking algorithm. In most cases, the performance of a data hiding system are expressed in terms of bit error rate (BER), for a given watermark strength. In such a case, the performance improvement brought by channel coding may be given in terms of BER reduction. Alternatively, the bit error rate may be fixed, and the improvement expressed in term of reduced watermark power, or signal to noise ratio[20]. Performance improvement also depends on the way the coded bit sequence is decoded. With this regard two main approaches may be used: hard decoding and soft decoding. To formalize these concepts, we focus on a single coded block. More specifically, in this section, we will let $\mathbf{b} = \{b_1 \ldots b_{k_c}\}$ denote the k_c-bit long block with the to-be-coded bits, and $\mathbf{c} = \{c_1 \ldots c_{n_c}\}$ the corresponding block of coded bits. Finally, we will refer to the set of host features hosting $\mathbf{b}$ as $\mathbf{f} = \{f_1 \ldots f_{n_c}\}$. Note that we are assuming that each bit is tied to a single host feature. If this is not the case, each f_i must be treated as a vector of features.

Hard decoding

If hard decision decoding is used, an independent decision is taken for each bit of the received sequence, thus producing a received word. Then the received word is compared to all possible codewords and the codeword with minimum Hamming distance chosen. Hard decision decoding is not optimal in the ML sense, however decoding is very simple since very efficient algorithms exist to decode the received codeword. Non-optimality leads to a performance loss which, for a Gaussian additive channel, amounts to roughly 3dB, which means that a 3dB stronger watermark is needed to achieve the same performance achieved by ML decoding.

Soft decoding

In this case, decoding corresponds to the ML estimate of the transmitted bit sequence. In other words, the decoder seeks for the codeword $\mathbf{c}^* \in \mathcal{C}$, that maximizes the probability of receiving the host features at hand:

$$\mathbf{c}^* = \arg \max_{j=1\ldots 2^{k_c}} p(\mathbf{f}|\mathbf{c}_j), \tag{3.42}$$

where by $p(\mathbf{f}|\mathbf{c}_j)$ the probability density function of the received host features conditioned to the transmission of $\mathbf{c}_j$ is meant.

---

[20]In this context, the signal to noise ratio is a measure of watermark asset quality, since it is defined as the ratio between watermark and host asset power.

Table 3.6: Coding gain of some popular block codes.

Code	k	n	d_m	G_c
Repetition code	1	n	n	1
Hamming	$2^m - 1 - m$	$2^m - 1$	3	$\frac{3(2^m - 1 - m)}{2^m - 1}$
Golay	12	23	7	3.65
BCH	7	15	5	2.33
BCH	21	31	5	3.39
BCH	51	63	5	4.05
BCH	106	127	7	5.84
BCH	231	255	7	6.34
BCH	223	255	9	7.87

The exact computation of the performance improvement obtained by means of block coding coupled with soft decoding, depends on the actual watermarking algorithm, and will not be discussed. Anyway, just to give an idea of how channel coding impacts on system performance, we recall from digital communications theory that for a code with coding rate R_c and minimum Hamming distance d_m, the performance improvement is quantified by the coding gain G_c:

$$G_c = R_c d_m = \frac{k_c}{n_c} d_m, \qquad (3.43)$$

where by coding gain the power saving allowed by the code is meant. The coding gain for some of the most common block codes is given in table 3.6. Unfortunately, soft decoding of block codes is a very difficult task, which is affordable only in some simple cases such as repetition codes or orthogonal Walsh-Hadamard codes. Alternatively, suboptimal soft decoding algorithms may be used.

Note that unlike in AWGN digital communication, where G_c has an immediate interpretation as the power saving allowed by the code, in digital watermarking, the way G_c impacts on system performance largely depends on the characteristics of the watermarking algorithm.

3.4.2 Convolutional codes

As block codes, convolutional codes operate by dividing the to-be-coded sequence into k_c bit long blocks, then each block is mapped into n_c ($n_c > k_c$) bits. Unlike block codes, the bits associated to each input block do not depend solely on the bits of the current input block, but are a function of

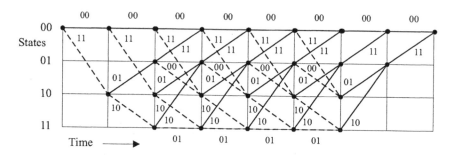

Figure 3.10: Trellis diagram for a convolutional code with $m = 3$, $n = 2$, $k = 1$. Solid lines correspond to a 0 input, whereas for solid lines the input is equal to 1.

$m - 1$ previous blocks as well (for this reason m is sometimes called the memory of the code). A convolutional code, then, acts as a finite state machine, whose state is defined by the $(m - 1)k_c$ bits preceding the current input.

A detailed description of convolutional codes is out of the scope of this book, nevertheless we will briefly recall the Viterbi decoding algorithm since we will often use it in the rest of the book. We will describe it by referring to a simple $C(2, 1)$ code with $m = 3$, the extension to more complex cases being straightforward. Definition of the Viterbi decoder goes through the trellis representation of the code, reported in figure 3.10. The horizontal axis represents time, while the states of the convolutional encoder are arranged vertically. Since we assumed that $m = 3$ we have $2^{m-1} = 4$ different states[21]. When a new bit arrives, the encoder state changes as it is shown in the figure, where for sake for clarity, transitions due to the arrival of a 0 are indicated by a solid line and transitions corresponding to 1 by a dashed line. Thus if the current state is 00 and a 0 arrives the encoder remains in the 00 state, whereas upon the arrival of an 1, the state 10 is entered. Each branch of the Trellis diagram is labelled with the bits output by the encoder in response to the particular input that caused the state transition. As it can be seen, the output does not depend only on the input bit, but on the encoder state as well (e.g. the label of a solid branch depends also on the state the branch originates from). Note that the Trellis diagram always starts from the 00 state since it is usually assumed that the initial state of the encoder is known and equal to 00. At the end of the transmission, the encoder is fed with a sequence of 0's to bring it back to the initial state, thus explaining why at the end of the trellis only solid branches are present.

---

[21]Remember that the state corresponds to the last two bits which entered the encoder.

Hard Viterbi decoding works as follows. Received bits are first decoded independently, then the decoded sequence is split into n_c by n_c blocks. By starting from the first block of the sequence (i.e. from the first step of the Trellis diagram), the received bits are compared with the labels of the branches in the Trellis and the (Hamming) distance between the received bits and the labels computed. In this way, the distance between the received sequence and all possible paths in the Trellis is computed. At this point we note that after some initial steps, the Trellis enters a stationary configuration in which 2 branches enter each node. At each step, i.e. for each group of n_c bits, Viterbi's decoder works by comparing the distance between the paths entering the same node and eliminating the one with highest distance. By referring to the example given in figure 3.10, we note that in this way only 4 (2^{m-1}) paths survive at each step, thus diminishing dramatically the algorithm complexity. At the end of the Trellis, the number of possible states halves at each step, until the 00 state is reached again, at this point only one path survives and a decision is taken as to which sequence was transmitted.

One of the main advantages of Viterbi's decoder is that it can be used for soft decoding as well. It only needs that the distance between the received signal and the paths in the Trellis are computed prior to hard, bit by bit, decoding. Soft decoding through Viterbi's algorithm achieves optimum decoding in the ML sense.

Assessing the performance of convolutional codes is rather a complex task, and will be not detailed here. We only recall that the performance of a convolutional code is mainly dictated by its free distance, i.e. the minimum distance between any two path in the Trellis diagram. It is also worth recalling that if a decoding error occurs, a burst of errors is produced in the decoded sequence, due to the memory of convolutional codes.

As a final remark we note that convolutional are usually preferred to block codes because of their superior performance, such a superiority being mainly due to encoder memory, leading to higher distances between coded sequences, and to the possibility of performing soft decoding at a very low computational cost.

3.4.3 Coding vs bit repetition

So far we did not treat bit repetition as part of channel coding, since we considered repetition as part of the overall binary channel depicted in figure 3.9. From this point of view, bit repetition is just a mean to improve the characteristics of the binary channel coding is applied to. In other words, it serves the scope of making the bit error probability in the absence of coding small enough.

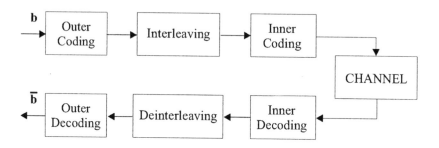

Figure 3.11: General scheme of a concatenated code.

If bit repetition is not included in the binary channel, we can easily realize that the binary channel becomes a very bad one, due to the necessity of keeping the watermark invisible. Then, in order to improve watermark reliability very long codes are needed. Unfortunately, designing very long codes ensuring high coding gains is not an easy task. Concatenated codes have been introduced just to overcome the above difficulty. The general form of a concatenated code is given in figure 3.11, where only one level of concatenation is shown for simplicity. Note the presence of the interleaver between the inner code and the outer code, whose presence serves to cope with possible burst errors introduced by the inner code[22]. By noting that bit repetition is nothing but a primitive form of channel coding, we can, then, look at bit repetition plus channel coding as a particular concatenated code. Nevertheless, much better concatenated codes can be built than the mere juxtaposition of a repetition code and a more performing code such a BCH or a convolutional code. It is known, in fact, from deep space communication theory, that much better results are obtained by using a convolutional code as the inner code and a block code, usually a multilevel Reed-Solomon code, as the outer code.

Turbo codes represents an alternative way of building codes with large minimum distance by starting from simpler, less performing codes. More specifically, in their basic form, turbo codes are built by the parallel concatenation of two (or more) convolutional codes. The main advantage of turbo codes is the possibility of decoding them iteratively at a computational cost which is roughly the same as the decoding cost of the constituent codes. The use of turbo codes for digital watermarking has been suggested only recently, and it is one of the current research areas in the field of robust

---

[22]In digital communication systems, the presence of the interleaver is sometimes a problem due to the delays it introduces in the transmission, and to the memory requirements. Usually this is not a problem in a watermarking system, since in most cases the watermarking process does not need to be performed in real time.

watermarking.

3.4.4   Channel coding vs orthogonal signaling

Though we have presented it from a different perspective, the use of orthogonal watermarking signals (e.g. Walsh sequences) may be seen as a particular kind of code in which all the codewords are orthogonal. In particular, it is known from communication theory that orthogonal coding are capacity-achieving codes, in that it is possible to show that the capacity limit is reached when the number of codes, namely n, tends to infinity. Nevertheless, it is also known that other channel coding schemes, such as convolutional codes, have superior performance, since they are closer to the capacity limit than orthogonal codes.

In some cases, though, orthogonal codes may be preferable to more powerful codes, since they allow to simultaneously embed more than one codeword virtually without interference. hence, their use is particularly indicated in all the applications calling for multiple watermarking of the same asset.

3.4.5   Informed coding

In traditional communication theory channel coding is independent on channel noise, for the very simple reason that the encoder has not means to know in advance the noise that will affect the communication. It is only assumed that such a noise is statistically known, in that the statistical properties of the communication channel are assumed to be available at the encoder. The encoder, then, uses such a knowledge to define a particular coding strategy, e.g. choosing a proper code length.

In data hiding applications the situation is rather different. By referring to blind watermark recovery, we have to consider the simple scheme reported in figure 3.12.

As it can be noted, the transmitted signal, namely the to-be-hidden information, is affected by two kinds of noise: the host features, that being unknown at the detector/decoder side must be treated as disturbing noise, and signal degradation introduced as a consequence of possible manipulations of the host asset[23]. Whether both kinds of disturbs are unknown by the detector/decoder, the first source of noise, namely the host features, is known by the encoder. This is a particular channel in which side information is available at the encoder. It is better for the encoder, then,

---

[23]In figure 3.12 both effects are indicated by an addition, however it should be kept in mind that actually the channel is much more complicated, since attacks can rarely be modelled as the addition of disturbing noise, and the hidden signal and the host features need not be mixed by means of a simple addition.

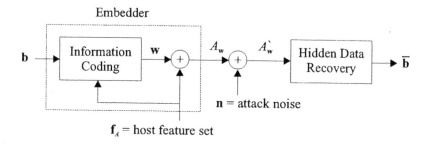

Figure 3.12: Digital watermarking with blind detection/decoding is better modelled by a communication system with side information at the encoder.

to exploit such an information to increase the reliability of the transmission. Actually the theoretical properties of the channel depicted in figure 3.12 can be investigated by resorting to classical information theory. The, somewhat surprising, result is that, in the presence of additive Gaussian noise and Gaussian features, channel capacity does not depend on the first source of noise, since its presence can be accounted for by the transmitter. More specifically, channel capacity is achieved by using a particular form of channel coding, namely informed channel coding, in which the codeword used to transmit a particular information sequence depends on channel condition, that is on the host asset itself.

To be more specific, let $\mathcal{U}$ denote the set with all possible codewords. Instead of associating each message to a single codeword, $\mathcal{U}$ is partitioned into a number of cosets $\mathcal{U}_i$, then each information message is associated to an entire coset. Before transmitting a message $\mathbf{b}_i$, the encoder searches the coset $\mathcal{U}_i$ associated to $\mathbf{b}_i$ for the code $\mathbf{u}$ which better fits the channel state, i.e. the host feature set, and then transmits $\mathbf{u}$. Note that in many cases, fitting the host features simply means to be close to them, nevertheless more complicated fitness measures may be used.

Upon receiving a signal $\mathbf{y}$, the decoder looks for the codeword $\mathbf{u}^* \in \mathcal{U}$ which is closest to $\mathbf{y}$, then it identifies the coset $\mathcal{U}_{i^*}$ containing $\mathbf{u}^*$ and outputs the message $\mathbf{b}^*$ associated to $\mathbf{u}^*$.

In the original work by Costa[24], which first studied the channel depicted in figure 3.12 for the case of Gaussian noise, transmission of the information sequence was paralleled to writing on a dirty piece of paper. The writer knows where dirt is and thus can take the proper countermeasures. Note that conventional channels may be paralleled to a situation in which the sender writes his message on a white paper, which is dirtied at a later

---

[24]M. Costa, "Writing on dirty paper", *IEEE Trans. Inform. Theory*, vol. 29, pp. 439-441, 1983.

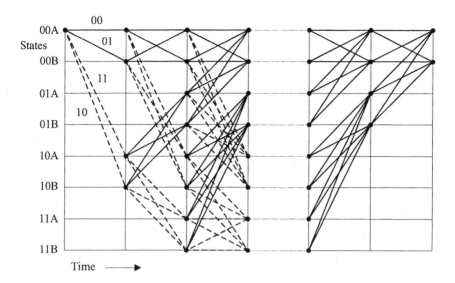

Figure 3.13: 8-states, redundant Trellis diagram for a convolutional code with $m = 3$, $n = 2$, $k = 1$, and 2 branches for each input bit. Solid lines correspond to a 0 input, whereas for solid lines the input is equal to 1. For simplicity output labels are given only for the first stage.

time. For this reason, a code obeying the general informed coding approach described above is sometimes referred to as a dirty-paper code.

Unfortunately, the information theoretic analysis of communication with side information is not a constructive one, in that it only demonstrates the existence of capacity-achieving codes, but no practical hints are given on how to construct them. Something similar happened with Shannon's work on channel capacity, the existence of good codes was proved, without giving any indications on how they could be built. In the years following the seminal work by Shannon, researchers developed a number of theories allowing to construct good codes. It is likely that digital watermarking research will follow the same path, thus we can expect that a number of alternatives will appear in the years to come, as to how design good dirty-paper codes. So far, only few solutions have been proposed. They comprise binary as well as non binary codes (quantization-based schemes). In the following, we will give an example of how a binary dirty paper code can be built, and leave the description of, quantization-based, informed coding to the next chapter.

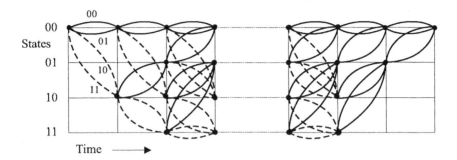

Figure 3.14: 4 state, redundant Trellis diagram for a convolutional code with $m = 3$, $n = 2$, $k = 1$, and 2 branches for each input bit. Solid lines correspond to a 0 input, whereas for solid lines the input is equal to 1. For simplicity output labels are given only for the first stage.

Dirty-paper trellis coding

In figure 3.10 a the trellis representation of a traditional convolutional code is given. Since we assumed that bits were coded one at a time, two arcs exit from each state, corresponding to the two possible inputs: 0's are associated to solid arcs whereas 1's are associated to dashed arcs. This causes each different message to be coded into a different coded sequence. In addition, the coded sequence is unique for each message, since at each step no choice is possible, the encoder follows a solid or a dashed arc according to the bit currently at its input. Suppose now that the number of arcs exiting each node is larger than 2, i.e. two or more arcs exit each node for the same input. Such a concept is exemplified in figure 3.13, where 2 solid arcs and 2 dashed arcs exit each node. Arc labels are designed so that several alternative bit sequences exist for each transition, as it is exemplified in the first stage of the Trellis depicted in the figure. For example, if the coder in its initial $00A$ state is fed with a 0 bit, it may decide to remain in the $00A$ state and output the sequence 00 or to pass into the $00B$ state and output the 01 sequence. Note that both the output sequence and the state transition itself are not completely defined by the input bit. It is clear that many alternative paths exist in the trellis to encode the same input sequence.

It is worth observing that parallel arcs may exist as well, as illustrated in figure 3.14, where alternative paths differ only for the output bit sequence, while the nodes (states) touched by alternative paths associated to the same message are the same.

Given an input message, the problem is now to select one of the possible alternative paths encoding it. According to the informed coding principle,

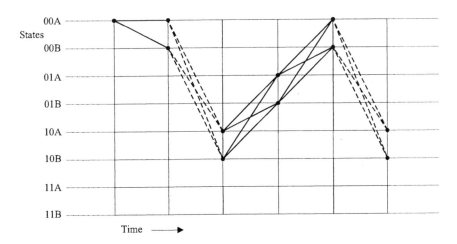

Figure 3.15: Pruned Trellis for coding the sequence 01001 (the original Trellis is given in figure 3.13).

in doing so the particular host asset the message has to be embedded in must be taken into account. This can be done in a very simple and elegant way. First the trellis is pruned so that all the paths that do not encode the to-be-coded-message are removed. This amounts to eliminating all the wrong arcs, i.e. solid arcs for the steps whose input is a 1 and dashed arcs when the input is 0. By referring again to figure 3.13 and by assuming that the sequence 01001 has to be coded, the pruned trellis assumes the form reported in figure 3.15 (it is assumed that coding starts from the 00A state). In order to select one of the possible paths, the trellis decoder is applied to the non-marked host asset, and the resulting path chosen to encode the input message. In this way watermarking distortion is minimized, since the path which is closest to the host asset is chosen.

Decoding is now straightforward. The traditional Viterbi algorithm is applied to the entire trellis, thus identifying the path which is closest to the marked host asset features. Then the sequence is decoded by looking at the bits of the arcs in the extracted path. Note that the metric used to compute distances between host features and trellis arcs, depends on the particular embedding algorithm used to hide the watermark.

3.5   Further reading

The brief description of pseudo-random sequence generation given in section 3.1.1 only serves the limited scope of this book. For a more comprehensive introduction to such a topic, readers are referred to [123].

Though continuous valued and binary pseudo-random sequences share many common properties, they are usually used in different contexts. More specifically the bulk of theory for binary pseudo-random sequences generation have been developed in the context of channel coding for digital communication. An excellent introduction to such topic may be found in [87, 194, 136]. A more detailed discussion of the properties of Gold sequences is contained in [86], whereas the class of Kasami sequences, which constitute a valid alternative to Gold sequences, is described in [117].

The comparison between systems based on spread-spectrum (PN-CDMA) and orthogonal signalling (OW-CDMA) has long been a research subject in digital communication theory. Though the extension to the watermarking case is not straightforward, such a comparison can give some hints on the potentialities of these competing technologies even in data-hiding scenarios. For an-easy-to-read survey of the pro's and con's of PN-CDMA and OW-CDMA in a communication framework, readers are referred to the introductory work by H. Sari, F. Vanhaverbeke and M. Moeneclaey [193].

As it will be detailed in section 7.6.5, the adoption of a periodic, self-synchronizing, watermarking signal is one of the most effective countermeasures to cope with geometric transformations of the host signal (an image, in this case). Readers interested to study such a topic in more depth may refer to the seminal PhD thesis by M. Kutter [128].

The use of chaotic signals for digital watermarking has mainly been explored by I. Pitas et al.. The theory underlying this particular class of watermarking systems, and the some experimental results showing the potentialities of such systems is contained in [227, 218, 219]. For an introduction to the use of chaotic signals for spread spectrum communication, readers may refer to [147, 189], whereas a more general introduction to chaotic signals analysis may be found in [101].

Position encoding watermarking has received only a little attention from watermarking researchers, the only noticeable works on this matter are those by M. Maes, T. Kalker, J. Haitsma and G. Depovere [140] and by R. Baitello, M. Barni, F. Bartolini and V. Cappellini [5].

Though many works have been written on the use of channel coding for digital watermarking, the most comprehensive work on such a subject is the one by F. Perez-Gonzalez, J. R. Hernandez and F. Balado [171], where the performance improvement achievable through conventional channel coding techniques is discussed for several realistic cases [25].

For a general, yet simple, introduction to channel coding theory readers may refer to a number of classical books, e.g. [185], [136] and [223] just to

---

[25] Dirty paper coding is not treated in [171]

mention some. For an introduction to turbo codes, more recent texts are needed, for example [200].

The first work in which the watermarking channel is explicitly modelled as a communication channel with side information is a paper by I. J. Cox, M. L. Miller and A. L. McKellips [57]. At the same time, in a paper by B. Chen and G. Wornell [37], a first implicit algorithm in which such principles were exploited was given. Indeed, the theoretical analysis of a communication channel with side information at the encoder was developed many years before by and by S. I. Gelf'and and M. S. Pinsker [84] and by M. H. M. Costa for the additive Gaussian case [50]. Since then the so-called informed embedding approach has been largely studied leading to innovative watermarking algorithms, which in many cases largely outperform conventional SS techniques (see next chapters.)

Dirty paper coding is a relatively new research field, yet some interesting and insightful works have already been published showing the potentiality of this kind of techniques. In addition to the works by Miller et al. [151, 155], already described in section 3.4.5, it is worth mentioning the papers by J. Chou, S. S. Pradhan and K. Ramchandran [42, 41, 43], where the design of dirty paper codes is framed in an information theory context.

4

Data embedding

Having discussed how the to-be-hidden information can be coded, we must now describe the possible procedures to hide it within the host asset. This is a crucial task, since watermark properties highly depend on the way the hidden information is inserted within the host asset. From a very general point of view, embedding is achieved by first extracting a set of features (host features) from the host data, and by modifying them according to the watermark content. Two steps are, then, required in order to define the embedding process: choice of host features, and definition of the embedding rule. Several solutions have been proposed, leading to different classes of watermarking systems. In this chapter, we review the main approaches proposed so far, paying attention to discuss the advantages and the drawbacks of systems operating in different feature domains and adopting different embedding rules[1].

4.1   Feature selection

In designing an effective data hiding system, it is important to determine the feature set which will convey the hidden information.

Many applications require a scheme where the hidden information does not alter the perceptual quality of the host signal, thus host features should be chosen so that the watermarked asset is identical to the non-watermarked asset in terms of visibility, audibility, intelligibility or some other relevant perceptual criterion. Another requirement heavily affecting the choice of the host features is robustness to signal processing alterations, that inten-

---

[1] It is worth noting that a clear distinction between information coding and embedding can not always be made, thus the division of the material between this chapter and chapter 3 may sometimes appear arbitrary.

tionally or unintentionally attempt to remove or alter the watermark. The choice of the feature set (and the embedding rule) should provide a watermark that is difficult to remove or alter without severely degrading the integrity of the original host signal. For other applications, capacity rather than robustness is a critical requirement, thus privileging features that can accommodate a large payload. According to the application scenario served by the watermarking system, the importance of the above requirements changes thus calling for the design of a wide a variety of algorithms, without that any of them prevails on the others.

Generally speaking, data hiding techniques can be divided into four main categories: those operating in the asset domain, be it the spatial or the time domain, those operating in a transformed domain, often the DCT or DFT domain, those operating in a hybrid domain retaining both spatial/temporal and frequency characterization of the host asset, and those operating in a compressed domain. In the last case, host features may correspond to frequency, spatial or temporal features, however the peculiarities of techniques operating directly in a compressed bit stream justify their separate treatment.

Another distinction can be made between systems in which the cardinality of the feature space is lower than the size of A, and those for which the host feature set contains as many samples as the asset samples. In the former case, the feature extraction operator is not strictly invertible, therefore either the missed data is retrieved by accessing the original non-marked image, or feature modification is performed by operating in the asset domain, without actually passing in the feature domain (see figures 1.2 and 1.3 and the corresponding discussion in section 1.1.2).

4.1.1   Watermarking in the asset domain

The most straightforward way to hide a signal within a host asset is to directly embed it in the original signal space, i.e. by letting the feature set correspond to signal samples. For audio signal, this amounts to embedding the watermark in the time domain, whereas for still images this corresponds to spatial domain watermarking.

In many cases, the choice to embed the watermark in the asset domain is the only possible, such a necessity being dictated by low complexity, low cost, low delay or some other system requirements. Another advantage of operating in the asset domain is that in this way temporal/spatial localization of the watermark is automatically achieved, thus permitting a better characterization of the distortion introduced by the watermark and its possible annoying effects. Additionally, an exact control on the maximum difference between the original and the marked asset is possible, thus permitting the design of near-lossless watermarking schemes, as required by certain applications such as protection of remote sensing or medical images (more details about near-lossless watermarking are given in section 5.5.2).

Still images

In the case of still images $(A = I)$, the asset domain corresponds to the spatial domain, and the host feature set coincides with pixel values. Let I be an $M \times N$ host image. By adopting the notation introduced in section 1.1.2, we have:

$$\mathcal{F}(I) = \mathbf{f}_I = \{I(i, k)\}_{i,k=(0,0)}^{(M-1,N-1)} \in \mathbb{F}^{M \times N}. \qquad (4.1)$$

Note that according to the particular pixel representation, $\mathbb{F}$ may correspond to the set of real numbers $\mathbb{R}$, to the set of integer numbers $\mathbb{N}$, or to a subset of integers, e.g. $\mathbb{F} = [0, 255] \cap \mathbb{N}$ as it is usual if pixel values are represented through 8 bits. Of course, the feature set needs not to correspond to the whole set of image pixels. On the contrary, in many cases only a subset of image pixels is marked.

When considering color images, the host features may correspond to the pixel values of a single image component, e.g. the blue component, or to triplets of RGB values:

$$\mathcal{F}(I) = \mathbf{f}_I = \{R(i, k), G(i, k), B(i, k)\}_{i,k=(0,0)}^{(M-1,N-1)}, \qquad (4.2)$$

where by $R(i, k)$, $G(i, k)$ and $B(i, k)$ the red, green and blue components of the color image are meant. Alternatively, any combination of the RGB values maybe used. For example, a common approach to the watermarking of color images consists in embedding the hidden information in the luminance component only. In this case we have:

$$\mathcal{F}(I) = \mathbf{f}_I = \{L(i, k)\}_{i,k=(0,0)}^{(M-1,N-1)}, \qquad (4.3)$$

where the exact definition of the luminance component $L(i, k)$ depends on the color space adopted to represent images, e.g. we can have:

$$L(i, k) = \frac{R(i, k) + G(i, k) + B(i, k)}{3}. \qquad (4.4)$$

A problem with separate watermarking of RGB color bands, is that the correlation between such bands somewhat complicates the design of the watermark detector. The dependence between color bands, in fact, is very difficult to model, hence making it difficult to take it into account when retrieving the hidden information from the host features. A possible way to cope with this problem, consists in de-correlating the image color bands through Karhunen-Loeve Transform (KLT).

Given a real $N \times 1$ random vector $\mathbf{u}$, its Karhunen-Loeve transform is defined as:

$$\mathbf{v} = \mathbf{u}\mathbf{\Phi}, \tag{4.5}$$

where $\mathbf{\Phi}$ is a matrix whose columns are the eigenvectors of the covariance matrix of $\mathbf{u}$, $\mathbf{v}$ is the KL-transformed random vector, and $*$ and T denote conjugation and transposition respectively. In other words:

$$\mathbf{\Phi}^{*T}\mathbf{C_u}\mathbf{\Phi} = Diag(\lambda_k), \tag{4.6}$$

where

$$\mathbf{C_u} = E\{(\mathbf{u} - \mu_{\mathbf{u}})^{*T}(\mathbf{u} - \mu_{\mathbf{u}})\} \tag{4.7}$$

is the covariance matrix of $\mathbf{u}$, $\mu_{\mathbf{u}}$ is the expected value of $\mathbf{u}$, and λ_k's are the eigenvalues of $\mathbf{C_u}$.

As stated above, the most important property of KLT is that transformed coefficients are mutually uncorrelated. Hence, by applying the KLT to the RGB triplets of the pixels in a color image, three uncorrelated bands K_1, K_2 and K_3 are obtained. Once pixels have been expressed in the KLT domain, they may be directly marked (spatial domain watermarking) or transformed in a frequency domain (frequency domain watermarking).

More details about the choice of a feature set for spatial watermarking of color images are given in chapter 5, since such a choice mainly depends on visibility issues.

Another approach to spatial domain watermarking consists in letting the host feature set correspond to pixel differences rather than pixel values. By assuming the image is raster scanned from left to right and from top to bottom, and by letting $e(i, k)$ denote the difference between pixel at position (i, k) and the previous pixel in the raster scan, we have:

$$\mathcal{F}(I) = \mathbf{f}_I = \{e(i, k)\}_{i,k=(1,1)}^{(M-1, N-1)}. \tag{4.8}$$

The above choice is not very popular, since watermark distortion propagates through the image, thus making it difficult to predict the actual effect of watermark insertion on image quality.

We conclude this paragraph by observing that in the still image case the number of host features is limited by the image size, whereas in the case of an audio or video signal, the number of available host features depends on signal duration (or on the size of the minimum watermarked segment). In many cases, it is the small number of available host features that limits the capacity of the watermark channel[2].

---

[2]We refer here to the overall capacity of the watermark channel, rather than to per-host-feature capacity.

Audio signals

Temporal domain techniques are probably the most used for audio. Similarly to what happens for images, in the case of audio signals $(A = S)$ watermarking in the asset domain implies the modification of the audio samples themselves, i.e., if S is an audio track of length M samples we have that

$$\mathcal{F}(S) = \mathbf{f}_S = \{S(i)\}_{i=0}^{M-1} \in \mathbb{F}^M. \qquad (4.9)$$

As opposed to still images the sampling rate needs to be known for having a correct reproduction (only slight sampling rate fluctuations can be tolerated). The most common sampling rate for high quality audio is 44.1 kHz (the value used in audio CD), although the 48 kHz value (originally used by audio DAT) can also be found. Samples are usually quantized with 16 bits per channel (again as in audio CD)[3]. The number of samples of an audio track is also very different with respect to images, for representing a track of only a few minutes, in fact, some tens of millions of samples are needed.

High quality audio tracks are usually described by multiple simultaneous channels, at least 2 for stereo representation, but more can also be used, as for example the 6 (5 normal wideband channels plus 1 lowpass channel) of 5.1-channel surround format. Considerations similar to those drawn when dealing with color images watermarking can be repeated here: multichannel watermarking requires that the correlation among channels is taken into account during the recovery phase or compensated for during encoding.

The video case

A video signal consists of two synchronized signals, an audio signal and an image sequence. Generally speaking, then, the host feature set for asset domain watermarking should consist of all audio and image sequence samples, or a suitable subset. However, by video watermarking the watermarking of the sole image sequence is usually meant. In this case, most of what has been told regarding still images also holds for the case of video, however video watermarking presents a number of peculiarities, the most important of which are highlighted below.

First of all it has to be decided if the sequence is watermarked on a frame by frame basis, for example by using a known still image watermarking tool, or if the watermark should be spread over more than one frame. Given that some particular frame-based attacks (frame dropping, frame

---

[3]These are the most widely used values for high quality audio, although some experts complain that for obtaining real transparency with respect to analog audio a sampling frequency of 50 kHz, and at least 20 bits per sample should be used.

exchanging, frame rate variation) can occur, it seems that frame by frame techniques are preferable, since, in this case, each frame contains the entire watermark, and time synchronization is not needed. Nevertheless, it is sure that watermark recovery can greatly benefit from exploiting the information which is contained in a sequence of frames, i.e. watermark recovery should be performed on a sequence basis.

Another important issue regarding video is related to the possibility of embedding the same watermark in every frame, thus obtaining a system which is sensible to statistical attacks , or to make the watermark changing from frame to frame, thus risking to produce visible temporal artifacts.

In any case, when watermarking is carried out at the asset level, the host feature set is defined as:

$$\mathcal{F}(V) = \mathbf{f}_V = \{V(i,k,t)\}_{i,k,t=(0,0,0)}^{(M-1,N-1,T-1)}, \qquad (4.10)$$

where T indicates the number of frames the image sequence consists of. Note that the number of host features is now much larger than in the still image case, thus allowing for higher capacity or robustness.

4.1.2 Watermarking in a transformed domain

In transformed domain techniques, the watermark is inserted into the co-efficients of a digital transform of the host asset. The most common choice consists in embedding the watermark in the frequency domain, usually the DFT (Digital Fourier Transform) or DCT (Digital Cosine Transform) domain[4]. However other solutions are possible, including the usage of the Mellin, Radon or Fresnell transforms.

Usually, transformed domain techniques exhibit a higher robustness to attacks. In particular, by spreading the watermark over the whole asset, they are intrinsically more resistant to cropping than asset domain techniques, where resistance to cropping can only be granted by repeating the watermark across the asset. Also robustness against other types of geometric transformations, e.g. scaling, or shifting, is more easily achieved in a transformed domain, since such a domain can be expressly designed so to be invariant under a particular set of transformations. For instance, techniques operating in the magnitude-of-DFT domain are intrinsically robust against shifting, since a shift in the time/space domain does not have any impact on DFT magnitude.

Perceptual constraints aiming at ensuring invisibility can also be readily incorporated into frequency domain representations, e.g. by avoiding modifying low spatial frequencies where alterations may produce very visible

---

[4]Watermarking in the wavelet domain is treated separately in section 4.1.3.

distortions. On the other side, frequency domain techniques do not allow to localize precisely the watermarking disturb in the asset space, thus making it difficult to tune it to the HVS or HAS characteristics[5].

Another drawback of transformed domain techniques is computational complexity. As a matter of fact, many applications can not afford the extra time necessary to pass from the asset to the transformed domain and backward.

Still images

Many frequency domain image watermarking systems have been developed since the early days of watermarking research. At the beginning, the DCT was preferred to DFT, mainly for its assonance with JPEG coding standard.

The DCT transform of an $N \times N$ image $I(i, k)$ is defined by the following set of equations:

$$C(u,v) = c(u)c(v)\frac{2}{N} \sum_{i=0}^{N-1}\sum_{k=0}^{N-1} I(i,k)\cos\left(\frac{\pi}{N}u(i+\frac{1}{2})\right)\cos\left(\frac{\pi}{N}v(k+\frac{1}{2})\right),$$

(4.11)

$$I(i,k) = \frac{2}{N} \sum_{u=0}^{N-1}\sum_{v=0}^{N-1} c(u)c(v)C(u,v)\cos\left(\frac{\pi}{N}u(i+\frac{1}{2})\right)\cos\left(\frac{\pi}{N}v(k+\frac{1}{2})\right),$$

(4.12)

with

$$\begin{cases} c(w) = 2^{-\frac{1}{2}} & w = 0 \\ c(w) = 1 & w > 0 \end{cases}$$

(4.13)

In DCT domain watermarking, then, we have:

$$\mathcal{F}(I) = \mathbf{f}_I = \{C(u,v)\}_{u,v=(0,0)}^{(N-1,N-1)}.$$

(4.14)

Note that the above equations define a full-frame DCT, i.e. the transformation is applied to the image as a whole. Such an approach should be contrasted to block-based techniques described in section 4.1.3. Note that the computational burden associated to full-frame DCT calculation is alleviated by the availability of fast algorithms to compute it.

Through full-frame DCT, $N \times N$ host features are made available. However, it is a common practice not to use all the $N \times N$ DCT coefficients to embed the watermark, thus restricting the embedding area to a subregion of the DCT spectrum. The reason for the above choice is twofold.

---

[5] Actually, an accurate analysis of watermark perceptibility requires that both the spatial (temporal) and frequency domains are considered, since different phenomena underlying the human visual (auditory) system are better analyzed in the frequency or the spatial (temporal) domain. See chapter 5 for more details.

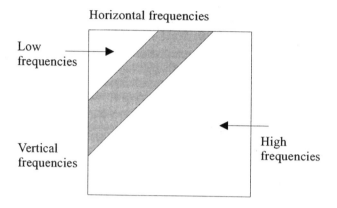

Figure 4.1: Watermarking in the DCT domain usually interests the medium frequency portion of the frequency spectrum (grey region).

Low frequency components are discarded since a low frequency watermark tends to be too visible, whereas high frequency components are discarded because a high frequency watermark would be too vulnerable to attacks such as JPEG compression or low pass filtering. The set of host features in DCT watermarking, then, is usually chosen to lie in the medium portion of the spectrum, as depicted in figure 4.1.

Another possibility consists in marking the M highest DCT coefficients[6]. However, this may lead to some problems in the recovery phase, since it is not sure that after the watermarked image has been attacked (or even as a consequence of watermark embedding), the highest DCT coefficients remain the same.

The subset of host DCT coefficients may also be chosen at random, by letting it depend on a secret key known to both the encoder and authorized decoders.

The most obvious alternative to DCT is the Discrete Fourier Transform. Such a transform is sometimes preferred to DCT since the division of the frequency domain into a phase and a magnitude spectrum, permits to isolate the effect of spatial translations. It is well known, in fact, that a circular spatial translation only affects the phase of the DFT spectrum, leaving the magnitude unchanged. It is, then, sufficient to embed the watermark in the magnitude of DFT coefficients to obtain a watermark which is invariant to circular spatial translations[7]. The full-frame DFT of a still

---

[6]Values near the DC coefficient are usually not taken into account.

[7]Spatial translations are mainly important because they are always associated to image cropping, one of the most common operations carried out on images.

image is defined by the following equations[8]:

$$F(u, v) = \frac{1}{N} \sum_{i=0}^{N-1} \sum_{k=0}^{N-1} I(i, k) \exp \left\{ \frac{-2\pi j (ui + vk)}{N} \right\}, \qquad (4.15)$$

$$I(i, k) = \frac{1}{N} \sum_{u=0}^{N-1} \sum_{v=0}^{N-1} F(u, v) \exp \left\{ \frac{+2\pi j (ui + vk)}{N} \right\}. \qquad (4.16)$$

If watermark embedding is achieved by operating in the magnitude of the DFT domain, we have:

$$\mathcal{F}(I) = \mathbf{f}_I = \{ \| F(u, v) \| \}_{u, v = (0,0)}^{(N-1, N-1)}. \qquad (4.17)$$

Note that due to the symmetry properties of the DFT spectrum, the effective number of DFT coefficients is less that N^2. By noting that the magnitude of coefficients in the first and the second quadrants is equal to that of the third and fourth quadrants, we obtain $N^2/2$ independent coefficients that can be used to insert the watermark. In addition, as for the DCT case, in order to find a trade off between visibility and robustness only medium frequency coefficients are usually exploited. As a rule, then, the subset of coefficients conveying the watermark assumes the form depicted in figure 4.2.

As for the DCT domain, alternative solutions to select a subset of DFT coefficients have been proposed including random selection, and choice of highest magnitude coefficients.

Some authors also proposed to mark the phase of DFT coefficients, however this approach has only had a limited diffusion, thus we will not consider it any further.

Frequency domain watermarking of color images is usually achieved by operating on the DFT/DCT of image luminance. Anyway some different approaches have been proposed as well. Among them, the independent watermarking of the DFT/DCT coefficients of each color band, and the watermarking of the DFT/DCT coefficients of the Karhunen-Loeve Transform of the image, have proved to give good results in term of robustness and invisibility.

Robustness against a selected set of global geometric transformations may be achieved by letting the host feature set belong to a transformed domain which is invariant to the selected transformations. An example of this approach has already been presented, when we noted that by watermarking the magnitude of DFT coefficients we automatically achieve invariance

---

[8]As for the DCT, the availability of fast algorithms to compute the DFT greatly diminishes the computational complexity of equations (4.15) and (4.16).

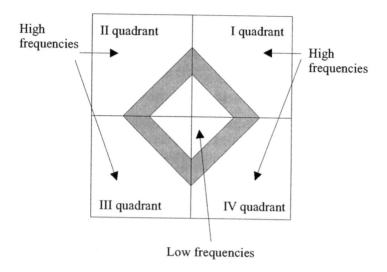

Figure 4.2: Watermarking in the DFT domain usually interests the medium frequency portion of the frequency spectrum (grey region).

against circular shifts in the spatial domain. An interesting generalization of such a concept may be used to achieve simultaneous invariance against spatial shifts, isotropic scaling and rotation. Let us assume we want to watermark a host image $I(i, k)$. We start by taking the magnitude of the DFT of $I(i, k)$. Let such a magnitude be $\| F(u, v) \|$. For the translation property of the Fourier transform we already got rid of circular spatial translations, since they only affect the phase of $F(u, v)$ (ordinary translations can be seen as cropped circular translations). In order to gain invariance against rotation and scaling, let us note that scaling the image axes in the spatial domain causes an inverse scaling in the frequency domain. In addition, rotating the image through an angle θ in the spatial domain corresponds to the same rotation in the frequency domain[9]. If we log-polar map the DFT magnitude spectrum, then, rotations and isotropic scalings are mapped into translations of the DFT spectrum. To be more specific, consider a point $(u, v) \in \mathbb{R}^2$, and let:

$$u = e^\rho \cos \theta,$$
$$v = e^\rho \sin \theta, \tag{4.18}$$

with $\rho \in \mathbb{R}^+$ and $0 \le \theta \le 2\pi$. It can be readily seen that in the (ρ, θ) coordinate system, scaling and rotation are converted to a translation of

---

[9]This is exactly true with continuous Fourier transform, however the same property approximately holds for the Discrete Fourier transform.

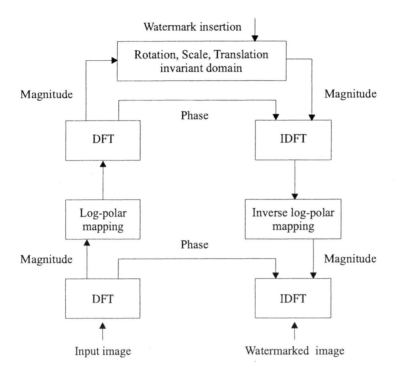

Figure 4.3: A sketch of Fourier-Mellin-based watermarking for RST invariance.

ρ and θ respectively. In order to get rid of translations in the ρ and θ axes one can apply again the DFT and retain the coefficients magnitude only. Taking the Fourier transform of a log-polar map is known as the Fourier-Mellin Transform. Watermarking in the Fourier-Mellin domain is summarized in figure 4.3.

Video signals

In contrast to the image case, transform domain watermarking of video is not feasible, if we mean that the whole video sequence is transformed. Due to the typical time duration of a video sequence, in fact, the dimension of the transformation would be prohibitive. Thus transform domain watermarking is only feasible on a segment by segment basis. This approach is quite similar to block based transform methods that we will present later as hybrid techniques. A distinction can anyway be made: in general block based transform methods are classified as hybrid if they operate on blocks of data which can be assumed to be quasi stationary; thus we classify as transform domain methods all those techniques that apply the transformation on a segment of asset data so long that the quasi stationarity hypothesis fails to hold. According to this definition a video watermarking method can be classified as a transform domain technique if it embeds the watermarking signal in the coefficients of a 3D transform of the video which have a spatial or temporal extent exceeding the limits over which the video signal can be considered quasi stationary. Thus, for example, a video watermarking system, embedding the watermark on a frame by frame basis, by using a full frame DCT or DFT, may be considered as a transform domain method.

The most used transforms for video watermarking are the same already presented for the still images case, and their 3D extensions for considering the time dimension.

Audio signals

As for video signals, transform domain methods are those using as embedding features the coefficients of a transform whose temporal duration exceeds the length over which an audio signal can be considered quasi stationary. Usually an audio signal can not be considered to be quasi stationary over a time period longer than a few tens of milliseconds (typically 50ms)[10].

In the case of audio too, the transforms presented for still images (in particular their 1D versions) are the most suitable.

4.1.3 Hybrid techniques

In the attempt to trade off between the advantages of asset domain techniques in term of localization of the watermarking disturb, and the good resistance to attacks of transformed domain techniques, several hybrid techniques have been proposed[11]. Despite the wide variety of techniques proposed, all of them share the same basic property: they keep trace of the spatial/temporal characterization of the host signal, while at the same time exploiting the richness of the frequency interpretation. Among hybrid techniques, those based on block DCT/DFT and those relying on wavelet decomposition of signals have received a particular attention. An interest which is also motivated by the close connection of such representations

---

[10]This is approximately the time frame used in compression standards, although in some time instant a shorter segment would be needed for granting quasi-stationarity.

[11]One may argue that instead of sharing the advantages of asset domain and transformed techniques, hybrid watermarking inherits the drawbacks of both approaches. Actually, this is sometimes the case, however, as we said above, no proven superiority of a set of host features with respect to the others has ever been demonstrated.

with the most popular signal coding standards, namely the JPEG, MPEG and H.26X coding standard families.

Still images

Block-based DCT is one of the most popular choices for the watermarking of image data because it is a basic component of image and video compression standards such as JPEG and the MPEG and ITU H.26x families of coders. By choosing a framework that matches compression standards, in fact, watermark embedding schemes can be designed to avoid hiding information into the coefficients that are typically discarded or coarsely quantized, resulting in a scheme that is robust to compression.

Let $I(i, k)$ be the host image. Block DCT watermarking operates by first splitting $I(i, k)$ into squared, non overlapping, blocks of size $n_b \times n_b$ (usually $n_b = 8$ in accordance with JPEG standard). Let us indicate by $B_l(i, k)$ the l-th block the image is decomposed into. Each block is DCT-transformed producing a transformed block $B_l(u, v)$, containing n_b^2 DCT coefficients, which are used to embed the watermark within $I(i, k)$. With block-DCT watermarking, the set of host features is defined by:

$$\mathcal{F}(I) = \mathbf{f}_I = \{B_l(u, v)\}_{u,v=(0,0);l=1}^{(n_b-1,n_b-1),N_b}, \qquad (4.19)$$

where N_b indicates the number of blocks $I(i, k)$ consists of. As for full-frame methods, not all the coefficients in a block are suitable for embedding: low and high frequency coefficients are usually discarded to avoid visible artifacts and to increase robustness. The decomposition of the host image into $n_b \times n_n$ sized blocks and the choice of mid-frequency coefficients for watermark embedding are illustrated in figure 4.4. Though mid-frequency watermarking is the most common solution, some methods have also been proposed which insert the watermark in different portions of the block-DCT spectrum, e.g. the block DC value, or high frequency coefficients. Such solutions, in fact, are still of interest for applications where robustness or visibility are not the main concerns.

Given its suitability to characterize both the spatial and frequency properties of signals, the Digital Wavelet Transform (DWT) has gained more and more popularity in image processing and coding applications. Indeed, wavelet decomposition provides a natural framework for time/frequency, spatial/frequency analysis of signals. In addition, it provides a multiresolution approach to signal representation which is particularly suited to describe the impact of disturbs on the HAS and HVS. The above considerations make the DWT domain an ideal candidate for image watermarking applications. Without pretending to be exhaustive, a goal which is by far outside the scope of this book, we give now some introductory notions of

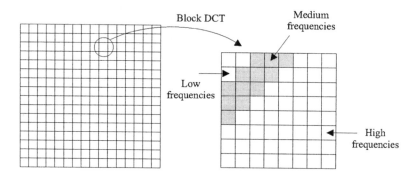

Figure 4.4: Watermarking into the medium frequency portion of block (8 × 8) DCT domain.

wavelet analysis which should permit readers to understand the basics of wavelet-domain watermarking. For sake of simplicity we will consider only 1D signals, the extension to the 2D case being straightforward.

Let us indicate by $L^2(\mathbb{R})$ the Hilbert space of real square summable functions, with a scalar product

$$< f, g > = \int f(x)g(x)dx. \qquad (4.20)$$

A multiresolution analysis with J levels of a continuous signal f is a projection of f on a basis $\{\phi_{J,k}, \{\psi_{j,k}\}_{j \leq J}\}_{k \in \mathbb{Z}}$. The basis functions defining the multiresolution framework, i.e.

$$\phi_{j,k}(x) = \sqrt{2^{-j}}\phi(2^{-j}x - k) \qquad (4.21)$$

result from translations and dilations of the same function $\phi(x)$. Such a function is called *scaling* function, and it verifies the property $\int \phi(x)dx = 1$. The set of functions $\{\phi_{j,k}\}_{k \in \mathbb{Z}}$ spans a subspace $V_j \subset L^2(\mathbb{R})$. The projection of f on V_j gives an *approximation* $\{a_{j,k} = < f, \phi_{j,k} >\}_{k \in \mathbb{Z}}$ of f at the scale 2^j. In the same way, the functions

$$\psi_{j,k}(x) = \sqrt{2^{-j}}\psi(2^{-j}x - k) \qquad (4.22)$$

result from dilations and translation of the same function $\psi(x)$, which is called *wavelet* function, and verifies the property $\int \psi(x)dx = 0$. The set of functions $\{\psi_{j,k}\}_{k \in \mathbb{Z}}$ spans a subspace $W_j \subset L^2(\mathbb{R})$. The projection of f onto W_j yields the wavelet coefficients $\{w_{j,k} = < f, \psi_{j,k} >\}_{k \in \mathbb{Z}}$ of f representing the *details* between two successive approximations. For the above reason, W_{j+1} is the complement of V_{j+1} in V_j:

$$V_j = V_{j+1} \oplus W_{j+1}. \qquad (4.23)$$

Subspaces V_j represent a multiresolution framework for signal analysis.

Eventually, through multiresolution analysis any function $f \in L^2(\mathbb{R})$ can be decomposed as:

$$f(x) = \sum_k a_{J,k} \tilde{\phi}_{J,k}(x) + \sum_{j \leq J} \sum_k w_{j,k} \tilde{\psi}_{j,k}(x). \qquad (4.24)$$

The functions $\tilde{\phi}_{J,k}(x)$ and $\{\tilde{\psi}_{j,k}(x)\}_{j \leq J}$ are generated from translations and dilations of dual functions, $\tilde{\phi}(x)$ and $\tilde{\psi}(x)$, that are to be defined in order to ensure a perfect reconstruction.

The above multiresolution framework is closely related to filter bank decomposition of signals. As a matter of fact, a multiresolution analysis of a signal f can be performed with a filter bank composed of a low-pass analysis filter $\{h_i\}$ and a high-pass analysis filter $\{g_i\}$

$$a_{j+1,k} = < f, \phi_{j+1,k} > = \sum_i h_{i-2k} a_{j,i}$$

$$\qquad (4.25)$$

$$w_{j+1,k} = < f, \psi_{j+1,k} > = \sum_i g_{i-2k} a_{j,i}.$$

As a result, successive coarser approximations of f at scale 2^j are provided by successive low-pass filters, with a downsampling operation applied on each filter output. Wavelet coefficients at scale 2^j are obtained by high-pass filtering, and downsampling, an approximation of f at the scale 2^{j-1}.

Signal reconstruction is derived from (4.23)

$$\begin{aligned} a_{j,k} &= \ < f, \phi_{j,k} > \\ &= \ \sum_i \tilde{h}_{k-2i} a_{j+1,i} + \sum_i \tilde{g}_{k-2i} w_{j+1,i} \end{aligned} \qquad (4.26)$$

where the coefficients $\{\tilde{h}_i\}$ and $\{\tilde{g}_i\}$ define the synthesis filters.

When the wavelet framework is applied to a discrete sequence, the original signal samples, $\{f_n = f(nX)\}$, with $X = 1$, are regarded as the coefficients of the projection of a continuous function $f(x)$ onto V_0. The coefficients relative to the lower resolution subspace and to its orthogonal complement can be obtained by subsampling the discrete convolution of f_n with the coefficients of the impulse response of the two digital filters $\{h_i\}$ (low-pass) and $\{g_i\}$ (high-pass). The two output sequences represent a smoothed version of $\{f_n\}$, and the rapid changes occurring within the signal. Such sequences are usually referred to as the *approximation* and the *detail* signals.

To achieve reconstruction of the original signal, the coefficients of the approximation and detail signals are upsampled and filtered by the dual

$f(n)$ 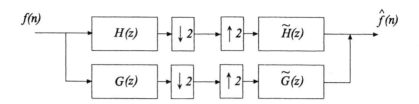 $\hat{f}(n)$

Figure 4.5: Wavelet decomposition of 1D signals.

filter of $\{h_i\}$ and $\{g_i\}$, or *synthesis* filters, $\{\tilde{h}_i\}$ (low-pass) and $\{\tilde{g}_i\}$ (high-pass). The scheme of a wavelet coefficient decomposition and reconstruction is depicted in figure 4.5, in which $\{f_n\}$ is a discrete 1D sequence and $\{\hat{f}_n\}$ the sequence reconstructed after the analysis/synthesis stages.

Extension to the 2D case is achieved by applying the above 1D analysis to the rows and columns of images. At each stage the image is decomposed into four, half-sized, sub-images, called image sub-bands: a low-pass (LL) sub-band resulting in the application of low-pass filtering in both horizontal and vertical directions, two detail sub-bands obtained by applying a low (high) pass filter in the horizontal direction and a high (low) pass one in the vertical direction (LH and HL sub-bands), and a high pass sub-band (HH) obtained by applying a high pass filter in both horizontal and vertical directions. This procedure is then applied again to the low-pass sub-band and iterated until the desired level of decomposition is obtained. An example of image decomposition through wavelet filtering is shown in figure 4.6.

By turning the attention to image watermarking, we can summarize the data embedding process as follows. The image to be watermarked is first decomposed through DWT in n_l levels: let us call I_l^θ the sub-band at resolution level $l = 0, 1 \ldots n_l$ and with orientation $\theta \in \{0, 1, 2, 3\}$ (an example with $n_l = 4$ is given in figure 4.7). Then, the sub-bands the watermark has to be embedded in, must be chosen. Here a tradeoff which is similar to that encountered in frequency domain watermarking must be reached. More specifically, by embedding the watermark in largest detail (high frequency) sub-bands some robustness is lost, since these sub-bands are more sensitive to image processing operations such as filtering and coding. At the same time, they ensure a higher invisibility than sub-bands containing the coarsest image details. Note that the number of host coefficients in large detail sub-bands is considerably larger than the number of coarse-detail coefficients. This permits to counterbalance the poor robustness of high frequency coefficients by inserting redundant information in the form of repeated bit insertion or channel coding. This is the reason why some of

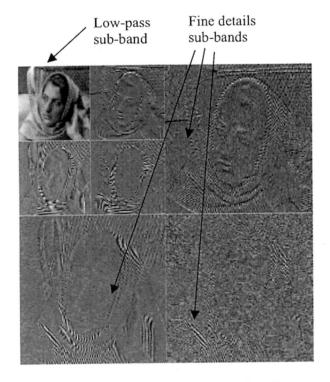

Low-pass
sub-band

Fine details
sub-bands

Figure 4.6: 2-level wavelet decomposition of a still image.

the most popular DWT watermarking methods only insert the watermark in the highest detail sub-bands (level 0 in figure 4.7).

Two drawbacks with DWT are the lack of shift invariance, which means small shifts in input signal can cause big changes in the wavelet coefficients, and poor directional selectivity for diagonal features. Possible alternatives include the Undecimated Discrete Wavelet Transform (UDWT), which is shift invariant but it is highly redundant and still has poor selectivity for diagonal features, and the Complex Wavelet Transform (CWT) which, at the price of a moderate redundancy, offers approximately shift invariance.

Video signals

Hybrid techniques are often used for video watermarking. In principle any hybrid still image watermarking method results in a video hybrid water-marking method when applied on a frame by frame basis. More generally, by hybrid techniques we mean all those methods that embed the water-

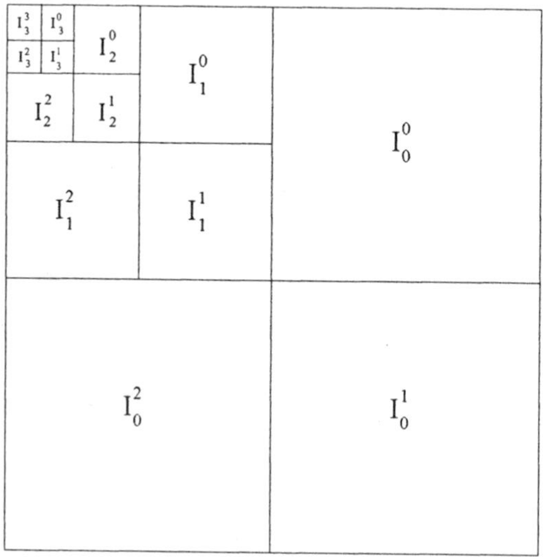

Figure 4.7: Wavelet decomposition of image I $(n_l = 4)$.

mark into the coefficients of a transform of a data block over which the video can be assumed to be quasi stationary. In particular, a commonly used approach employs a spatial block based DCT transform that has the advantage of well adapting to the most common video compression algorithms (e.g. ISO MPEG and ITU-T H.26X), thus allowing to exploit the knowledge developed in the field of video compression for concealing disturbs, and the technological achievements for real time transform implementation. Sometimes the transform is applied to 3D blocks although this approach results to be sensitive to frame dropping and exchanging.

The DWT can also be used on a frame by frame basis, as for still images. A mixed approach has also been proposed [210] which applies DWT along the temporal axis and then embeds a watermark into the obtained transformed frames with a block DCT based method. The application of the DWT along the temporal axis allows to obtain a multiresolution temporal representation of the video, i.e. to separate its static and dynamic components. Basically, embedding the watermark in the low frequency DWT components allows to spread it over all frames.

Audio signals

Hybrid techniques are also often used for audio watermarking. In addition to block based transforms, subband decomposition algorithms are em-

$S_r(i)$

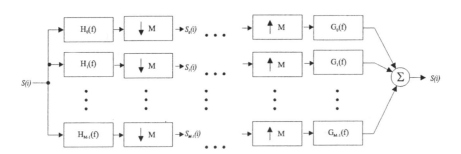

Figure 4.8: General scheme of a uniform filter bank, with analysis $(H_l(f))$ and synthesis $(G_l(f))$ filters.

ployed. Basically subband decomposition systems use a bank of M equal bandwidth passband analysis filters to process the signal $S(i)$ (see figure 4.8): the output of each filter $H_u(f)$ is subsampled by a factor M to obtain the contribution $S_u(i)$ in a given subband. Every M input samples, 1 sample is obtained in each subband: subband decomposition can thus be seen as a block transform. Under some given conditions, the original signal can be exactly recovered from the subsampled signals (we speak in this case of perfect reconstruction), by interpolating them with a bank of synthesis filter $G_u(f)$. Many forms have been proposed for the analysis and synthesis filters, sometime only approximately respecting the perfect reconstruction constraint[12]. Here we present the Modified Discrete Cosine Transform (MDCT) which is widely used for perceptual audio coding[13], and has perfect reconstruction characteristic. Although the MDCT can be well modelled as a filter bank, it is common to describe it as a block transform: in particular it is a transform which processes partially overlapping blocks of $2M$ samples (with M being the overlap length)[14]. Transform coefficients are then obtained as:

$$S(u) = \sum_{i=0}^{2M-1} h_u(i)S(i) \qquad u = 0, \ldots M - 1, \qquad (4.27)$$

---

[12]As an example the bank used in the MPEG1 standard does not ensure perfect reconstruction.

[13]This is the filter bank standardized by MPEG2 for the Non Backward Compatible Advance Audio Coding (NBC/AAC) mode.

[14]For his reason the MDCT is also sometimes referred to as Modulated Lapped Transform (MLT).

where $h_u(i)$ is the impulse response of the analysis filter and is given by:

$$h_u(i) = w(i)\sqrt{\frac{2}{M}} \cos\left(\frac{(2i + M + 1)(2u + 1)\pi}{4M}\right) \tag{4.28}$$

and $w(i)$ is the prototype filter whose design can ensure perfect reconstruction. In particular a popular choice of $w(i)$ is given by:

$$w(i) = \sin\left(\left(i + \frac{1}{2}\right)\frac{\pi}{2M}\right). \tag{4.29}$$

The first M samples of the original signal in the present block can be reconstructed as:

$$S(i) = \sum_{u=0}^{M-1} [S(u)h_u(i) + S^p(u)h_u(i + M)], \tag{4.30}$$

where $S^p(u)$ are the transformed coefficients of the previous block. Usually, for watermarking, only a subset of the transformed coefficients, belonging to the mid frequency range is modified.

4.1.4 Watermarking in the compressed domain

Regardless of the particular domain the watermark is embedded into, an important distinction can be made between techniques which embed the watermark directly in a compressed bit-stream and those operating on the, so to say, *baseband*, non-compressed signal. In order to highlight the advantages of compressed-domain watermarking, we note that by choosing a framework that matches a compression standard, we can avoid adding watermark information to the coefficients that are typically discarded or coarsely quantized, resulting in a scheme that is robust to compression. Another important reason to choose compressed-domain watermarking, is that for many applications, direct embedding in a compressed bitstream is a feature which is desirable (or necessary) to keep the computational burden as low as possible. This is especially true for some video applications where the video will most likely be in some compressed format such as an MPEG2 bitstream, and it is desirable to insert the watermark information directly in the MPEG2 bitstream with only a partial decode.

A problem with compressed-domain techniques is that often they are sensitive to transcoding, thus making it possible that the watermark is lost when the representation of the host signal changes.

Still images

Virtually all block-DCT techniques are suitable for operating in the JPEG domain[15], thus making block-DCT watermarking methods very popular. A difference between conventional block-DCT watermarking and watermarking in the JPEG domain, is that in the latter case the host features correspond to quantized DCT coefficients. Any modification of the host coefficients, then, will be magnified when the image is decompressed and quantized coefficients multiplied by the proper de-quantization values. This makes an exact control of watermarking visibility more difficult, since watermark energies are, so to say, quantized at a coarser level with respect to techniques operating on non-quantized coefficients. It is also worth noting that such a phenomenon is more evident at high frequencies, since for high frequency coefficients JPEG quantization is heavier, and for heavy compressed JPEG images, since in this case a coarser quantization matrix is used.

As to color images, usually only the luminance component is watermarked, since chrominance coefficients are quantized too heavily.

Video signals

Watermarking in the compressed domain is more popular in the case of video data, first because most of the video material is available only (and since the origin) in compressed format, second because the computational cost of completely decoding and re-encoding the video for adding the watermark in some other domain can be really prohibitive. Apart from this, considerations similar to those drawn for still images are also, in general, valid for video.

Another particularity is worth to be mentioned with regard to video watermarking in the compressed domain. This is related to the availability, in the video compressed domain, of a slightly larger amount of information with respect to base band video. More specifically the information about scene motion is available in the form of macro block motion vectors: this information can thus be considered as a possible feature for hiding data. This is just what has been proposed in [203] where the video compression process is slightly modified in such a way to constrain the motion estimation procedure to choose the motion vectors among those belonging to a subset of all possible vectors, based on the watermark message. More precisely, 2 bits for macro block can be hidden by imposing the motion estimation algorithm to choose the macro block motion vector as having

---

[15]Actually a partial decode is necessary since in the JPEG bit-stream block DCT coefficients are entropy coded.

both components of integer pixel precision or both half pixel precision, the first integer and the second half pixel precision or vice versa, according to the 2 bits configuration to hide. The reduction in coding efficiency is quite small (a loss of a few dBs for a given bit rate is obtained).

Audio signals

Compressed domain watermarking of audio signals is not very common. In this case the computational burden of decoding and re-encoding is not, in fact, very heavy and can be easily sustained also by low cost devices. The loss of flexibility of compressed domain techniques is then not adequately compensated by some other advantages.

4.1.5 Miscellaneous non-conventional choices of the feature set

Though encompassing most of the data hiding methods developed so far, the classification given up to here does not completely account for the huge variety of techniques that has been investigated by watermarking researchers. As a matter of fact, the number of possible choices of the host feature set is virtually unlimited, as it is witnessed by a number of ingenious systems described in the literature. It is the goal of this section to revise a bunch of, so to say, non-conventional approaches to feature selection. We selected them either because they answer to problems which turned out to be particularly hard to solve with more classical methods, or because of their theoretical appeal.

Before proceeding, it is worth noting that some of the algorithms discussed in the following could be equivalently classified as spatial, temporal or transformed domain techniques. This because, even if the features actually bearing the watermark are defined in a different domain, their modification is carried out in a conventional domain, e.g. the spatial or time domain. Let us consider, for example, image watermarking through histogram modification. In this case, data are hidden within histogram bins, however, embedding is achieved by modifying pixel values in the spatial domain[16], thus such a technique could be considered as belonging to the class of spatial domain algorithms. Nevertheless, we find it more natural to think as histogram bins as the true host feature set, and considering pixel modification in the spatial domain just as a convenient way of modifying histogram bins. Stated in another way, given a watermarking technique, we identify the host feature set, by considering the set of features truly

---

[16]As a matter of fact, it is impossible to embed the watermark directly in the histogram domain, since the transformation between the image and the corresponding histogram is not invertible.

conveying the watermark, even if the actual insertion is performed in a different domain.

Histogram watermarking

An interesting approach to cope with geometric transformations relies on manipulations of the image histogram. It is in fact evident, that by letting the set of host feature coincide with histogram values, geometric invariance is more easily achieved. This is exactly true for geometric transformations preserving pixel values, such as zooming by repetition, flipping, or rotation by angles multiple of 90 degrees. However, invariance approximately holds for other transformations as well, e.g. scaling or rotation through interpolation.

Histogram-based watermarking can be seen as a classical histogram specification problem: modify the host image so that the watermarked image has a desired target histogram. The degrees of freedom implicit in the histogram specification procedure[17] are used to match the invisibility constraint as much as possible. As to the choice of the target histogram the following considerations must be taken into account:

- The target histogram must ensure that the watermarked image is identical to the original one, or an enhanced version of it. For instance, by letting the target histogram be roughly flat, the effect of watermarking on the host image may even be a pleasant one.

- The set of possible target histograms is virtually unlimited, nevertheless if the watermark has to be robust, target histograms must be far apart, e.g. they must be easily recognizable, either by visual inspection or by an automatic detection procedure. A solution which is known to ensure good robustness is reducing the presence of certain groups of grey levels, i.e. introducing holes in the histogram.

- The overall image brightness must remain the same.

A commonly adopted solution, consists in letting the target histogram be a periodic variation of the uniform histogram, where periodicity may be used both to enable visual watermark detection and to preserve certain intermediate gray level averages with respect to the uniformly distributed histogram (equalized image). An example of two such target histograms is given in figure 4.9.

Despite some good properties, the use of histogram watermarking seems to be limited to fragile watermarking applications, mainly for lack of robustness and security. For instance, the watermark may be removed by

---

[17]The number of images having the same target histogram is virtually infinite.

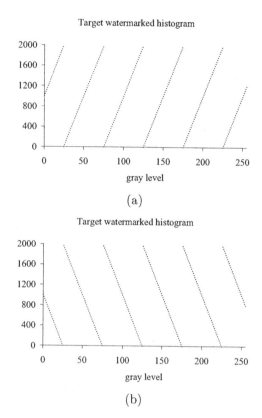

Figure 4.9: Example of target histograms for histogram-based watermarking.

simply inserting within the same image a different watermark, thus forcing the image histogram to assume a different form.

Fractal watermarking of still images

Fractal analysis of still images has received a considerable attention as a powerful tool to discover and code the redundancy present in all natural images. All fractal image analysis algorithms are based on the same basic principle: discover self-similarities between the whole image and its subparts, i.e. discover smaller copies of the entire image buried in it at every scale. Alternatively, local self-similarities can be considered, where similarities are searched between small image subparts.

Among the algorithms developed so far to decompose an image into self-similar subparts, the most popular one works by constructing a so-called *Collage Map* (CM) of the image. The image is first partitioned into two kinds of blocks: the range and the domain blocks that are respectively extracted from a range partition R and a domain partition D. More specifically, range partitioning is usually achieved by means of squared blocks of size $B \times B$, whereas domain blocks have a larger size, typically $2B \times 2B$, (a common choice consists in choosing $B = 4$ or 8).

The Collage Map is built by associating to each block $R_i \in R$, the block D_j which is more similar to R_i (except itself). The test of self-similarity may vary from an algorithm to another. A common choice consists in finding the couple of reals s and b minimizing a quadratic error between the block R_i and the affine transformed block $D_j = s \cdot R_i + b$. To each image corresponds a collage map composed of a range partition R, the indices I_j of the blocks associated to the blocks in D, the scale s_j and the offset b_j:

$$\text{CM} = \{R; I_1 \ldots I_n, s_1 \ldots s_n, b_1 \ldots b_n\}. \tag{4.31}$$

Note that map representing similarities in the image can also be considered in the frequency domain, as it has been proposed for some fractal image compression schemes.

In fractal image watermarking, the mark is embedded by altering the original CM of the host image. Because it is statistically rare to find a block similar to another in an ordinary image, adding similarities permits to obtain singular information in the image. Modification of the CM may be performed by substituting a range block R with a new block $\hat{R} = sD + b$. By this way, the CM can be given the desired form specified by the watermark. For example, new range blocks may be introduced so that certain indexes I_i are present in the CM. Alternatively, the watermark may be tied to the presence of certain scale and offset values s_i and b_i. As to the actual algorithm used to embed a new range block, a bunch of solutions have been proposed. For example, given a domain block D and a to-be-replaced range block R, the new block may be defined as follows:

$$\hat{R} = c \cdot \gamma \cdot \frac{D}{\max(D)} + \bar{R}, \tag{4.32}$$

where $\bar{R}$ is the mean value of R, $\hat{R}$ is the replaced block, γ is a scaling parameter controlling the watermark strength and c may be set to ± 1 depending on the bit to be inserted.

Of course, to make fractal watermarking feasible several problems have to be solved, e.g. choice of range and domain blocks whose modification does not impair image quality, or definition of CM modifications which are as robust as possible to image manipulations such as coding or geometric transformations.

Second generation techniques

Most of the watermarking schemes described so far use a set of host features which is not directly related to the semantic content of the host asset. This is the case, for example, of pixel or audio sample values, and frequency coefficients. A drawback with such schemes is that the watermark is not tied to the significant content of the asset, and as such is more prone to attacks attempting to remove it without destroying the value of the host asset. These kind of techniques are often referred to as as first generation schemes.

An improvement with respect to first generation schemes can be obtained by means of so-called second generation watermarking. The concept of second generation watermarking (sometimes called object-based watermarking) involves the notion of perceptually significant features in the data. By considering the example of still images, such features may correspond to edges, corners or textured areas. For audio signals, the relationship between harmonics may be considered. Features suitable for watermarking should have the following properties:

- Invariance to signal processing attacks (lossy compression, additive, multiplicative noise). This property can be achieved by ensuring that only salient features are chosen, since attacks are likely not to alter them because otherwise the commercial value of the data would be lost.

- Covariance to geometrical transformations (rotation, translation, subsampling, resizing). Features should be chosen in such a way that a moderate amount of geometrical modification should not alter significantly the feature set.

- Robustness to cropping. This is, perhaps, the most difficult property to achieve. It states that cropping the data should not alter the remaining feature points, and should not prevent the exact recovery of the watermark.

Salient features may be used within the watermarking process in two different ways. According to the former, the features serve as reference for standard watermarking techniques. For example, the features may be used to provide a reference orientation for a standard watermark scheme. The goal of this kind of schemes is basically to increase the robustness against geometrical modifications. The latter scheme uses the features directly in the embedding process. That is, the extracted features are directly modified to embed the watermark information.

Whereas the use of salient feature extraction for improved robustness against geometrical transformation will be detailed in chapter 6, we will now

Figure 4.10: In the example reported in the figure, the watermark consists of a bunch of lines, and image watermarking corresponds to moving image corners so that they lie on the watermark lines.

give an example of a second generation scheme for image watermarking where salient features are directly used to embed the watermark. The algorithm described here follows the original algorithm described in [142].

Watermark embedding is a three-step process. First a set S of key points in the host image is selected. Key points should be linked at the semantic content of the image, e.g. they may correspond to edge corners. Then, a pattern P is generated by picking out a subset of pixel positions. The pattern is generated by starting from the to-be-embedded watermark and should be dense enough, i.e. it should cover almost uniformly the whole image space: as an example, the watermark pattern could consists of a bunch of lines spread across the image (figure 4.10). Lines are generated so that a significant percentage p of the pixels in the image lies in the vicinity of one of the lines (a pixel lies in the vicinity of a line if the distance from that pixel to the line is less than a threshold value δ). Finally the image is watermarked by introducing small, local, geometrical changes to the image (warping) such that a significantly high percentage of pixels in S lies in the vicinity of P. The geometric changes introduced by warping affect neighborhoods of the pixels in S in a smooth way, such that at some distance of a given pixel the geometric change decays to zero. Of course some of the points are already in the vicinity of P, whereas for some others warping must be applied. If the amount of warping necessary to bring a pixel in the vicinity of P is too large, then such a pixel is said to be un-warpable and left as is. Note that the number of points in P must be carefully set. Having too much points, in fact, may result in an excessively high false detection rate. At the opposite side, if P contains only a few points, most of the key points in S will be un-warpable thus making it impossible to watermark the image.

Watermarking the picture type sequence for compressed video [137]

In this case the feature conveying the message is the sequence of picture types (PTY) of an MPEG compressed video. It is well known that the frames of a video sequence can be compressed by an MPEG encoder in one of the three modalities, I, P or B. By considering a GOP of 12 frames the most common sequences are IBBPBBPBBPBB or IBPBPBPBPBPB. Linnartz and Talstra suggest to modulate this sequence of picture types in such a way to carry a symbol taken from an alphabet of 60. Basically a binary block code of length 11 is built having minimum Hamming distance of 4; then a B picture type is associated to each 1 bit and a P to each 0 bit. If 1 of the 60 symbols has to be embedded into a GOP, its representative code word is translated into the corresponding sequence of picture types, that is then used for MPEG encoding that GOP. The code is built in such a way that all code words have an equal number of 1s (e.g. of B type frames) set to 6, in order to grant that coding complexity does not change too much with respect to common encoders; furthermore having an Hamming minimum distance of 4 grants a certain degree of flexibility to the modified encoder, which can handle particular situations (e.g. scene changes) that can require a frame to be encoded as P instead of as B. This system requires the complete decoding and re-encoding of the video sequence for the watermark to be removed.

Echo hiding for audio signals

An original approach to data hiding for audio signals exploits the inability of the ear to perceive echoes that are very near in time to the corresponding original sounds. The method, firstly proposed by Bender *et al.* [29] partitions an audio file into small segments, each segment is then delayed by an offset whose value depends on the bit to be embedded (although being always below the threshold of distinguishability of the ear), scaled down not to be audible, and added back to the original signal. A transition region is defined between each pair of subsequent segments, where both the informative bits are embedded with different and smoothly varying strength, in order to avoid annoying blocking artifacts caused by the sudden change of the echo offsets. The information is conveyed by the delay of the added echo signals.

4.2 Blind embedding

Having said how the host feature set can be chosen, we are now in the position of dealing with watermark embedding, that is how the information string b (or the watermark signal w) is hidden within the host asset A. Techniques for watermark embedding can be divided in two main categories: those using always the same embedding rule regardless of the particular host asset to be watermarked, and those adapting the embedding strategy to the host asset. We will refer to techniques belonging to the first category as blind embedding techniques, whereas we will refer to the others as informed embedding techniques. Blind embedding techniques almost exclusively apply to cases in which information hiding passes through the definition of a watermark signal $\dot{w}$. In this case, embedding is defined by means of a mathematical operator which is responsible of *mixing* w and the host feature sequence f. Conversely, informed embedding techniques are more frequently used in conjunction with direct watermark embedding, even if the informed embedding paradigm may be successfully applied to other cases as well. This section is devoted to the analysis of blind embedding techniques only, while informed embedding will be discussed in the next section.

4.2.1 Additive watermarking

The most common approach to blind embedding is the additive one, for which:

$$f_{A_{\mathbf{w}},i} = f_{w,i} = f_i + \gamma w_i, \tag{4.33}$$

where f_i is the i-th component of the original feature vector, w_i the i-th sample of the watermark signal, γ is a scaling factor controlling the watermark strength, and $f_{w,i}$ is the i-th component of the watermarked feature vector[18].

The main reason for the popularity of additive watermarking is its simplicity. Additive watermarks are mainly used in the asset domain, since in this case watermark concealment is achieved very simply by adapting the watermark strength γ to the local characteristics of the cover asset (equations (4.35) through (4.37)). Another advantage of additive watermarking is that under the assumption that the host features follow a Gaussian distribution and that attacks are limited to the addition of white Gaussian noise

---

[18]Hereafter, we will adopt the simplified symbolism $f_{w,i}$, instead of the more exact, but more cumbersome, symbolism $f_{A_{\mathbf{w}},i}$.

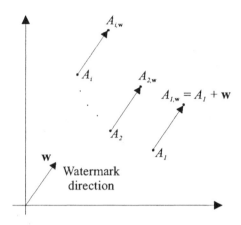

Figure 4.11: Blind, additive watermarking may be seen as the addition of the same watermark vector **w** to the host feature vector (indicated here by A_i.)

(AWGN model), correlation-based decoding is optimum, in that either the overall error probability, or the probability of missing the watermark given a false detection rate, is minimized. The adoption of correlation decoding, in turn, permits to cope with temporal or spatial shifts due, for example, to asset cropping. The exhaustive search of the watermark by looking at all possible spatial/temporal location, in fact, can be accomplished efficiently in the transformed domain, since signal correlation in the asset domain corresponds to a multiplication in the Fourier domain. It is also worth noting that, if the additive approach is used, the bulk of theory addressing digital communications through additive channels may be applied, getting many benefits in terms of system insight and availability of techniques to improve system reliability

A geometrical interpretation of additive watermarking in the feature space is given in figure 4.11. As it can be seen, watermarking always corresponds to moving the original host asset in the direction of the watermark signal. Note that the displacement between the original asset and the marked one does not depend on the original asset itself, in accordance with the blind embedding principle.

A deviation from the blind additive embedding paradigm, is obtained when the watermark strength γ is allowed to vary with i, that is:

$$f_{w,i} = f_i + \gamma_i w_i. \tag{4.34}$$

The main reason for letting γ_i depend on i, is that in this way the watermark strength can be adapted to each host feature in **f** to better match the imperceptibility constraint. To be more precise, equation (4.34) is rewritten as

$$f_{w,i} = f_i + \gamma_i(\mathbf{f})w_i, \tag{4.35}$$

where the dependence between γ_i and the host feature set is explicitly indicated. Note that γ_i may depend on the feature set as a whole, even if in many cases it depends only on a small neighborhood of f_i. If γ_i depends only on f_i we have:

$$f_{w,i} = f_i + \gamma_i(f_i)w_i. \tag{4.36}$$

To make the exact role of γ_i clear, equation (4.34) is usually given the form:

$$f_{w,i} = f_i + \gamma m_i w_i. \tag{4.37}$$

where the dependence of the watermark strength from the index i is incorporated in the sequence of parameters m_i. The sequence m_i is usually referred to as masking sequence. Now γ only accounts for the global watermark strength. Equations (4.34) through (4.37) are usually adopted in conjunction with asset domain watermarking, since in this case the parameter m_i can be directly related to the perceptibility of the watermark disturb at a given position within A. For example, in spatial domain image watermarking, m_i usually depends on a measure of the local image energy, e.g. the local image variance, since it is known that disturbs in high activity regions are less perceivable. More details about the possible choices of m_i are given in chapter 5.

Due to the dependence of m_i upon the host feature set, one may argue that the embedding rule expressed by equation (4.37) is no more a truly additive one. At the same time, m_i could be interpreted as part of the watermark itself, i.e.:

$$w_i' = w_i m_i, \tag{4.38}$$

thus yielding:

$$f_{w,i} = f_i + \gamma w_i', \tag{4.39}$$

where γ does not depend on $\mathbf{f}$ anymore. However, in this way, the watermark is no more independent on the host asset, thus complicating considerably the analysis of the watermarking system. Moreover, the statistical properties of $\mathbf{w}$ would change, thus making the efforts to properly design the watermark signal useless. The approach usually adopted to deal with the presence of the masking sequence within the additive embedding framework, is to assume that m_i's are constant over a small subset of $\mathbf{f}$. In other words, it is usually assumed that the frequency content of m_i is much lower than that of the watermark sequence, which is often assumed to be spectrally white. In this way, equation (4.37) can be approximated to a truly additive embedding rule, and watermarking analysis carried out accordingly.

In the following we will give some examples of additive watermarking schemes. The choice of the algorithms described is by far incomplete, since they are only included for demonstrative purpose. For a more comprehensive covering of the wide variety of watermarking algorithms proposed up to date, readers may refer to a number of excellent tutorials giving a thorough overview of the methods developed so far. A list of such tutorials is given in the further reading section at the end of the chapter.

Example: Patchwork [30]

The patchwork algorithm is one of the earliest watermarking algorithms which appeared in the scientific literature. Here we describe its original implementation, however several modifications have been proposed to overcome some of the limits of early implementation. The original algorithm operated in still images, nevertheless the extension to other media types is straightforward.

Patchwork is a typical spatial domain, additive algorithm. Embedding is achieved by randomly selecting a subset S of image pixels, and then dividing S into two equal subparts S_1 and S_2. In order to increase security selection of S, S_1 and S_2 may depend on a secret key K. Then, pixels belonging to S_1 are increased by a small quantity d, whereas pixels in S_2 are decreased by the same amount (see figure 4.12 for a graphic sketch of Patchwork behavior). It is clear that for a non-watermarked image the average difference between pixels in S_1 and S_2 should be close to zero, whereas for a watermarked image such a difference should approach $2d$, thus making it possible the distinction between watermarked and non-watermarked images. Note also that watermark recovery is not possible if the exact composition of S, S_1 and S_2 is not known.

It is instructive observing that the Patchwork algorithm can be seen as the addition of a pseudo-random watermarking signal to the host image. To be more specific, let us consider the following signal:

$$w(i,k) = \begin{cases} +1 & if \ \ I(i,k) \in S_1 \\ -1 & if \ \ I(i,k) \in S_2 \\ 0 & if \ \ I(i,k) \notin S \end{cases} \qquad (4.40)$$

where by $I(i,k)$ the image pixel at position (i,k) is meant. According to the Patchwork approach, watermark embedding can be expressed as:

$$I_w(i,k) = I(i,k) + d \cdot w(i,k), \qquad (4.41)$$

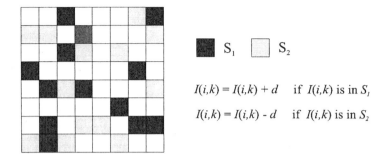

$I(i,k) = I(i,k) + d$ if $I(i,k)$ is in S_1

$I(i,k) = I(i,k) - d$ if $I(i,k)$ is in S_2

Figure 4.12: Patchwork watermarking algorithm.

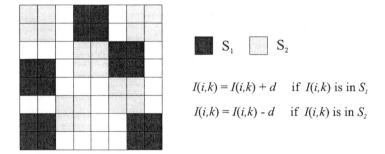

$I(i,k) = I(i,k) + d$ if $I(i,k)$ is in S_1

$I(i,k) = I(i,k) - d$ if $I(i,k)$ is in S_2

Figure 4.13: By increasing the size of blocks defining S_1 and S_2, the robustness of Patchwork against low-pass filtering is considerably increased.

where, according to equation (4.40), $w(i,k)$ is a white, pseudo-random signal taking values $+1$ or -1 with equal probability.

A possible modification of the basic patchwork approach consists in introducing a certain amount of correlation between the samples in $w(i,k)$, e.g. by increasing the size of the blocks in S_1 and S_2 (figure 4.13). In this way the robustness of the watermark to low-pass attacks, such as filtering and JPEG compression is considerably increased.

Example: additive watermarking in the wavelet domain [18]

Though widely used in the asset domain, additive watermarking can be conveniently adopted in other domains as well. One of these is the wavelet domain. The example given here refers to the algorithm described by Barni et al. in [18], which has been proved to ensure an excellent robustness against many common manipulations, including cropping and JPEG coding.

Watermarking follows the general steps described in section 4.1.3. The host image is decomposed through DWT in four levels (figure 4.7). The watermark is inserted by modifying the wavelet coefficients belonging to the three detail bands at level 0, i.e. I_0^0, I_0^1 and I_0^2. The choice of embedding the watermark only into the three largest detail sub-bands is motivated by experimental tests, as the one offering the best compromise between robustness and invisibility. Actually, inserting the watermark into these sub-bands may result in a lower robustness against recompression or low-pass filtering, however, given the low visibility of disturbs added to these frequencies, a higher level of watermark strength is allowed, thus compensating for its fragility.

The watermark consists of a pseudo-random binary sequence $w_i = \pm 1$ which is arranged in 2D, in such a way to scan the host sub-bands, thus leading to a 2D watermark:

$$\omega^\theta(i,j) = w_{(\theta N^2 + iN + j)}, \tag{4.42}$$

where $2N \times 2N$ is the size of the, supposedly square, host image and $\theta \in \{0, 1, 2\}$; DWT coefficients are modified according to the rule:

$$\tilde{I}_0^\theta(i,j) = I_0^\theta(i,j) + \gamma m^\theta(i,j)\omega^\theta(i,j), \tag{4.43}$$

where, as usual, γ is a global adjustable parameter accounting for watermark strength, and $m^\theta(i,j)$ is a masking function considering the local sensitivity of image to noise (see chapter 5 at page 203 for more details).

Example: spread spectrum watermarking of video [91, 115]

Additive spread spectrum watermarking has also been used for the watermarking of raw image sequences. In the simplest case, the pseudo-noise sequence defining the watermark is repeatedly inserted within the luminance component of the image sequence.

The pseudo-noise sequence may either span more than one video frame, or be smaller than it. As a limit case, the pseudo-noise sequence may have the same size of video frames, thus letting each frame contain exactly one copy of the watermark signal. In this case, image sequence watermarking reduces to the independent marking of video frames. Independent frame watermarking provides some advantages with respect to sequence-based techniques, including simplicity, and invariance to sequence-based attacks such as frame removal, frame shuffling or temporal shifting. As a drawback, frame-based watermarking tends to be less robust than sequence-based watermarking, since the available size of the host feature set is limited by the frame size, thus limiting the spreading gain. In order to provide a specific example of frame-based additive watermarking, we briefly describe

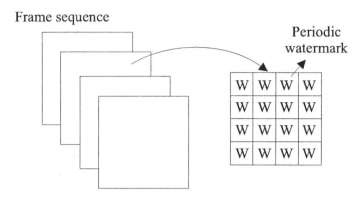

Figure 4.14: The adoption of a periodic, self-synchronizing, watermark is exploited to recover the original frame size and orientation.

an image sequence watermarking scheme developed for DVD protection [115, 215]. The watermark signal consists of a periodic, normally distributed, pseudo-random signal which is directly added to each video frame (figure 4.14). As discussed in section 3.1.4, the adoption of a periodic, self-synchronizing, watermark permits to cope with geometrical manipulations, since periodicity can be exploited to recover the original image size and orientation. More specifically, in the scheme described in [115, 215] the period of the watermark is set to 256 × 256 (or 128 × 128), thus reaching a good compromise between reliable detection (calling for a large period), robustness against geometric attacks, and payload. It has also been proposed to increase the payload of this basic scheme through multiple watermarking associated to position encoding (see section 3.2.2). For example, in [140] Maes et al. demonstrated that by inserting four patterns of $\frac{1}{4}$ image a payload of 36 bits per second can be obtained while at the same time providing a satisfactory degree of robustness.

Example: video watermarking through luminance mean modification [88]

The system described above treats the video signal as a sequence of still images which are marked independently thus ignoring the temporal dimension of the signal. At the opposite extreme, Haitsma and Kalker [88] proposed a scheme in which only the temporal axis is retained by ignoring the spatial dimension of video. In this way, robustness against virtually all kinds of geometric manipulations affecting the image sequence is obtained.

To be specific, watermarking is achieved by varying the mean luminance of video frames. The watermark is a pseudo-random sequence **w** of length

n, where $w_i \in [-1, 1]$. In its simplest form, the watermark is embedded by increasing the luminance of every pixel of frame i by 1 if the watermark sample $w_i = 1$ and decreased by 1 if $w_i = -1$. For example, if n is 1024, then the watermark is repeated every 1024 frames. Since the Human Visual System (HVS) is sensitive to flicker, this simplistic approach suffers from artifacts, especially in non-moving flat areas. Then, to improve the quality of the marked sequence, the watermark strength must be modulated both spatially and temporally. A systematic description of the basic principles to be used for watermark concealment is given in chapter 5.

Example: temporal perceptually masked audio watermarking [211]

This is one of the most classical approaches to audio watermarking. The watermark signal consists of a pseudo random sequence which is added to the samples of the audio file. Before being added, the watermark is perceptually weighted, on a block by block basis, in the frequency domain, in such a way to adapt it to the local frequency characteristics of the host audio. In particular, both the audio and the watermark signals are partitioned into blocks, the DFT is then applied to each block of both signals: the DFT of the audio signal is used for estimating a perceptual audibility threshold (in particular the Psychoacoustic Model 1 of MPEG1[19] is used) which scales the DFT of the watermark signal. The shaped DFT of the watermark signal is then inverse transformed and added to the audio signal. An effect of frequency domain shaping is that the pseudo noise sequence (originally uncorrelated) becomes correlated in the temporal domain: in practice the watermarking process can be modelled as

$$f_{w,i} = f_i + \gamma_i(\mathbf{f}) \otimes w_i \tag{4.44}$$

where the embedding features f_i are the audio samples, and the masking filter $\gamma_i(\mathbf{f})$ is the IDFT of a scaled version of the perceptual audibility threshold.

4.2.2 Multiplicative watermarking

In the attempt to match the characteristics of the watermark to those of the host asset, it may be desirable that larger host features bear a larger watermark. Stated in another way, we may let the energy of watermark samples be proportional to the corresponding host feature samples. The simplest way to implement the above principle is by means of multiplicative

---

[19]More details about this model will be given in section 5.4.5

watermarking[20]. A multiplicative watermark is one obeying the following embedding rule:

$$f_{A_w,i} = f_{w,i} = f_i + \gamma w_i f_i, \tag{4.45}$$

where the symbols have the same meaning as in equation (4.33). Depending on the characteristics of the host feature set, equation (4.45) may be modified as follows:

$$f_{w,i} = f_i + \gamma w_i |f_i|, \tag{4.46}$$

where the absolute value of f_i is used instead of f_i to let the watermark depend only on the magnitude of the host features rather than on their signed values.

Multiplicative watermarking is often used together with full-frame frequency domain watermarking. More specifically, equations (4.45) and (4.46) are used with DCT domain watermarking, whereas for DFT-based schemes, where the watermark is inserted in the magnitude of DFT coefficients, only equation (4.45) may be used.

The main reason for the success of multiplicative embedding coupled with frequency domain watermarking relies in the masking properties of the HVS and the HAS. It is known, in fact, that it is more difficult to perceive a disturb at a given frequency, if the host asset already contains such a frequency component. Let us consider, for example, the still image case. For a better match of the invisibility constraint, it is preferable to embed a watermark whose energy at a given frequency is proportional to the energy of the image at that frequency. Another advantage of multiplicative watermarking is that, according to equation (4.45), we obtain an image-dependent watermark, thus increasing system security, since in this case it is more difficult to estimate the watermark by averaging a set of watermarked images. Finally, we note that multiplicative watermarking obeys a fundamental rule, which says that in order to simultaneously match the invisibility and the robustness constraints, the watermark should be inserted in the most important parts of the host asset (see chapter 5 for more details). Equation (4.45) provides a very simple way to obey the above principle.

In contrast to additive watermarking, in the multiplicative case, the introduction of a masking sequence as in equation (4.37) is rarely used. This is because the proportionality of the watermark with the host feature, already takes masking into account. In addition, as we already noticed, multiplicative watermarking is often used by systems operating in frequency domain, where masking as in (4.37) loses most of its meaning.

---

[20]For historical reasons, we will refer to watermarking schemes obeying the rule expressed in (4.45) as multiplicative, even if the terms proportional or additive/multiplicative watermarking would be closer to the nature of equation (4.45).

Drawbacks of multiplicative watermarking are better understood by looking at the benefits of additive schemes. More specifically, the multiplicative framework is much more difficult to analyze, thus making it hard the optimization of all the steps watermarking consists of, e.g. detection or decoding. In addition, classical results derived from digital communications and information theory cannot be used, since they are usually derived under the assumption of additive noise. As an example let us consider correlation-based detection. Though widely used even in the multiplicative case, the deviation from optimality of correlation detection is much more evident in the multiplicative than in the additive case. Abandoning correlation detection, however, is not harmless, since, in most cases, this prevents the possibility of exploiting fast detection algorithms based on FFT.

Example: spread spectrum watermarking in the DCT domain [53]

Among frequency domain watermarking algorithms, the one proposed by Cox et al. [53] deserves particular attention for the influence it has exerted on subsequent watermarking research. The algorithm was originally proposed for still images, however the underlying principles were adopted for other signal types as well. The algorithm belongs to the category of non-blind, detectable systems. The to-be-marked image is first transformed through full-frame DCT, then the watermark is embedded in the 1000 largest-magnitude DCT coefficients. In so doing, very low frequency coefficients are discarded in order to preserve invisibility. The embedding rule is a multiplicative one, as in equation (4.45). The watermark strength γ is about 0.1. The watermark signal w_i is white and normally distributed with zero mean and unitary variance. In the original paper by Cox et al. other possible embedding rules were proposed, for example:

$$f_{w,i} = f_i \cdot e^{\gamma w_i}, \qquad (4.47)$$

which for small values of γ is approximately equivalent to the multiplicative rule given in (4.45). However the effectiveness of the above exponential embedding strategy has never been tested thoroughly.

Example: making Cox's algorithm blind [9]

A problem with Cox's scheme is that blind detection is not possible. At the detection side, the original image is needed for two main reasons: i) to retrieve the 1000 largest DCT coefficients; ii) to allow correlation detection even in the presence of zero mean host features (see chapter 6 for more details). In order to make blind detection possible, the original scheme by Cox et al. may be modified as follows. Instead of embedding the watermark in the largest DCT coefficients, always embed it in the same set of coefficients belonging to the mid-frequency portion of the frequency spectrum (figure 4.1). In addition, the absolute value of DCT coefficients is used to define the embedding rule, as specified in equation (4.46).

A problem with blind detection is that the original coefficients are not known by the detector and must be treated as disturbing noise. In order to compensate for this lack of knowledge a larger number of coefficients is marked with respect to the original algorithm by Cox. For example, given a 512 × 512 image, DCT coefficients ranging from the 180-th to the 250-th diagonal are watermarked, for a total of about 16.000 coefficients.

A further modification of the above approach consists in inserting the watermark in the DFT instead than in the DCT domain. In this case, the set of host features is chosen as in figure 4.2. In addition, only the magnitude of DFT coefficients is marked. The main advantage of magnitude DFT marking is that in this way robustness against circular spatial translations is automatically achieved. As to the number of marked coefficients, given a 512 × 512 image the portion of the frequency spectrum interested by the watermark ranges from the 80-th to the 150-th diagonal.

Example: Multiplicative temporal domain audio watermarking [27]

In this case the watermarking signal, which consists of a chaotic sequence, is first multiplied, sample by sample, by the absolute value of the audio signal, then it is shaped by means of a Hamming low pass filter, and finally added to the cover audio, thus resulting in the following watermarking relation:

$$f_{w,i} = f_i + h_i \otimes [|f_i|w_i] \tag{4.48}$$

where the embedding features are the audio samples, and h_i is the impulse response of the low pass Hamming filter. Low pass filtering is proposed as a simple way to reduce the audibility of the watermark.

4.3 Informed embedding

According to the blind embedding paradigm described in the previous section, watermark insertion is looked at as the mixing of a watermark signal $\mathbf{w}$ and the host feature set $\mathbf{f}$. As such, watermarking reduces to a classical communication problem, where $\mathbf{w}$ plays the role of the to-be-transmitted signal, and $\mathbf{f}$ plays the role of channel noise. Depending on the particular embedding rule, the channel may be additive or multiplicative, with the pdf of noise samples determined by the pdf of the host features.

Since the very beginning of watermarking research, a different approach to watermark embedding was also used: modify the set of host features so

Figure 4.15: Numbering of DCT blocks in Zhao-Koch's algorithm [237].

that they satisfy a particular relationship. It is easily understood that, according to this formulation, the definition of a watermark signal **w** to be mixed with **f** is no more necessary. It is only necessary to define a rule and modify **f** so that the desired relationship is verified. Due to the absence of an intermediate watermark signal, these kind of techniques may be referred to as direct embedding methods. In some cases, direct embedding leads to watermarking schemes where the host features are removed and replaced with a different feature set, thus justifying the name of substitutive watermarking which is sometimes used for this class of algorithms.

Example: the Zhao-Koch's algorithm [237]

The algorithm developed by Zhao and Koch in 1995 [237] is one of the first image watermarking algorithm published in the scientific literature. It is a hybrid algorithm embedding the watermark in the block-DCT coefficients of the host image. To this aim, the host image is first split into 8 × 8 non overlapping blocks. Blocks are then DCT transformed and medium frequency coefficients selected and numerated as in figure 4.15.

By starting from these 8 medium frequency coefficients, 18 subsets each containing 3 different coefficients are formed as exemplified in table 4.1.

Watermark insertion corresponds to letting the magnitudes of coefficients in one of these 18 subset having a predefined order. The exact subset actually conveying the watermark is determined according to a secret key K. A possible example of the mapping between information bits and coefficients order is given in table 4.2 (L = lowest magnitude, H = highest magnitude, M = medium magnitude).

When trying to embed a bit, three situations are possible. If the selected coefficients already exhibit the desired order, then the embedder leaves the coefficients unaltered. If the coefficients are not in the desired order they

Table 4.1: DCT coefficients subsets for watermarking embedding in the Zhao-Koch's scheme. Pixel position refers to numbering in figure 4.15.

Set no.	P_1	P_2	P_3
1	2	9	10
2	9	2	10
3	3	10	1
. . .	. . .	. . .	. . .
18	17	10	18

Table 4.2: Mapping bits into order of DCT coefficients in the Zhao-Koch's algorithm (L=lowest, M=medium, H=highest).

P_1	P_2	P_3	bit	P_1	P_2	P_3	bit
H	M	L	1	L	L	H	0
M	H	L	1	H	L	M	*invalid*
H	H	L	1	L	H	M	*invalid*
M	L	H	0	M	M	M	*invalid*
L	M	H	0				

are modified so that such an order is obtained. If the modification needed to reach the desired order is too large the embedder may decide to modify the coefficients so that one of the invalid combinations is reached. In this last case, no information bit is hidden within the DCT block. In order to increase the robustness of the watermark, the possibility of adding some redundancy bits to the information string is envisaged.

Watermark decoding is straightforward, since it only needs to recover the sequences of marked coefficients according to K and look for the order of DCT coefficients. Note also that, unlike with blind embedding schemes, in the absence of attacks the error probability is null.

The informed embedding principle

By comparing direct embedding watermarking and blind embedding schemes, we can observe that they exhibit an opposite behavior. In the blind case, the embedding rule is first defined, then the decoder/detector is designed so to recover the information from the watermarked asset. The direct embedding paradigm acts in the opposite way. First the detector/decoder is defined, then the host asset is modified so that the desired information is correctly recovered. In other words, first the detection (decoding) region in the feature space is defined, then the host asset is mapped into a point inside it (figure 3.1).

Whereas the analogy between blind embedding algorithms and communication theory is straightforward, the parallelism between direct embedding algorithms and digital communication was not recognized until late nineties, when the communication-with-side-information paradigm was first used to model the general watermarking problem. The basic observation the communication-with-side-information model relies on is that, though the original host feature set is unknown at the decoder, which must treat it as disturbing noise, such feature set is known at the encoder side (figure 3.12). Stated in another way, the encoder knows in advance part of the noise[21] that will affect the communication, and hence it can take some proper countermeasures to reduce the impact of decoder blindness on watermarking reliability. For this reason, watermarking algorithms exploiting the knowledge of the host feature set to embed the watermark are referred to as informed embedding algorithms.

In the attempt to clarify, and somewhat oversimplify, the informed embedding concept, let us consider the example reported in figure 4.16, where it is assumed that the transmitter has to send over the channel depicted in figure 3.12 one out of four signals (s_1 through s_4 in the figure). We also assume that a maximum transmitting power is set, thus compelling the transmitter to send a signal lying on, or within, the circle depicted in the figure. The decoding regions for each of the signals are also indicated. Let assume, now, that the signal s_1 has to be transmitted, and that the first source of noise affecting the transmitted signal is known by the encoder; let such a noise be denoted by n_1. A blind embedder would always transmit the same signal, namely s_1, since in the absence of any knowledge about noise this choice ensures the maximum transmission reliability. On the contrary, and informed embedder would exploit its knowledge about n_1 to decide the embedding strategy. Instead of transmitting s_1, it would transmit s_{if}, so that after noise addition the transmitted signal is still well inside the correct decoding region.

Let us consider now a typical detectable watermarking scheme where the watermark is simply added to the host feature set. Also assume that the detection region assumes the form given in figure 4.17[22]. If the watermark signal w is to be defined regardless of the host asset, a suitable choice would be to let it lie on the x-axis, as shown in the figure, in the hope that

---

[21]Noise due to attacks or processing of the watermarked asset is unknown both by the encoder and the decoder.

[22]As we will see in chapter 6, this is the typical detection region of a normalized correlation detector.

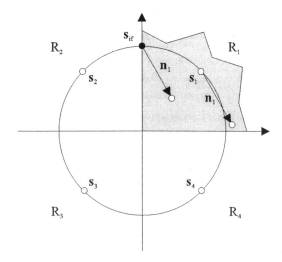

Figure 4.16: Since noise $\mathbf{n}_1$ is known in advance, the embedder transmits $\mathbf{s}_{if}$ instead of $\mathbf{s}_1$, so that after noise addition the transmitted signal is still well inside the correct decoding region.

after host asset addition, the watermarked asset A_w will still lie inside the detection region. The behavior of an informed embedder would be rather different: it would choose $\mathbf{w}$ so that after addition of A the watermarked asset lies as inside as possible within the detection region, e.g., by referring to figure 4.17, it would let $\mathbf{w} = \mathbf{w}_{if}$.

To further exemplify the informed embedding approach, let us revisit the example of figure 4.17 as in figure 4.18. A blind embedder would add the same watermarking signal $\mathbf{w}$ to all the host assets. Note that sometimes this may even lead to a watermarked asset which is outside the detection region (A_2 in the figure). On the contrary, an informed embedder would adapt the watermarking signal to the host asset, thus falling deeper inside the detection region.

As it can be readily seen, direct embedding watermarking is easily modelled by the informed embedding paradigm, with the advantage that the watermarking process can now be grounded on a strong theoretical basis[23], thus avoiding the heuristic flavor typical of early research in the field. By following the communication-with-side-information model, many useful insights about the ultimate achievable performance of any watermarking algorithm may be obtained. For example, it may be come as a surprise that, at least in the Gaussian additive case, decoder blindness does not affect

---

[23]See chapter 9 for more details.

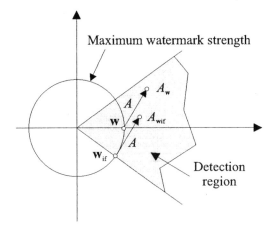

Figure 4.17: Since noise (the original asset, black arrow) affects mainly the second host feature, the informed embedder reinforces the second watermark component to increase immunity to noise.

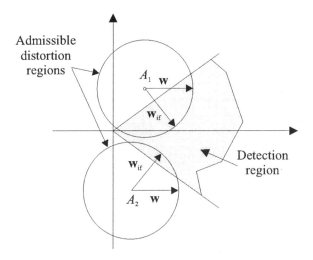

Figure 4.18: According to the informed embedding paradigm, the watermark signal is adapted to the host asset, thus falling deeper inside the detection region.

channel capacity at all, thus supporting the idea that, at least asymptotically, no loss of robustness has to be expected by denying the decoder the access to the original, non-marked, asset.

Actually, to get full advantage of the possibilities allowed by the informed embedding paradigm, knowledge of the host feature set must not only be used to properly map the host asset within the detection/decoding region. On the contrary, the detection (decoding) region itself should be properly designed so to allow the embedder to take full advantage of knowledge of A; thi can be done, for example, through dirty paper coding (section 3.4.5).

In this chapter we treat informed embedding at a rather heuristic level, without studying in depth the theoretical aspects of transmission through channels with side information at the encoder. Such a theory will be detailed in chapter 9, where an attempt will be done to give an information-theoretic background to digital watermarking.

In the following, we consider first the application of the informed embedding paradigm to the case of detectable watermarking, then we discuss the readable case.

4.3.1 Detectable watermarking

A general discussion of informed embedding algorithms is rather difficult for their strong dependence on the detection strategy used to assess watermark presence. In order to clarify the informed embedding principles in a practical case, then, we consider a case study in which the watermark is simply *added* to the host asset and detection is based on the normalized correlation between the watermark and the, possibly marked, asset[24]

Informed embedding

Let us begin by noting that, for the sake of watermark detection, any asset, be it watermarked or not, can be seen as a point in an n dimensional space, where n is the cardinality of the host feature set. With respect to a given watermark code, watermark detection and watermark invisibility can be defined via two regions in the feature space, namely the region of acceptable distortion and the detection region, i.e. the set of all points that the detector will classify as containing the watermark.

The exact shape of the region of acceptable distortion depends on the perceptual distortion metric used to model the perceptual appearance of the degradation introduced by the watermark. Unfortunately, perceptual metrics are very difficult to define, and, often, leading to results which are too cumbersome to be treated analytically. It is customary, then, to adopt the square Euclidean metric, thus resulting in regions of acceptable distortion having a (hyper)-spherical form in the feature space.

---

[24]The analysis presented in this section relies on the work by M.Miller et al. at NEC research Institute [153, 151].

At the same time, definition of the detection region requires that watermark detector is specified. In the sequel we assume that a detector based on normalized correlation is adopted. The possible rationale for adopting such a detector will be discussed extensively in chapter 6. To be specific, let $\mathbf{f}'$ be the vector with the host features and let $\mathbf{w}$ be the watermark we are looking for. Note that we can either have $\mathbf{f}' = \mathbf{f}_w$, if $\mathbf{f}'$ contains $\mathbf{w}$, or $\mathbf{f}' = \mathbf{f}$, if $\mathbf{f}'$ does not contain $\mathbf{w}$. Let us also consider, for simplicity, that $\mathbf{f}'$ and $\mathbf{w}$ have zero mean. In order to assess the presence of $\mathbf{w}$ within $\mathbf{f}'$, the normalized correlation ρ_n between such two vectors is computed:

$$\rho_n = \frac{\mathbf{f}' \cdot \mathbf{w}}{\|\mathbf{f}'\|\|\mathbf{w}\|} = \frac{\sum_{i=1}^{n} f_i' w_i}{\sqrt{\sum_{i=1}^{n} f_i'^2 \sum_{i=1}^{n} w_i^2}}, \qquad (4.49)$$

then ρ_n is compared against a detection threshold. If ρ_n is above the threshold the detector decides for the watermark presence, otherwise a negative answer is output. Thresholding equation (4.49) amounts to thresholding the angle formed by $\mathbf{w}$ and $\mathbf{f}'$. This leads to a hyperconic detection region R_w. In spite of the n-dimensional nature of the problem, a simple sketch of the detection region can be obtained by projecting it on the plane defined by $\mathbf{w}$, $\mathbf{f}'$ and the origin of the feature space. Under this assumption, R_w assumes the form of a triangular area pointing in the direction of the watermarking signal $\mathbf{w}$, as in figure 4.19, where, for simplicity, the x-axis has been taken in the same direction as $\mathbf{w}$. Along with the detection region, the point representing the or.ginal asset and the region of acceptable distortion are also given.

A blind embedder simply adds to $\mathbf{f}$ a scaled version of $\mathbf{w}$, thus moving the host asset in the direction of $\mathbf{w}$. In figure 4.19, the behavior of a blind embedder is indicated by the arrow labelled "BE" (Blind Embedding). As it can be seen, this strategy, which maximizes the linear correlation between $\mathbf{w}$ and $\mathbf{f}_w$, may even fail to map A to a point which is inside the detection region, thus leading to a non-null probability of missing the watermark even in the absence of attacks. This is only rarely the case, when a detection threshold which is much lower than that used in the figure is adopted, however, it clearly appears that more clever strategies may be used to obtain a marked asset which can be more easily detected.

An immediate way to exploit the knowledge of R_w consists in embedding a watermark which maximizes the answer of the detector. In our case, this amounts to minimizing the angle between the marked asset and $\mathbf{w}$, as indicated by the vector labelled "MNC" (Maximum Normalized Correlation) in figure 4.19. It is readily understood that, as soon as the detection region and the region of acceptable distortion overlap, this strategy always results in a watermarked asset which is inside the detection region. In this

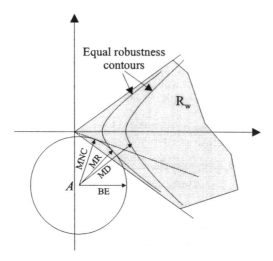

Figure 4.19: Different embedding strategies: blind embedding (BE), maximum detector answer (MNC), maximum robustness (MR), minimum distortion (MD). The figure refers to a detector based on the correlation coefficient between the host features and the to-be-looked-for watermark.

way, the probability of missing the watermark in the absence of attacks is zero. A problem with the MNC strategy is that it may result in a marked asset which is very close to the space origin, thus compromising watermark robustness, since even the addition of a small amount of noise may succeed in moving the marked asset outside R_w. To get around the problem one may choose to maximize robustness instead than detector answer. To do this, it is first necessary to define a measure of watermark robustness. This, in turn, depends heavily on the model used to describe attacks, especially on the correlation between attack noise and host data. In figure 4.19 an example is given in which constant robustness contours assume a hyperbolic form. In this case, maximizing robustness given a maximum admissible distortion (vector with "MR" label in the figure) may result in a watermarked asset which does not coincide with the one obtained by applying the MNC criterion. Interestingly, one may also decide to minimize distortion for a given level of robustness, thus obtaining the watermarked asset indicated by the "MD" (minimum distortion) label.

Example: Watermarking in the Fourier-Mellin domain [135]

A good example of how the informed embedding can be used to actually watermark an host asset is given by the image watermarking scheme proposed by Lin et al. in [135]. This scheme belongs to the class of detectable, transformed domain techniques, since, in the attempt to achieve automatic resilience against rotation, scale and translations, it embeds the watermark in the magnitude of Fourier spectrum expressed in log-polar coordinates. To be specific, let $I(x, y)$ denote the host image[25], and let $I'(x, y)$ indicate a rotated, scaled and translated version of $I(x, y)$, that is:

$$I'(x, y) = I(\sigma(x \cos \alpha + y \sin \alpha) - x_0, \sigma(-x \sin \alpha + y \cos \alpha) - y_0), \quad (4.50)$$

where α, σ and (x_0, y_0) indicate rotation angle, scale factor and translation amount, respectively. By taking the Fourier transform of $I'(x, y)$, we have:

$$\|F'(u, v)\| = \frac{1}{\sigma^2} \|F(\frac{1}{\sigma}(u \cos \alpha + v \sin \alpha), \frac{1}{\sigma}(-u \sin \alpha + v \cos \alpha)\| \quad (4.51)$$

where by $\|F(u, v)\|$ the magnitude of the Fourier Transform of $I(x, y)$ is meant. As it can be seen, $\|F'(u, v)\|$ does not depend on (x_0, y_0), thus ensuring translation invariance of the watermark. In order to achieve invariance against rotation and scaling, $\|F'(u, v)\|$ is resampled using log-polar coordinates, defined as:

$$u = e^\rho \cos \theta; \quad (4.52)$$

$$v = e^\rho \sin \theta. \quad (4.53)$$

By substituting the above equation in (4.51), we obtain:

$$\|F'(u, v)\| = \frac{1}{\sigma^2} \|F(\frac{1}{\sigma}e^\rho \cos(\theta - \alpha), \frac{1}{\sigma}e^\rho \sin(\theta - \alpha))\|, \quad (4.54)$$

which can be easily put in the following form:

$$\|F'(u, v)\| = \frac{1}{\sigma^2} \|F(e^{(\rho - \log \sigma)} \cos(\theta - \alpha), e^{(\rho - \log \sigma)} \sin(\theta - \alpha)\|; \quad (4.55)$$

$$\|F'(\rho, \theta)\| = \frac{1}{\sigma^2} \|F(\rho - \log \sigma, \theta - \alpha)\|, \quad (4.56)$$

which clearly shows how scaling and rotations are mapped into translations in the log-polar Fourier domain. As anticipated at the end of section 4.1.2, we can eventually get rid of scale and rotation by considering the magnitude of the Fourier transform of $\|F'(\rho, \theta)\|$. The approach propose by Lin et al. [135] is a different one: embed the watermark in the projection of $\|F'(\rho, \theta)\|$ along the ρ axis. To do so, let $g(\theta)$ be defined as follows:

$$g(\theta) = \int \log(\|F'(\rho, \theta)\|)d\rho, \quad (4.57)$$

---

[25]For sake of simplicity we initially ignore the problems stemming from the discrete nature of I.

or, by considering the discrete nature of the host image:

$$g(\theta) = \sum_k \log(\| F'(\rho_k, \theta) \|), \tag{4.58}$$

where log-magnitudes are used in order to equalize the dynamic range of Fourier coefficients. In addition, in order to take into account the symmetry properties of Fourier transform, and not to bias the scheme towards vertical or horizontal directions, the portions of $g(\theta)$ ranging between $0°$ and $90°$ are summed together, yielding:

$$\hat{g}(\theta) = g(\theta) + g(\theta + 90°), \tag{4.59}$$

with $\theta \in [0°, 90°)$. It is readily seen that $\hat{g}(\theta)$ is invariant to both translations and scaling, whereas rotations result in a shifting of $\hat{g}(\theta)$. In the algorithm by Lin et al. shifts of $\hat{g}(\theta)$ are dealt with through exhaustive search.

According to the informed embedding principle, before defining how the watermark is embedded within the host feature samples, namely $\hat{g}(\theta)$, we have to define how detection works. For the case at hand, we assume that detection relies on the correlation coefficient between the host feature sample $\mathbf{f} = \hat{g}(\theta), \theta = 0°, 1° \ldots 89°$ and the watermark $\mathbf{w}$:

$$D = \frac{\mathbf{w} \cdot \mathbf{f}}{\sqrt{(\mathbf{w} \cdot \mathbf{w})(\mathbf{f} \cdot \mathbf{f})}}, \tag{4.60}$$

where we used the symbol D to indicate correlation, not to get confused with the polar coordinate ρ. As usual, if D is above a detection threshold T, then the detector indicates that the watermark is present, otherwise that it is absent.

Having defined how the detector works, we can now establish a suitable embedding procedure. To do so, we start by applying the detection process to the non-watermarked original image I, that is we consider the set of host features $\mathbf{f} = g(\theta)$, with $g(\theta)$ directly computed on $F(\rho, \theta)$. Of course we would like that $\mathbf{f} = \mathbf{w}$, but we can not simply replace $\mathbf{f}$ with $\mathbf{w}$ since this would result in a highly visible watermark. On the contrary, let us compute a new signal $\mathbf{s}$ which is a suitable mixture of $\mathbf{f}$ and $\mathbf{w}$, e.g. a weighted average of $\mathbf{f}$ and $\mathbf{w}$. Now replace $\mathbf{f}$ with $\mathbf{s}$. This can be easily accomplished by operating on the rows of $F(\rho, \theta)$, e.g. by adding the value $(s_i - f_i)/K$ to each of the K values in the i-th row of $F(\rho, \theta)$.

A problem with the procedure described above is that inversion of log-polar sampling is an ill-conditioned problem, thus it is not feasible to embed the watermark in the log-polar Fourier domain and then go back to cartesian coordinates. The solution suggested in [135] exploits again the informed embedding principle. The watermark is inserted by modifying Fourier samples in the cartesian coordinate system in such a way that by applying the detector a proper answer is obtained. More specifically, let F be a column vector containing all the elements of the log-polar array, and C a column vector with the elements of the cartesian array of frequencies, we have:

$$F = MC, \tag{4.61}$$

where M is an interpolation matrix containing the weights to perform interpolation. We can assume that each interpolated sample only depends on four neighbor samples in the cartesian plane, thus each row of M contains only 4 nonzero values, moreover such values need to sum to 1. Then, instead of adding the quantity $(s_i - f_i)/K$ to samples in F, we can add the same quantity to the 4 non-zero samples in C which contribute to forming the $i - th$ sample in F. Unfortunately, such a procedure is not an exact one, since the same sample in C may contribute to form more than one sample in F, thus a set of conflicting modifications would be needed. To get around the problem, each sample in C is modified by weighting all the desired changes affecting it. For example, by assuming that $M_{j,i}$ and $M_{j,k}$ are the only non zero elements of row j, and that $F_i' - F_i$ and $F_k' - F_k$ are the changes to be applied to the i-th and k-th elements of F, we would apply the following change:

$$\frac{M_{i,j}(F_i' - F_i) + M_{k,j}(F_k' - F_k)}{M_{i,j} + M_{k,j}}. \tag{4.62}$$

Of course, the above procedure provides only an approximation of the desired inversion problem, since in general there will be more than two non zero elements. Hence, if after applying it the detector answer is not sufficiently high, the same procedure is iterated as many times as necessary (usually three of four iterations are enough).

In order to make watermarking in the log-polar Fourier domain effective, many other implementation problems need to be considered, including the influence of DFT tiling on rotated images, the unreliability of very low and very high frequencies, difficulty in applying a truly white watermark. For a comprehensive treatment of all this aspects, the reader is referred to the original work by Lin et al.

Informed coding

Full application of the side information available at the embedder is not limited to the definition of the embedding rule, e.g. to the mapping of the host asset into a point within the detection region. On the contrary, the detection region itself can be designed so to improve system performance.

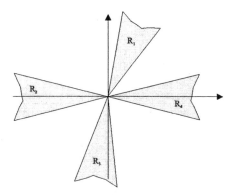

Figure 4.20: Informed coding for detectable watermarking: 4 different detection regions are associated to the same message.

In section 3.4.5, this led us to the definition of the dirty-paper coding principle, in which a redundant set of codewords is used for each to-be-transmitted message. Here we will adapt the same principle to fit the needs of detectable watermarking.

Let us assume, again, that watermark detection is carried out via normalized correlation (as in equation (4.49)). As opposed to classical watermark embedding, in which a single codeword is associated to each message, now the encoder chooses one among several watermark signals, on the basis of the particular host asset to be marked. The set of signals associated to each message can be chosen at random, so that the entire asset space is almost equally covered. Given a message, the choice of the signal to be transmitted, can be made by looking at the signal which is closest to the host asset. First of all the watermark signals associated to the to-be-transmitted message are looked for in the original non-marked asset. Then the signal resulting in the highest response is chosen for embedding. Suppose, for example, that the MD strategy described previously is used. Given the desired level of robustness, by choosing a signal for which the detector answer is already high even without actually embedding it, permits to achieve the desired robustness with a lower distortion. Alternatively, given the maximum allowed distortion, by choosing a proper watermarking signal, it is possible to maximize robustness, i.e. getting deeper into the detection region. The detector needs only to be modified to search for more than one watermark signal for each possible message. The above principles are exemplified in figure 4.20, where the detection regions corresponding to 4 watermark signals associated to the same message are depicted.

The performance improvement achievable through informed coding is quite substantial. For example, by considering the watermarking of still images, by increasing the number of signals associated to each message from 1 to 1000, the signal to noise ratio[26] of the watermarked image may be improved up to 4.5dB.

4.3.2   Readable watermarking

Exploitation of side information (host asset knowledge), in readable watermarking, goes the same way followed for the detectable case. To be specific, for each message $\mathbf{b}_i$ let us indicate by R_i the set of all the points in the feature space for which the decoder results in

$$\mathcal{D}(A, K) = \hat{\mathbf{b}} = \mathbf{b}_i, \qquad (4.63)$$

where A is the cover asset under examination and K accounts for the presence of a secret key ensuring privacy of the system. At a first, simpler level, the informed embedding principle may be used to map the host asset into a point within R_i. As opposed to blind embedding, this results in a watermark signal which adapts itself to the host asset at hand (see figure 3.1). At a more sophisticated level, the form of decoding regions is modified so that they are composed by a set of disjoint sets $R_{i,j}$ (decoding subregions) uniformly spread over the whole feature space. As an example, in figure 4.21 the detection regions corresponding to two different messages $\mathbf{b}_1$ and $\mathbf{b}_2$ are shown. Informed watermark embedding, then, consists of two separate steps. Choice of a suitable decoding subregion the host asset should be mapped in, and informed embedding of the host asset within such a sub-region. By referring again to figure 4.21, we can see how two different subregions are chosen for two different assets A_1 and A_2 conveying the same message $\mathbf{b}_1$. In this case, choice of subregions was done based on a minimum distance criterion.

A difficulty with informed coding, is that definition of decoding subregions may be a difficult task. Theoretical analysis suggests adoption of a random coding approach, however, as soon as the watermark payload increases, this requires a huge computational effort to memorize all the decoding subregions and, consequently, to extract the hidden message from the host asset. Possible solutions, include the adoption of structured codes such as the dirty paper codes described in section 3.4.5, or lattice quantization of the feature space.

---

[26]In this context the original image plays the role of the signal and the watermark signal that of noise.

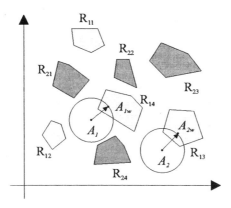

Figure 4.21: Informed coding for readable watermarking. Different assets are marked by mapping them into different decoding sub-regions.

A general informed embedding scheme for readable watermarking

In this paragraph, we present a general approach to informed embedding that can be used virtually with any watermarking scheme for which a measure of robustness is available. Let us start by noting that in the readable case, measuring watermark robustness means measuring the amount of distortion the host asset may undergo before the decoder outputs a wrong sequence.

Without losing, generality we can start by assuming that we want to embed the watermark code $\mathbf{b}_1$, and that channel coding is not used. A decoding error occurs when the decoded bit sequence is not equal to $\mathbf{b}_1$, i.e. when

$$\mathcal{D}(A) = \mathbf{b}_i \qquad i \neq 1, \tag{4.64}$$

where we have neglected the possible dependence of $\mathcal{D}$ on a secret key K, or

$$\mathcal{D}(\mathbf{f}) = \mathbf{b}_i \qquad i \neq 1, \tag{4.65}$$

where the dependence of $\mathcal{D}$ upon the host feature set $\mathbf{f}$ is explicitly shown. Assume, then, that embedding $\mathbf{b}_1$ results in a marked feature vector equal to $\mathbf{f}_w$. We start by considering a simplified situation in which we take into account only two elements of $\mathbf{B}$ at a time, say $\mathbf{b}_1$ and $\mathbf{b}_i$. We indicate by

$$R_2(\mathbf{f}_w, \mathbf{b}_1, \mathbf{b}_i), \qquad i \neq 1, \tag{4.66}$$

a generic measure of the probability that when the host asset is attacked the watermark code $\mathbf{b}_1$ is correctly decoded. Note that, in order to define

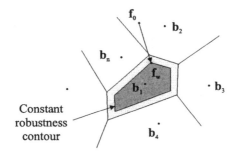

Figure 4.22: Embedding a watermark ensuring a certain degree of robustness amounts to mapping the host asset into a point which is sufficiently inside the right decoding region.

$R_2(\mathbf{f}_w, \mathbf{b}_1, \mathbf{b}_i)$, a statistical characterization of the attack is needed. Clearly, a global measure of the robustness of $\mathbf{b}_1$ when embedded in $\mathbf{f}_w$ is given by

$$R(\mathbf{f}_w, \mathbf{b}_1) = \min_{i \neq 1} R_2(\mathbf{f}_w, \mathbf{b}_1, \mathbf{b}_i). \tag{4.67}$$

As illustrated in figure 4.22, embedding a watermark ensuring a certain degree of robustness, i.e. a sufficiently low probability of decoding error, amounts to mapping the host asset into a point which is sufficiently inside the right decoding region. Constant robustness contours are reported as well.

If the robustness measure $R_2()$ is available at the encoder, informed watermark embedding may be accomplished through the following iterative algorithm:

1. Let R_t be the desired level of robustness,

2. Let the initial host feature set be equal to the non-marked features: $\mathbf{f}_w = \mathbf{f}_0$,

3. Find $\mathbf{b}_i \neq \mathbf{b}_1$ such that $R_2(\mathbf{f}_w, \mathbf{b}_1, \mathbf{b}_i)$ is minimum,

4. If $R_2(\mathbf{f}_w, \mathbf{b}_1, \mathbf{b}_i) \geq R_t$, then stop,

5. Else modify $\mathbf{f}_w$ in such a way that $R_2(\mathbf{f}_w, \mathbf{b}_1, \mathbf{b}_i) \geq R_t$, and go to 3.

The above algorithm is a very general one and can be applied to a variety of situations. However, in order to be actually used some points need to be specified. First of all, it is necessary to devise a modification strategy that permits to satisfy the constraint $R_2(\mathbf{f}_w, \mathbf{b}_1, \mathbf{b}_i) \geq R_t$. This strongly depends on the robustness measure. Let us assume, for example, that attacks can be modelled as addition of Gaussian white noise, and that decoding is based on correlation, i.e.:

$$\mathcal{D}(\mathbf{f}) = \arg\max_i(\mathbf{f} \cdot \mathbf{b}_i), \tag{4.68}$$

which, in the case of watermarks with equal energy corresponds to minimum distance decoding. Then, by letting $\mathbf{f}'$ indicate the set of attacked host features we have

$$P\{\mathbf{f}' \cdot \mathbf{b}_1 > \mathbf{f}' \cdot \mathbf{b}_i\} = P\{(\mathbf{f}_w + \mathbf{n}) \cdot \mathbf{b}_1 > (\mathbf{f}_w + \mathbf{n}) \cdot \mathbf{b}_i\}, \tag{4.69}$$

where $\mathbf{n}$ indicates the white noise added by the attacker. By further manipulating equation (4.69), we have:

$$P\{(\mathbf{b}_1 - \mathbf{b}_i) \cdot \mathbf{f}_w > (\mathbf{b}_i - \mathbf{b}_1) \cdot \mathbf{n}\} = P\left\{\frac{(\mathbf{b}_1 - \mathbf{b}_i) \cdot \mathbf{f}_w}{\|\mathbf{b}_1 - \mathbf{b}_i\|} > \frac{(\mathbf{b}_i - \mathbf{b}_1)}{\|\mathbf{b}_i - \mathbf{b}_1\|} \cdot \mathbf{n}\right\}. \tag{4.70}$$

By noting that the right term of the above inequality is nothing but the projection of noise on the direction defined by $(\mathbf{b}_1 - \mathbf{b}_i)$, we can easily argue that

$$R_2(\mathbf{f}_w, \mathbf{b}_1, \mathbf{b}_i) = \frac{(\mathbf{b}_1 - \mathbf{b}_i) \cdot \mathbf{f}_w}{\|\mathbf{b}_1 - \mathbf{b}_i\|} \tag{4.71}$$

is a suitable measure of watermark robustness when the analysis is limited to codewords $\mathbf{b}_1$ and $\mathbf{b}_i$. In this case, a feasible way to implement the last step of the watermark embedding procedure described above, consists in moving $\mathbf{f}_w$ in the direction connecting $\mathbf{b}_1$ and $\mathbf{b}_i$, by an amount which is proportional to $R_t - R_2(\mathbf{f}_w, \mathbf{b}_1, \mathbf{b}_i)$, i.e. the marked features at step $n + 1$ are computed as follows:

$$\mathbf{f}_w(n + 1) = \mathbf{f}_w(n) + \alpha\frac{\mathbf{b}_1 - \mathbf{b}_i}{\|\mathbf{b}_1 - \mathbf{b}_i\|}, \tag{4.72}$$

with

$$\alpha = R_t - R_2(\mathbf{f}_w, \mathbf{b}_1, \mathbf{b}_i). \tag{4.73}$$

A geometric interpretation of the above equations is given in figure 4.23. Note that the final position of the vector with the watermarked features is not an optimum one ($\mathbf{f}_w$ in figure 4.22), even if the robustness constraint:

$$R(\mathbf{f}_w, \mathbf{b}_1) \geq R_t \tag{4.74}$$

is clearly satisfied.

A second problem with the general embedding scheme outlined in this section is that step 3 may be extremely demanding from a computational point of view, especially when the watermark payload is high, and when

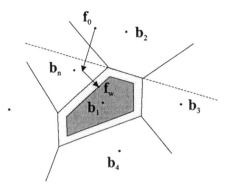

Figure 4.23: Iterative informed embedding. Note that the final position of the watermarked asset does not necessarily coincide with the optimum one (f_w in figure 4.22).

channel coding is used. For a solution to the above problem which couples the informed embedding principles exposed here, and informed coding through dirty-paper trellis coding as described in section 3.4.5, the reader is referred to [155].

Another example in which the informed embedding approach is exploited to optimize watermark embedding, while also taking into account perceptual considerations is given below.

Optimum embedding in block DCT domain [167]

A problem with the general scheme described above is the lack of perceptual modelling, a problem that can be solved only by loosing in generality and considering a specific watermarking scheme. A good, yet simple, example of how perceptual modelling can be accommodated within an informed embedding framework is given by the still image watermarking system developed by Pereira et al. [167]. This system operates in the block DCT domain[27], with each block accommodating up to 2 bits. Each bit is embedded in the sign of a host DCT coefficient. In order to change the sign of host coefficients (if necessary), or to increase watermark robustness, DCT coefficients are increased or decreased. Of course, during embedding it is desirable to increase or decrease the DCT coefficients as much as possible for maximum robustness. However such a distortion is limited by the invisibility constraint. According to the formulation by Pereira et al., it is assumed that such a constraint can be given the form of a Noise Visibility

---

[27]A version of the same system operating in the wavelet domain has also been proposed by the same authors in [167].

Function (NVF) giving, pixel by pixel, the maximum positive and negative distortion that can be applied to such a pixel without introducing any visible distortion.

In order to define the optimum embedding strategy, let $\mathbf{x} = \{x_{1,1} \ldots x_{8,1}, x_{1,2} \ldots x_{8,2}, x_{1,8} \ldots x_{8,8}\}^t$ be a vector with the amount of modification applied to each DCT coefficient in the block, and let $\mathbf{v}$ be a vector giving the maximum admissible positive and negative distortion that can be applied to the pixels in the block. Let such distortions be denoted by $v_{+,i}$ and $v_{-,i}$ respectively, where i indicates pixel position within the 8×8 block. Note that $v_{+,i}$ and $v_{-,i}$ need not be equal since truncation effects are also taken into account. The problem can now be formulated as a standard constrained optimization problem. For each block two mid-frequency coefficients are selected, in which the information bits will be embedded. Optimum embedding reduces to solving the following problem:

$$\min_{\mathbf{x}}(\mathbf{p} \cdot \mathbf{x}), \qquad (4.75)$$

where $\mathbf{p}$ is a vector of zero's except in the positions of the two selected coefficients. More specifically, we insert in such positions a -1 or a 1 depending on whether the DCT coefficient needs to be increased or decreased respectively. Minimization of the expression in equation (4.75) is performed subject to the constraint

$$\mathbf{Tx} \le \mathbf{v}, \qquad (4.76)$$

with

$$\mathbf{T} = \begin{bmatrix} IDCT \\ \hline -IDCT \end{bmatrix}; \quad \mathbf{v} = \begin{bmatrix} \mathbf{v}_+ \\ \hline \mathbf{v}_- \end{bmatrix}; \qquad (4.77)$$

where $IDCT$ is the matrix yielding the 2D inverse DCT transform of $\mathbf{x}$, and $\mathbf{v}_+$ and $\mathbf{v}_-$ are column vectors with the maximum positive and negative distortions applicable to the pixels in the block. Stated in the form expressed by equations (4.75) through (4.77), the embedding problem can be easily solved through linear programming methods, e.g. the Simplex method. Note that additional constraints on the maximum admissible distortion in the frequency domain can be easily considered by imposing that:

$$\mathbf{L} \le \mathbf{x} \le \mathbf{U}, \qquad (4.78)$$

where $\mathbf{L}$ and $\mathbf{U}$ are two vectors with the lower and upper bounds on the admissible distortion affecting each DCT coefficient.

Whereas the above method is a very simple one, it can be regarded as a truly informed embedding scheme since in order to increase watermark

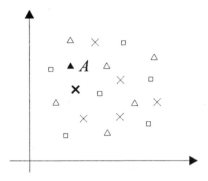

Figure 4.24: Geometrical interpretation of QIM watermarking. Based on the to-be-transmitted message a quantizer is chosen (× in this figure), then the quantized value closest to the host asset (▲) is selected (indicated here by boldface ×).

robustness, the embedder does not only increase the amount of distortion applied to the coefficients actually bearing the information bits. On the contrary, all DCT coefficients in a given block may be modified in order to *make room* for the hidden information.

QIM and Scalar QIM watermarking

As we already noted, full exploitation of the side information available at the encoder demands that decoding regions are properly designed, e.g. split into several subregions spread uniformly over the feature space. The embedder first decides into which subregion the host asset should be mapped (informed coding), then it applies the informed embedding principles to actually move the host asset within the selected subregion. Effective design of decoding subregion, though, is a very complex problem. In line of principle, it can be demonstrated that optimal results are obtained by random coding, i.e. random selection of subregions. However such an approach is computationally unfeasible, thus calling for the use of structured codebooks such as those provided by means of dirty-paper trellis coding. A very simple way to simultaneously achieve informed coding and embedding is through Dither Modulation (DM), a watermarking scheme belonging to the wider class of Quantization Index Modulation (QIM) algorithms.

In QIM schemes, watermarking is achieved through the quantization of the host asset, namely the host feature vector, according to a set of predefined quantizers, where the particular quantizer used for the case at hand depends on the to-be-hidden message b. Stated in another way, the to-be-hidden message modulates the quantizer index, hence justifying the QIM appellative. In order to exemplify the QIM concept, let us assume

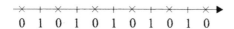

Figure 4.25: Geometrical interpretation of 1-bit SQIM. The host (scalar) feature is mapped into the closest × or | according to the to-be-hidden bit.

that only three possible messages are possible. In figure 4.24 the quantized vectors associated to each quantizer are depicted, where points marked with ×'s, □'s and △'s are associated with the first, second and third message respectively. Let us assume, now, that the first message, say b_1 has to be hidden within a host asset A, whose position in the feature space is indicated by ▲. The embedder looks for the × which is closest to ▲ and maps the host asset into such a point. The decoder looks for the quantized value which is closest to the asset at hand, and output the bit sequence corresponding to the quantizer the quantized value belongs to. Note that the decoder does not know which quantizer was used, thus it extends its search to all the quantized points in the asset space, i.e. the points marked with either with a ×, a □ or a △.

Of course, in order not introduce a perceptually significant distortion, each quantizer must be fine enough. On the contrary, to increase robustness quantized vectors belonging to different quantizers should be placed as far as possible. Note that as the host asset varies, the marked asset varies from one × point to another, but it never moves from a × point to a □ or a △ point. Thus, in the absence of attacks, the decoder never makes a mistake, regardless of the energy of the host signal.

Dither Modulation (DM) watermarking, hereafter referred to as Scalar QIM (SQIM), is a QIM scheme in which the ensemble of all quantizers forms a rectangular n-dimensional lattice, and single quantizers correspond to rectangular sub-lattices. In this way, both watermark encoding and decoding are straightforward, since they amount to componentwise feature quantization. In order to be specific, let us consider a simple situation in which a 2-bit message has to be hidden within a feature vector consisting of two features only. Also assume that the first bit is hidden in the first feature and the second bit in the second feature. Hiding of a single bit, then, corresponds to defining two different scalar quantizers, producing the quantized values marked by ×'s and |'s in figure 4.25. When the effect of the quantizers acting on each feature separately is visualized in a 2-dimensional space, we obtain four 2D quantizers, corresponding to the four possible input sequences 00, 01, 11, and 11. The position of these four quantizers is depicted in figure 4.26. SQIM watermarking corresponds to quantizing the input asset according to one of the four 2D quantizers reported in the figure. Note that vector quantization is not actually required since, due to

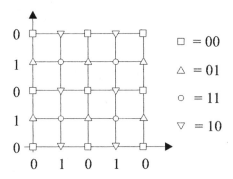

Figure 4.26: Geometrical interpretation of 2-bit SQIM. The host feature vector is mapped into the closest quantized values belonging to the quantizer associated to the to-be-hidden message.

the rectangular nature of quantization lattices, scalar quantization can be carried carried out on each host feature separately.

The simple quantization-based algorithm described above, is the progenitor of a wide class of powerful algorithms stemming from an information theoretic analysis of the watermarking problem. It is not possible, however, to describe them without delving into the details of the theoretical analysis of communication systems with side information at the encoder, thus we will postpone their description to chapter 9.

DM watermarking with bit repetition and ST-DM [39]

Hiding each bit into a single feature leads to a very unreliable watermark, since, in order to match the imperceptibility constraint, a very low energy, i.e. a very small quantization step, has to be used.

A first possibility to overcome this problem, consists in hiding the same bit into r consecutive host features. Let us start by formalizing the scalar DM scheme described in the previous paragraph. We first define the two codebooks associated, respectively to $b = 0$ and $b = 1$ (b is the to-be-hidden bit):

$$\mathcal{U}_0 = \{k\Delta + d, k \in \mathbb{Z}\}, \tag{4.79}$$

$$\mathcal{U}_1 = \{k\Delta + \Delta/2] + d, k \in \mathbb{Z}\}, \tag{4.80}$$

where d is an arbitrary parameter, possibly depending on a secret key to improve security. In the sequel, we will assume $d = 0$ for simplicity, however, as it will be argued in section 7.3.2, by letting $d = \Delta/4$ a less obtrusive watermark is obtained, while keeping the system performance

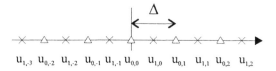

Figure 4.27: Codebook entries for scalar DM watermarking.

constant. Watermark embedding amounts to the application of either the quantizer $\mathcal{Q}_0$ associated to $\mathcal{U}_0$:

$$\mathcal{Q}_0(f) = \arg \min_{u_{0,i} \in \mathcal{U}_0} |u_{0,i} - f|, \tag{4.81}$$

where f is the feature hosting b and $u_{0,i}$ are the elements of codebook $\mathcal{U}_0$, or the quantizer corresponding to $b = 1$:

$$\mathcal{Q}_1(f) = \arg \min_{u_{1,i} \in \mathcal{U}_1} |u_{1,i} - f|. \tag{4.82}$$

By letting f_w indicate the marked feature, we then have:

$$f_w = \begin{cases} \mathcal{Q}_0(f) & b = 0 \\ \mathcal{Q}_1(f) & b = 1 \end{cases} \tag{4.83}$$

whose geometrical representation is given in figure 4.27.

When the same bit is embedded within r consecutive host features $\mathbf{f} = \{f_1, f_2 \ldots f_r\}$, two r-dimensional quantizer are defined by starting from the scalar quantizers defined above. In DM watermarking with bit repetition, the r-dimensional codebooks $\mathcal{U}_0^r$ and $\mathcal{U}_1^r$ are the product of the corresponding scalar codebooks, i.e.

$$\mathcal{U}_0^r = \underbrace{\mathcal{U}_0 \times \mathcal{U}_0 \cdots \times \mathcal{U}_0}_{r \text{ times}} \tag{4.84}$$

and

$$\mathcal{U}_1^r = \underbrace{\mathcal{U}_1 \times \mathcal{U}_1 \cdots \times \mathcal{U}_1}_{r \text{ times}} \tag{4.85}$$

The quantizer associated to $b = 0$, then, works as follows:

$$\mathbf{f}_w = \mathcal{Q}_0^r(\mathbf{f}) = \arg \min_{\mathbf{u}_{0,i} \in \mathcal{U}_0^r} \|\mathbf{u}_{0,i} - \mathbf{f}\|^2. \tag{4.86}$$

Given the particular form of the codebook, this amounts to quantizing each component of $\mathbf{f}$ separately, i.e.:

$$f_{w,i} = \mathcal{Q}_0(f_i) = \arg \min_{u_{0,i} \in \mathcal{U}_0} |u_{0,i} - f_i|. \tag{4.87}$$

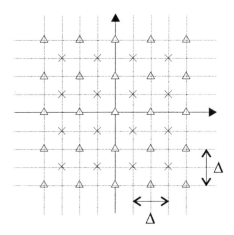

Figure 4.28: Codebook entries for DM watermarking with bit repetition ($r = 2$). Triangles refer to $\mathcal{U}_0$, whereas crosses correspond to $\mathcal{U}_1$.

Similar considerations hold for $b = 1$. The position of the codebook entries in the feature space in exemplified in figure 4.28 for $r = 2$.

As it will be detailed in chapter 6, bit repetition permits to improve significantly the performance of DM watermarking. A problem with bit repetition, which somewhat limits the improvement achievable through the usage of several host coefficients, is due to the fact that when r increases the number of nearby entries belonging to the other codebook increases too, thus limiting robustness. This is evident in figure 4.28 where each entry of $\mathcal{U}_0$ is surrounded by four elements of $\mathcal{U}_1$ (there were only two such neighbors for $r = 1$). Such a problem is addressed and solved by Spread Transform Dither Modulation (ST-DM). According to such a scheme, the correlation between the host feature set and a reference watermark signal **w** is quantized instead of the feature themselves. Let us assume that **w** is a normalized binary pseudorandom sequence. The watermarked feature vector $\mathbf{f}_w$ is derived from **f** as follows. First the correlation between **f** and **w** is calculated:

$$\rho_f = \mathbf{f} \cdot \mathbf{w}. \tag{4.88}$$

Then the projection of **f** over **w** is subtracted from **f** and a new vector component along the direction of **w** is added resulting in the desired quantized autocorrelation (ρ_w):

$$\mathbf{f}_w = \mathbf{f} - \rho_f \mathbf{w} + \rho_w \mathbf{w}, \tag{4.89}$$

as to ρ_w, it is calculated by applying a proper quantizer to ρ_f. In order to compute the admissible quantization step, let us remember that **w** is a

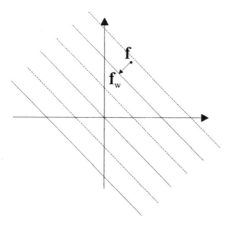

Figure 4.29: Geometrical representation of ST-DM watermarking. Points on solid lines form the $\mathcal{U}_0$ codebook (gray quantization regions, whereas dashed lines correspond to $\mathcal{U}_1$.

binary, normalized vector, hence its components take only values $\pm 1/\sqrt{r}$. If the maximum admissible quantization step along each feature component is Δ, then ρ_f may be quantized with step $\sqrt{r}\Delta$. A geometric interpretation of ST-DM watermarking for $r = 2$ is given in figure 4.29. Solid lines represent quantized values corresponding to $b = 0$, whereas dashed lines are relative to $b = 1$. For any host asset A, embedding $b = 0$ is obtained by projecting the host feature vector f over the closest solid line (dashed lines are used for $b = 1$). As it can be seen, quantization regions corresponding to $b = 0$ only border on two regions for which $b = 1$. In chapter 6, we will show that this property will result in a lower bit error probability with respect to DM with bit repetition.

4.4 Further reading

The use of the Fourier-Mellin transform to achieve invariance to the most common geometric attacks has been first proposed by J. J. K. Ó Ruanaidh and T. Pun in 1997 [162]. Since then the difficulties associated to the practical implementation of the ideas contained in [162] has been experimented by many researchers. A watermarking algorithm based on the Fourier-Mellin transform which can be actually applied to practical situations is described in [135].

The few notes about wavelet analysis we gave in this chapter can not be considered by no means as an introduction to multiresolution analysis based on the wavelet framework. For a comprehensive, yet simple, introduction to such a field, readers may refer to the seminal work by S. Mallat [143], or to one of the many book written on this subject. e.g. [222, 144]. As to the use of the complex wavelet transform to improve robustness against shifting in the asset domain, readers are referred to the work by P. Loo and N. Kingsbury [139].

A detailed description of the most popular audio compression standards can be found in the relevant publication of the standardization bodies, for instance [102] and [104]. With regard to subband decomposition of audio signals, a good introductory work has been written by T. Painter and A. Spanias [165]. Finally a good discussion of the requirements and limitations of audio watermarking techniques may be found in [122].

In section 4.1.5 we gave a brief overview of a set of techniques embedding the watermark in, so to say, non conventional host feature set. For a more detailed description of these techniques, readers are referred to [48, 47] for histogram watermarking, [204] for an algorithm using the Radon-Wigner distribution, and [24] for a fractal-based technique system (more details about the collage theorem, fractal-based methods rely on, nay be found in [111]). Finally a number of excellent surveys going through the most popular watermarking techniques developed so far have been published in the scientific literature in the last years [208, 92, 177, 131].

As we already noted at the end of chapter 3, the informed embedding paradigm was first introduced in [57]. The applications of such a to readable watermarking algorithms led to a class of algorithms based on dirty paper coding. In the context of detectable watermarking, informed embedding/coding principles have been applied successfully to improve the performance of conventional systems based on the spread spectrum paradigm, as it is carefully demonstrated in [153, 151, 135].

QIM watermarking algorithms have been introduced only recently, mainly thanks to the work by B. Chen and G. Wornell [37, 39] and J. Eggers, B. Girod, R. Bauml and R. Tzschoppe [70, 71].

5

Data concealment

To better understand the techniques commonly used to hide the watermark signal within the host asset, let us consider the exact meaning of the term *to hide*. According to the definition of the Oxford dictionary *to hide* means: *prevent something (or somebody, or oneself) from being seen; put or keep out of sight; prevent something from being known; keep something secret.* The hiding concept is, thus, mainly related to the possibility of perceiving the presence of an object by means of the most important human sense, the sight. Such an idea may be easily extended to the other human senses, or even to the *augmented* sensing capabilities achievable through the use of any other sensor, e.g. mechanical or electronic sensors. At the same time, to hide an object, or a piece of information, means to keep it secret, thus referring to the knowledge sphere of human capabilities. In this case, the existence itself of the object is kept secret.

In general the main approaches employed for hiding something are:

Keep it secret The hidden object is put in a place which is unknown to not authorized people. If an object location is unknown it is unlikely that it can be seen.

Make it small The hidden object is made so small that nobody is able to see it. The ability of people to perceive an object is limited by its dimension.

Make it similar The hidden object is made so similar to the surrounding environment that it is not possible to distinguish it. This is the most common approach used by animals for hiding themselves (e.g. think to chameleons).

Make it spread The object to be hidden is sub-divided into pieces which are spread around. In this case is the whole object that can not be perceived. This is also related to the second approach, if the object can not be made small, its pieces could be.

From what we have said about the meaning of *hiding*, the importance of having a good knowledge of the characteristics of the Human Visual System (HVS) and Human Auditory System (HAS) appears with evidence. Having a clear idea of the mechanisms underlying the perception of visual and auditory stimuli can help in fine tuning the watermark embedding phase, in particular for what regards making the embedded signal invisible. Over the past years the results of the studies of the HVS and the HAS have been largely exploited for developing effective image, video and audio compression techniques: in fact, data hiding has got many hints from compression research. One issue need to be addressed with this regard, that is the duality between multimedia data compression and multimedia data hiding, and, in particular the contrasting effects that these technologies tend to have on multimedia data.

The duality between the problem of data hiding and that of compression consists in the fact that, while in compression technology the aim is to remove from the multimedia document all those data which are perceptually less important, in data hiding technology the goal is, on the contrary, to add to the multimedia document some data in such a way that they result to be perceptually unimportant. Perceptual relevance has thus to be evaluated in both applications, in the former case for removing something, and in the latter for adding something. Tools for evaluating perceptual relevance are just those that can be developed based on the experimental results obtained by the researchers studying the HVS and the HAS. Given this duality a contradictory situation could be reached: an optimum compression technique would, in fact, be able to remove everything that is perceptual not relevant, and thus would also be able to completely remove (as perceptually not relevant) data possibly hidden inside the multimedia document. Although the presented situation have some elements of truth, in that, indeed, compression techniques will really reduce effectiveness of data hiding strategies, anyway reality is slightly different. To exploit in a highly effective way the perceptual characteristics of humans, in fact, two conditions have to be satisfied, the first regard the ability to embrace large computational costs, the second to strongly adapt to the data at hand. These two conditions can be satisfied with difficulty by compression tools, the former because it is almost always required that the compression process is performed very fast, the latter because a strong adaptation to the to be compressed data would require a large amount of side information to be transmitted to the decompression phase, thus partially nullifying the gain in compression performance. On the contrary data hiding tools not always have to satisfy stringent time requirements (only some applications, such as, for example, fingerprinting for infringement tracking, or copy control, have to), and never require side information to be transmitted (given that data extraction have to be performed only based on the watermarked data). Thus it is likely that data could be perceptually hidden inside multimedia documents more effectively than they could be removed by compression algorithms. Thanks to the absence of the above mentioned requirements (computational time limitation and side information minimization), it is even likely that data hiding could benefit from carefully exploiting the characteristics of the HVS and HAS more strongly than compression techniques do.

Of course the above considerations about the usefulness of exploiting the characteristics of the HVS or HAS are valid when hidden data concealment has to be granted with respect to human observers (or listeners). There are, anyway, situations in which the end users of the multimedia document is mainly not an human, but a machine. As two examples we can cite the case of video surveillance images, for which watermark invisibility should be granted with respect to automatic video surveillance algorithms, or the case of remote sensing multispectral images, for which watermark invisibility should be granted with respect to classification tools. When this happens the concealment criteria need to be adapted to the application at hand.

The first two section of this chapter are thus dedicated to describe the most popular models of, respectively, the HVS and the HAS from a general point of view. The following two sections will, on the contrary, concentrate on the approaches that can be followed for exploiting the human perception characteristics for data hiding. The last section will, finally, present how data concealment can depend on the application at hand, and thus, how the concealment criteria should be adapted in such cases.

5.1 The Human Visual System

The Human Visual System (HVS) is certainly one of the most complex biological devices. The processes that allow to our brain to build a model of the surrounding real world, based on the light which impacts the retina, are very far from being exactly understood. For trying to explain vision mechanisms, research is performed on several fronts spacing from psychology to physiology, from neurology to anatomy, from artificial intelligence to computational neurology. From the point of view of image/video data hiding (and similarly from that of image/video compression) only a particular aspect of human perception is of interest and it regards the ability of the HVS to perceive (or not perceive) certain stimuli. Let us consider for examples the two images in figure 5.1: on the left the popular Lenna

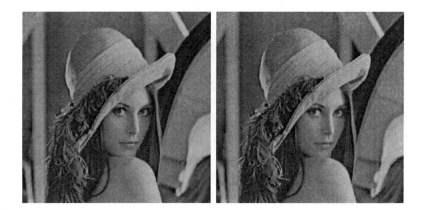

Figure 5.1: The popular Lenna image (Left) and a copy of it with added uniform noise with variance 81.

image is depicted, on the right a copy of it with added an uniform noise of variance 81 is shown. As it can be seen the added noise is completely imperceptible over the feathers of the hat of Lenna, it is, on the contrary, very easily perceived over the flat areas of the image (e.g. the background or the shoulders of Lenna) and slightly perceivable around object contours. Furthermore the noise is less visible on very dark and very bright regions.

These observations can be generalized, and we can list the following three rules of thumb:

Rule 1 Disturbs are much less visible on highly textured regions than on uniform areas.

Rule 2 Contours are more sensible to noise addition that highly textured regions but less than flat areas.

Rule 3 Disturbs are less visible over dark and bright regions

Although these rules of thumb are very comprehensive of the phenomenon, a mathematical formalization of the mechanisms underlying the perception of visual stimuli would be more manageable. Such a mathematical formulation has been developed over the past decades, mainly pushed by research in the field of image/video compression and on target recognition/sensing: the main results of this theory will be presented in the following.

Before going into the details it is anyway useful to overview the physical aspects of light perception. A simple model is constituted of an illumination source that produces light, this impacts an object and is partially reflected toward the eye (the retina). Given this simple model it is also useful to define some commonly used terms:

Illuminance is the amount of visible light incident on the surface of the object. It depends obviously on the source of illumination that will have a particular spectral content.

Reflectance is the proportion of visible light reflected by the object surface. Part of the light power is absorbed by the object, and part is reflected; dim objects absorb much of the light, while bright ones reflect much of it.

Luminance is the amount of light reflected by the object that is *recorded* by the retina. This is the physical phenomenon that influence perception. The eye is not able to perceive light radiation equally at all frequencies, on the contrary it is more sensitive to middle frequencies and less to low and high frequencies of the visible spectrum[1]: the function describing this different sensitivity is called Spectral Luminous Efficiency Function. Luminance is the measure of the radiometric power of light, per unit of emitting surface and steradian, weighted by the spectral luminous efficiency function of the eye. Luminance is measured in candles per square meter (cd/m^2), which is basically a power spatial density.

All these terms are basically referring to measures of power of the electromagnetic radiation. It has furthermore to be considered that the perception of the HVS is not linearly and directly correlated to the power of electromagnetic radiation (light) impacting the eye, for example it is not true that a light having double power with respect to another one, is perceived as doubly intense. Thus some perceptual terms are also useful to better understand the behaviour of the HVS:

Brightness is the perceived reflectance, i.e. it is related to the darkness or brightness of an object as it is perceived by the eye.

Lightness is the perceived luminance i.e. it is related to the sensation of light intensity as it is perceived by the eye.

These two perceptual phenomena are impossible to measure. One of the objectives of the following sections is just to understand how light intensity (in particular luminance) is related to perception.

---

[1]Let us remember that the visible spectrum extends approximately from 380 to 780 *nm*.

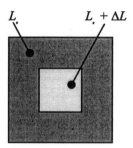

Figure 5.2: The Weber experiment consists of measuring when a small square of luminance $L_0 + \Delta L$ is perceived by a human observer as differing from a background of luminance L_0.

5.1.1 The Weber law and the contrast

First of all let us consider that images are perceived through the eye based on the amount of light impacting the retina. For correctly modeling this phenomenon it is thus needed to refer to images, as luminance arrays $L(x,y)$, where at each pixel location (x,y) we have a certain value of recorded light L. This is not the usual way images are stored: the topic of how to relate the image data stored in a file with the corresponding luminance values will be more deeply investigated later in this section.

A first simple experiment for quantifying the ability of the HVS to perceive luminance variations is depicted in figure 5.2. A small square of uniform luminance $L_0 + \Delta L$ is superimposed over a uniform background of luminance L_0: ΔL is then increased until the small square is perceived by an human observer, as differing from the background. Weber, who performed this experiment in the middle of 18th century, observed that the ratio of the just noticeable difference ΔL_{jn} and of the background luminance is almost constant, in particular:

$$\frac{\Delta L_{jn}}{L_0} = 0.02 \qquad (5.1)$$

This equation, known as Weber law, means that the higher is the luminance of the background the higher has to be the luminance difference of the stimulus for being perceived. Indeed it was soon also observed that the value of the ratio in equation (5.1) is not completely independent on the background luminance L_0, on the contrary, it increases for very low and very high luminance values. This last observation is completely in agreement with the **Rule 3.** that we have listed at the beginning of this section. Such a behaviour can be justified by the fact that the eye receptors have a response which varies in a non linear way with respect to the impacting light: the behaviour at the extrema of the luminance range is, on the other side, justified by the fact that receptors are not able to perceive luminance changes above and below a given range (saturation effect).

The main limitation of this experiment is that it only shows the response of the HVS to uniform stimuli; more complex experiments are needed to model the behaviour of the eye in the presence of textured patterns. Anyway the Weber law suggest that it is better to model HVS behaviour by referring to contrast measures, that is, to measures of the ratio between the increase in luminance caused by the stimulus and the mean luminance value of the background. For this reason in the following we will use the following definition of local contrast:

$$C(x,y) = \frac{L(x,y) - L_0}{L_0} \qquad (5.2)$$

where $L(x,y)$ is the luminance of the considered pixel and L_0 is the local mean background luminance.

5.1.2 The contrast sensitivity function

To deal with real images, where more complex texture patterns are present, the most straightforward approach is to refer to the harmonic decomposition of any signal that can be obtained by means of Fourier analysis. Given thus that every stimulus can be decomposed as the sum of sinusoidal stimuli, it needs first of all to investigate how sinusoidal stimuli are perceived. With this aim, experiments are carried out (see figure 5.3) where a sinusoidal stimulus of spatial frequency ν (measured in $cycles/m$), orientation θ and amplitude ΔL, superimposed to an uniform background of luminance L_0, is presented to a large set of viewers; the spatial luminance of the image can be modeled as:

$$L(x,y) = L_0 + \Delta L \cos\left(2\pi\nu\left(x\cos\theta + y\sin\theta\right)\right) \qquad (5.3)$$

The luminance of the sinusoidal stimulus is then increased until the observer perceives it: let us name the value of the luminance of the sinusoidal stimulus which is just noticeable as Just Noticeable Visibility Threshold (ΔL_{jn}). To obtain the independence on the viewing distance, the angular frequency $f = \frac{\pi d}{180}\nu$, where d is the distance between the observer and the monitor[2], has to be used which is expressed in $cycles/degree$. The qualita-

---

[2]If the distance d is expressed in meter, the spatial frequency ν is in $cycles/m$. By considering that images are sampled, it can be useful expressing the distance in pixel units, and consider that international norms have established that image quality should be judged from a distance of 4-6 times the image height; thus, for example, a 512×512 image should be viewed from a distance of 2048-3072 pixel units. In this case the spatial frequency is expressed in $cycles/pixel$.

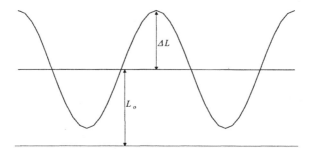

Figure 5.3: For measuring the CSF, a sinusoidal stimulus of increasing luminance amplitude, ΔL, is presented to a viewer, superimposed to an uniform background, until the viewer perceives it.

tive trend of ΔL_{jn} as a function of the angular frequency which is obtained by these experiments is exemplified in figure 5.4, its exact plot will depend on the luminance of the background (L_0), on the orientation of the stimulus (θ) and on the observer viewing angle[3] (W). Nevertheless the trend of figure 5.4 is already able to give us some useful indications regarding the dependence of the HVS sensibility to stimuli, with respect to the angular frequencies: in particular it is evident that the HVS has a peak of sensitivity at intermediate angular frequencies, where it is able to perceive stimuli of lower amplitude than in the low and high part of the spectrum. For large angular frequencies the sensibility is limited by the spatial distribution of the eye receptors which limits spatial resolution. On the other side, for very small angular frequencies the sensibility is mainly limited by the fact that the spatial scale of changes of the luminance pattern becomes larger than the fovea area[4].

In agreement to the Weber experiment it is usually preferred to consider

---

[3]The observer viewing angle in radiant is defined as the ratio between the square root of the area of the monitor and the distance between the monitor and the observer.

[4]The fovea is the central region of the retina, where visual receptors are much more dense than in the other retina regions: this area is the main responsible of our world detailed perception; surrounding regions are only devoted to preattentive tasks, i.e. to notice to brain interesting events occurring inside their viewing space, but not to perceive object details.

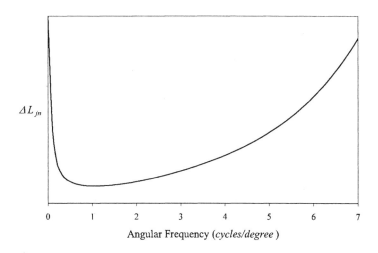

Figure 5.4: Qualitative trend of ΔL_{jn} as a function of the angular frequencies.

the Just Noticeable Contrast, defined as:

$$C_{jn} = \frac{\Delta L_{jn}}{L_0} \tag{5.4}$$

or its inverse, called Contrast Sensitivity Function (CSF)

$$S_c = \frac{1}{C_{jn}} = \frac{L_0}{\Delta L_{jn}} \tag{5.5}$$

As we have already outlined, this is a function of:

- the angular frequency of the stimulus,

- the orientation of the stimulus

- the observer viewing angle,

- the background luminance itself.

Many analytical expressions can be found in the literature for the CSF , we report here one of the most widely used, obtained by Barten (1990) by fitting data of psychophysical experiments:

$$S_c(f, L_0, W, \theta) = a(f, L_0, W) f e^{-\Gamma(\theta)b(L_0)f} \sqrt{1 + ce^{b(L_0)f}} \tag{5.6}$$

where:

$$a(f, L_0, W) = \frac{540 \left(1 + \frac{0.7}{L_0}\right)^{-0.2}}{1 + \frac{12}{W\left(1 + \frac{f}{3}\right)^2}}$$

$$b(L_0) = 0.3 \left(1 + \frac{100}{L_0}\right)^{0.15} \tag{5.7}$$

$$c = 0.06$$

$$\Gamma(\theta) = 1.08 - 0.08 \cos 4\theta$$

where f, the angular frequency of the stimulus, is measured in *cycles/degree*, W, the observer viewing angle, in *degree*, L_0, the mean local background, luminance in cd/m^2, and θ, the orientation of the stimulus, in *radiant*.

In figure 5.5 the plots of the CSF with respect to the angular frequency are reported for some values of background luminance, for an horizontal stimulus, and for an observer viewing angle $W = 180/(\pi\sqrt{12})$[5]. A trend analogous to that which has been observed in figure 5.4 for the Just Noticeable Visibility Threshold can be again evidenced for all values of background luminance: the maximum sensitivity is exhibited by the eye in the middle range of angular frequencies, while in the low and high part of the frequency range the HVS have a lower sensitivity to stimuli.

In figure 5.6 the plots of the CSF with respect to the background luminance are reported for some values of the angular frequency, for an horizontal stimulus and for an observer viewing angle $W = 180/(\pi\sqrt{12})$. In this case it is not immediately evident that the plots are consistent with **Rule 3.** (which states that disturbs are less visible over dark and bright regions), to verify that indeed they are, it is convenient plotting the Just Noticeable Visibility Threshold ΔL_{jn} versus luminance: this can be derived from figure 5.6 and by manipulating equation (5.5):

$$\Delta L_{jn} = C_{jn} L_0 = \frac{L_0}{S_c} \tag{5.8}$$

and it is plotted in figure 5.7 for an angular frequency of 15 *cycles/degree*.

Finally in figure 5.8 the plots of the CSF with respect to angular frequency are reported for two different stimulus orientation (namely 0° and 45°), for a background luminance of 50 cd/m^2 and for an observer viewing angle $W = 180/(\pi\sqrt{12})$. It can be observed that the sensitivity of the HVS

---

[5]For the observer viewing angle it is assumed that the monitor is viewed from a distance of four times its height (h), which brings to a value of $W = \frac{180}{2\pi} \frac{\sqrt{4/3h^2}}{4h} = 180/(\pi\sqrt{12})$

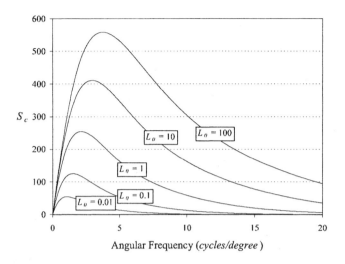

Figure 5.5: Plots of the CSF with respect to the angular frequency for values of background luminance of 0.01, 0.1, 1, 10, 100 cd/m^2.

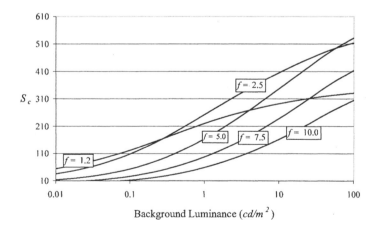

Figure 5.6: Plots of the CSF with respect to the background luminance for angular frequencies of 1.2, 2.5, 5.0, 7.5, 10 $cycles/degree$.

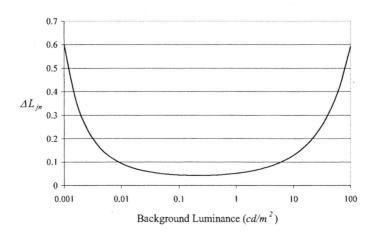

Figure 5.7: Plot of the Just Noticeable Visibility Threshold (ΔL_{jn}) versus image background luminance, for an angular frequency of 15 *cycles/degree*. It is evident how the amplitude of the just noticeable disturb increases for low and high background luminance values.

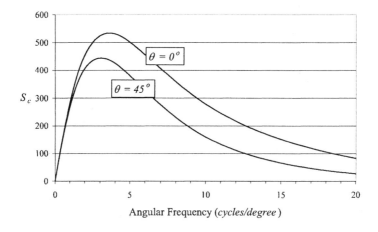

Figure 5.8: Plots of the CSF with respect to angular frequency for horizontal and diagonal stimuli and background luminance of 50 cd/m^2.

decreases by around 3 dB when the stimulus is diagonally oriented with respect to when it is horizontal (vertical stimuli exhibit the same behavior as horizontal ones).

5.1.3 The masking effect

The CSF it is very useful to model HVS capabilities, nevertheless it is still unable to consider some effects that can be observed: as an example the CSF alone is not able to justify the low sensitivity of the eye to disturbs which are superimposed to highly textured areas of the images. The Contrast Sensitivity function, in fact, has been derived by considering only sinusoidal stimuli superimposed to a uniform background. To go a step further it needs then to consider how a sinusoidal stimulus of amplitude ΔL, frequency f and orientation θ, superimposed to a spatially changing background composed by the sum of an uniform luminance value (L_0) plus a sinusoidal stimulus of amplitude ΔL_m, frequency f_m and orientation θ_m, is perceived:

$$
\begin{aligned}
L(x,y) = L_0 + & \\
& \Delta L_m \cos\left(2\pi f_m \left(x\cos\theta_m + y\sin\theta_m\right)\right) + \\
& \Delta L \cos\left(2\pi f \left(x\cos\theta + y\sin\theta\right)\right)
\end{aligned}
\tag{5.9}
$$

For simplicity we have here used directly the angular frequencies. The sinusoidal stimulus of frequency f_m is called masking stimulus. Let us first consider the case where $f_m = f$ and $\theta_m = \theta$, i.e. the iso-frequency masking case. From psychophysical experiments it has been observed that, in general, the presence of a masking stimulus increases the value of the Just Noticeable Threshold, i.e. higher values of ΔL are needed to perceive a stimulus when it is superimposed to an iso-frequency mask than when it is solely superimposed to an uniform background. Indeed this happens if the masking stimulus is already perceptible by itself: if it is not, the contrary effect (named pedestal effect) occurs, i.e. an increase in sensitivity is observed (in this case, in fact, the sum of the two stimuli make easier them to be perceived). This masking effect can be modeled with a Masked Just Noticeable Contrast function having the form:

$$
C_{jn}^m = \frac{\Delta L_{jn}^m}{L_0} = C_{jn}(f, L_0, W, \theta) F\left(\frac{\Delta L_m / L_0}{C_{jn}(f, L_0, W, \theta)}\right)
\tag{5.10}
$$

where ΔL_{jn}^m is the amplitude (ΔL in equation (5.9)) of the just noticeable disturb and the function $F(x)$ takes into account the effects due to the presence of the masking stimulus. In particular the function $F(x)$ will be such that:

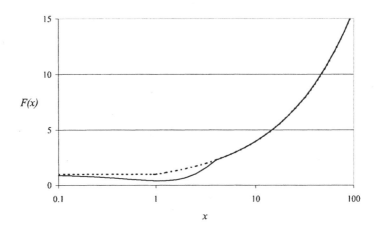

Figure 5.9: Plot of the masking function $F(x)$ (solid line) and of its approximation given by equation (5.11). It is assumed $w = 0.6$.

- when the masking stimulus is not present (i.e. $\Delta L_m = 0$ and thus $x = 0$) we have $C_{jn}^m = C_{jn}$ and thus $F(0) = 1$;

- when the masking stimulus is lower than the unmasked Just Noticeable Contrast (i.e. when $\Delta L_m/L_0 < C_{jn}$, the masking stimulus is not imperceptible by itself), we have $C_{jn}^m < C_{jn}$ which implies that $F(x) < 1$ for $0 < x < 1$;

- when the masking stimulus is higher than the unmasked Just Noticeable Contrast (i.e. when $\Delta L_m/L_0 > C_{jn}$, the masking stimulus is perceptible by itself), we have $C_{jn}^m > C_{jn}$ which implies that $F(x) > 1$ for $x > 1$;

This function if often approximated in practice as:

$$F(x) = \max\left(1, x^w\right) \tag{5.11}$$

where w can have values ranging from 0.5 to 0.8 depending on the masking frequency[6]. In figure 5.9 the masking function $F(x)$ having expression:

$$F(x) = \begin{cases} 1 - 1.1727x + 0.6445x^2 - 0.0676x^3 & 0 < x < 4 \\ x^{0.6} & x > 4 \end{cases} \tag{5.12}$$

---

[6] Usually this dependency is neglected in practical applications, and a constant value is chosen.

as obtained by fitting experimental results (solid line) and its approximation (5.11) (dashed line) are plotted.

The Masked Just Noticeable Contrast assumes thus the approximated form:

$$C_{jn}^m = C_{jn}(f, L_0, W, \theta) \max\left(1, \left[\frac{\Delta L_m/L_0}{C_{jn}(f, L_0, W, \theta)}\right]^w\right) \quad (5.13)$$

When the masking stimulus and the superimposed one have different frequencies (i.e. $f_m \neq f$ and $\theta_m \neq \theta$) the described effect decreases its influence, in particular if the frequencies are very different the presence of the masking stimulus become not influent, and the Just Noticeable Contrast for the superimposed signal takes the value that have when the background is uniform. In other words a disturb can only be masked by a background signal having similar angular frequency and similar orientation (near-frequency masking). In particular it has been observed that the decrease of the masking effect approximately depends on the ratio between the frequencies of the stimulus and of the mask (i.e. on f_m/f) and on the difference between the respective orientations (i.e. on $|\theta_m - \theta|$). This effect can be modeled by inserting a frequency weighting function into (5.13) as it follows:

$$C_{jn}^m = C_{jn}(f, L_0, W, \theta) \max\left(1, \left[S\left(\frac{f_m}{f}, |\theta_m - \theta|\right) \frac{\Delta L_m/L_0}{C_{jn}(f, L_0, W, \theta)}\right]^w\right)$$
$$(5.14)$$

where the weighting function $S(\ ,\ )$ is such that $S(1, 0) = 1$ and it decreases monotonically in all directions. A suitable function well fitting experimental data has a simple Gaussian shape:

$$S\left(\frac{f_m}{f}, |\theta_m - \theta|\right) = e^{-\left(\log_2^2\left(\frac{f_m}{f}\right)/\sigma_f^2 + (\theta_m - \theta)^2/\sigma_\theta^2\right)} \quad (5.15)$$

where the two parameters σ_f^2 and σ_θ^2 are related to the points, B_f and B_θ, where the weighting function amplitude gets smaller than half its maximum value as:

$$\sigma_f = 1.2 \log_2 B_f$$
$$\sigma_\theta = 1.2 B_\theta \quad (5.16)$$

where:

$$B_f = \sqrt{2}$$
$$B_\theta = 27 - 3 \log_2 f \quad (5.17)$$

In practice a masking stimulus of frequency f_m, farther from f than $1/2$ octave (i.e. $\log_2 f_m/f > 1/2$ or $< -1/2$), and of the same orientation ($\theta_m = \theta$) must have an amplitude more than double of an iso-frequency masking stimulus to cause the same increase in the Just Noticeable Contrast. Just based on this considerations, a simpler piecewise constant model is sometimes used, i.e. it is assumed that:

$$S\left(\frac{f_m}{f}, |\theta_m - \theta|\right) = \begin{cases} 1 & \text{if } 1/B_f < f_m/f < B_f \text{ and } |\theta_m - \theta| < B_\theta \\ 0 & \text{otherwise} \end{cases}$$

$$(5.18)$$

5.1.4   Mapping luminance to images

Until now the behaviour of the HVS has been described in terms of luminance measures, but digital images are often stored as grey-level values, and a watermarking system will directly affect grey-level values. In order to effectively exploit the acquired knowledge about HVS ability to perceive disturbs, it is thus very important to understand how grey-level values are related to the luminance perceived by the eye.

Of course such a dependence is connected to the way the image is shown to the observer. For a printed image, for example, grey-level pictures are basically mapped into much higher resolution binary images, having a spatially varying density of black dots (halftoning) which, thanks to the low pass nature of the eye[7], reproduce the spatially varying luminance of the picture itself (more dense black dots are put near darker pixels, and vice versa). Thus the luminance perceived by the eye will depend on: the grey-level value, the mapping function used to produce the spatially varying density of black dots, the amount of light illuminating the picture and reflected to the eye, the shape of the filter modeling the low pass behaviour of the eye. In this section we will mainly concentrate on the case of reproduction of pictures by means of a cathode ray tube (CRT), for which the dependence between grey-level values and luminance is better known and more easily modeled. Furthermore the models of the HVS behaviour presented until now have been mainly obtained in this case.

In general a non-linear relation exists between the grey level I of an image pixel and the luminance L produced by the corresponding CRT light emitting element, it is common to model such a relation as:

$$L = L(I) = q + (mI)^\gamma \qquad (5.19)$$

where q is the luminance produced by a CRT when displaying a black image, m is related to CRT contrast and γ to the intrinsic non-linearity of the CRT light emitting elements (the phosphors). While the γ parameter is characteristic of a given CRT, q and m depends on the regulations (of

---

[7]Indeed, we have seen that the CSF of the eye has not really a low pass characteristic; what is important in this case, anyway, is that high frequency are attenuated.

"brightness" and "contrast") that the observer can perform through the CRT electronics. Thus to apply the theoretical results presented in this section, a first possibility is to map the image grey-level values through (5.19) thus obtaining a luminance image, process this image based on the presented models, obtain the maximum values of luminance modifications that can be performed without being perceived, and finally mapping back these values to grey-level modifications through the inverse of (5.19).

Another solution is to try to directly approximate the Just Noticeable Contrast as a function of grey-level values. This can be obtained by considering a generic grey-level image composed, in analogy to equation (5.9), of an uniform background I_0, of a masking sinusoidal signal of amplitude ΔI_m and of a disturbing sinusoidal stimulus of amplitude ΔI:

$$
\begin{aligned}
I(x,y) = I_0 + & \\
& \Delta I_m \cos\left(2\pi f_m \left(x\cos\theta_m + y\sin\theta_m\right)\right) + \\
& \Delta I \cos\left(2\pi f \left(x\cos\theta + y\sin\theta\right)\right)
\end{aligned}
\tag{5.20}
$$

this is mapped to a luminance pattern as:

$$
\begin{aligned}
L(x,y) = L\left(I(x,y)\right) \approx L(I_0) + & \\
& L'(I_0)\Delta I_m \cos\left(2\pi f_m \left(x\cos\theta_m + y\sin\theta_m\right)\right) + \\
& L'(I_0)\Delta I \cos\left(2\pi f \left(x\cos\theta + y\sin\theta\right)\right)
\end{aligned}
\tag{5.21}
$$

where $L'(I_0)$ is the derivative of the luminance mapping function (for example of (5.19)) and a first order Taylor approximation has been used. By comparing (5.21) with (5.9) we can assume, as a first approximation, that $\Delta L_m = L'(I_0)\Delta I_m$ and $\Delta L_m = L'(I_0)\Delta I_m$. It is thus possible to define the Just Noticeable Grey-level Masked Contrast:

$$
\begin{aligned}
C^m_{Ijn} &= \frac{\Delta I^m_{jn}}{I_0} \\
&\approx \frac{\Delta L^m_{jn}}{I_0 L'(I_0)} = \frac{C^m_{jn} L_0}{I_0 L'(I_0)} = \\
&\frac{L_0}{I_0 L'(I_0)} C_{jn}(L_0) \max\left(1, \left[\frac{\Delta L_m / L_0}{C_{jn}(L_0)}\right]^w\right) \\
&\approx \frac{L(I_0) C_{jn}(L(I_0))}{I_0 L'(I_0)} \max\left(1, \left[\frac{L'(I_0)\Delta I_m}{L(I_0) C_{jn}(L(I_0))}\right]^w\right) = \\
&C_{Ijn}(I_0) \max\left(1, \left[\frac{\Delta I_m / I_0}{C_{Ijn}(I_0)}\right]^w\right)
\end{aligned}
\tag{5.22}
$$

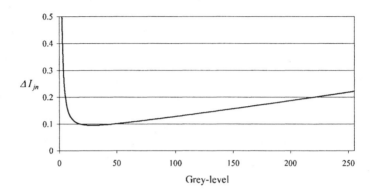

Figure 5.10: Plot of the Just Noticeable Visibility Grey-level Threshold (ΔI_{jn}) versus image background grey level, for an angular frequency of 5 *cycles/degree*. It is evident how the amplitude of the just noticeable disturb increases for low and high background grey level values.

where equations (5.8) and (5.13), have been used, and we have defined:

$$C_{Ijn}(I_0) = \frac{L(I_0)C_{jn}(L(I_0))}{I_0 L'(I_0)} \qquad (5.23)$$

In this derivation we have not made explicit, for simplicity, the dependence of C_{jn} (and as a consequence of C_{Ijn} too) by the frequency and orientation of the stimulus and by the image viewing angle (see section 5.1.2). Similarly we have assumed that the masking and the disturbing stimuli have the same angular frequency and orientation, if this is not the case, equation (5.22) can be easily generalized as equation (5.14).

By using the derived approximation (i.e. the generalized version of equation (5.22) and equation (5.23)) it is possible to process directly grey level images. Of course it needs to know the exact form of the $L(I)$ function, i.e. the q, m and γ parameters for a CRT[8].

In figure 5.10 the just noticeable grey level visibility threshold $(\Delta I_{jn} = IC_{Ijn})$ is reported with respect to grey level values for an angular frequency of 5 *cycles/degree*: the values of the parameters describing the CRT re-

---

[8]These can be estimated by means of some device able of measuring luminance, as for example a spectroradiometer.

sponse have been set to $q = 0.04$, $m = 0.03$ and $\gamma = 2.2$[9]. It is evident how this plot is in agreement with **Rule 3.** listed at the beginning of section 5.1.

5.1.5 Perception of color stimuli

In previous sections we have overview the main principles underlying the perception of grey scale images, but the real world is colored: we thus pass now to analyze how colour scenes are perceived by the HVS. Having this aim it is worth to better understand how the human eye is working. The eye is equipped with 2 types of photoreceptors: the rods and the cones. The rods are sensitive to very low level of luminance (approximately from 10^{-6} to 100 cd/m^2), these photoreceptors are spectrally undifferentiated (i.e. they all present the same spectral sensitivity) and they are responsible of human nightly vision (scotopic vision), in bright light conditions (e.g. during the day) their responses are almost saturated and do not contribute to visual perception. We will not investigate further this kind of photoreceptors. The cones, on the contrary, are sensitive to high levels of luminance (approximately from 10^{-2} to 10^6 cd/m^2)and, as such, they are responsible of the human daily vision (photopic vision), furthermore they are differentiated into three classes usually called Long (L), Mean (M) and (S) Short cones, according to the wavelength value for which they exhibit the maximum of the spectral response. In figure 5.11 the relative sensitivity of the three types of cones (L, M and S) are depicted: their respective range of prominent spectral response correspond respectively to the blue, green and yellow-green parts of the visual spectrum.

Similarly to what is done for grey-scale perception, some perceptual attributes of colors are useful for describing the way the HVS perceive them. These attributes are

Lightness (already defined at the beginning of this section) which is the perceived luminance i.e. it is related to the sensation of light intensity as it is perceived by the eye.

Hue refers to the color itself, regardless of its brightness or its purity, (e.g. a dark red and a bright red have the same hue). For monochromatic light sources differences of hue are related to differences of wavelength.

Saturation refers to how much a color is pure, i.e. to the amount of white light that is added to the color.

Let us now describe how color values can be represented. Given that color vision is due to the presence of three type of cones, every light ray,

---

[9]This set of parameters have been estimated on a Philips CRT monitor.

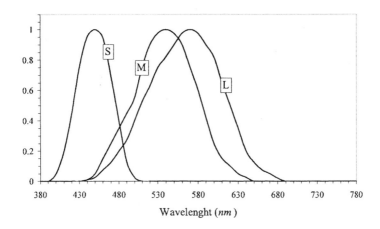

Figure 5.11: Relative spectral responses of the L, M and S cones. They concentrate respectively on the blue, green and yellow-green parts of the visual spectrum.

having spectral density $C(\lambda)$, impacting the retina, and producing a given color sensation, is basically represented by three values C_L, C_M and C_S which are the results of weighting the incident light spectral density $C(\lambda)$ with the spectral responses $L(\lambda)$, $M(\lambda)$ and $S(\lambda)$ of the three cones, i.e.

$$C_L = \int_{380}^{780} C(\lambda)L(\lambda)d\lambda$$

$$C_M = \int_{380}^{780} C(\lambda)M(\lambda)d\lambda \qquad (5.24)$$

$$C_S = \int_{380}^{780} C(\lambda)S(\lambda)d\lambda$$

A first observation is that we can have another type of light stimulus, with a different power spectral density $C'(\lambda)$, producing the same values C_L, C_M and C_S, and thus resulting to be perceived by the eye as the same color. This well known phenomenon is called metamerism. As a direct consequence we can assume that every colour having power spectral density $C(\lambda)$ can be reproduced by properly mixing three primary light sources having power spectral densities $P_1(\lambda)$, $P_2(\lambda)$ and $P_3(\lambda)$ with proportions

p_1, p_2 and p_3 [10] given that:

$$\int_{380}^{780} \left[p_1 P_1(\lambda) + p_2 P_2(\lambda) + p_3 P_3(\lambda) \right] C_L(\lambda) d\lambda = C_L$$

$$\int_{380}^{780} \left[p_1 P_1(\lambda) + p_2 P_2(\lambda) + p_3 P_3(\lambda) \right] C_M(\lambda) d\lambda = C_M \qquad (5.25)$$

$$\int_{380}^{780} \left[p_1 P_1(\lambda) + p_2 P_2(\lambda) + p_3 P_3(\lambda) \right] C_S(\lambda) d\lambda = C_S$$

By solving the resulting set of three linear equations, the proportions p_1, p_2 and p_3 (called primary co-ordinates) can be found. The proportions of each primary are also normalized with respect to a reference white stimulus having a given spectral density $W(\lambda)$: in practice if w_1, w_2 and w_3 are the primary co-ordinates of this reference white, the values

$$T_k = \frac{p_k}{w_k} \qquad k = 1, 2, 3 \qquad (5.26)$$

completely describe the colour on the basis of the chosen primaries. The T_k values are called tristimulus values. The tristimulus values of the reference white are always $(1,1,1)$. Of particular interest are the tristimulus values $T_k(\lambda_0)$ of a unit energy monochromatic light of wavelength λ_0, i.e. a light source having power spectral density $C(\lambda) = \delta(\lambda - \lambda_0)$ (for this kind of color it results that $C_L = L(\lambda_0)$, $C_M = M(\lambda_0)$ and $C_S = S(\lambda_0)$). The tristimulus values of a monochromatic light considered as a function of wavelength are called spectral matching functions, are usually indicated as $T_k(\lambda)$, and allow to completely describe the system of primaries, in that the tristimulus values of any color having power spectral density $C(\lambda)$ can be obtained as:

$$T_k = \int_{380}^{780} C(\lambda) T_k(\lambda) d\lambda \qquad k = 1, 2, 3 \qquad (5.27)$$

The chromaticity coordinates of a color are defined as:

$$t_k = \frac{T_k}{T_1 + T_2 + T3} \qquad k = 1, 2, 3 \qquad (5.28)$$

---

[10]This is basically the process by which CRT displays reproduce color pictures: on the surface of the screen three types of photoemitting elements (phosphors), each one characterized by a particular spectral density (peaked in the red, green and blue parts of the spectrum), are spread; every color can be produced by properly mixing these three components.

which are not independent given that $t_1 + t_2 + t_3 = 1$. Two chromaticity co-ordinates jointly describe the chrominance components (i.e. hue and saturation) of a color, and are usually represented in a plane (which is a plane of constant luminance). The chromaticity co-ordinates of the reference white are always $(1/3, 1/3)$,

Color spaces

A possible choice for the three primaries has been made by CIE[11] in 1931 and consists of three monochromatic light sources at wavelength of 700 nm (red), 546.1 nm (green) and 435.8 nm (blue). The reference white has a flat spectrum, which basically means that the three spectral matching functions are normalized to have unit area. The obtained tristimulus values are usually referred to as (R, G, B) values. In figure 5.12 the spectral matching functions $(R(\lambda), G(\lambda)$ and $B(\lambda))$ of this primary system are depicted. Given that one of these three functions can assume negative values, this primary system is not able to reproduce all visible colors (the mixing of the three primaries can in fact be performed only by non-negative proportions). Indeed any physical set of primaries can not be found which is capable to reproduce all colors.

To obviate this problem another set of primaries, which are indeed physically not realizable, has been proposed by CIE in 1931 which allows to obtain any visible color by a combination of the three primaries with non-negative proportions. This set of primaries, or color space, is called XYZ: its spectral matching functions are plotted in figure 5.13. Again the reference white has a flat spectrum. The CIE XYZ co-ordinates can be obtained by a linear combination of the CIE RGB values:

$$\begin{bmatrix} X \\ Y \\ Z \end{bmatrix} = \begin{bmatrix} 0.490 & 0.310 & 0.200 \\ 0.177 & 0.813 & 0.011 \\ 0.000 & 0.010 & 0.990 \end{bmatrix} \begin{bmatrix} R \\ G \\ B \end{bmatrix} \tag{5.29}$$

The co-ordinate Y is basically the luminance component of the color. In figure 5.14 the chromaticity diagram (i.e. a plane where each color is represented by a point having as spatial coordinates its chromaticity coordinates) of the XYZ color space is depicted. The horse-shoe shaped region represents all visible colors, that, as it is evident, can all be obtained by non-negative combinations of the XYZ primaries (the region completely lies in the first quadrant of the chromaticity diagram): all monochromatic colors have representative point in the boundary line. In figure 5.14 also some ellipses are depicted (MacAdam ellipses): these graphically represent

---

[11]CIE stands for Commission International d'Eclairage and is the international committee on color standards.

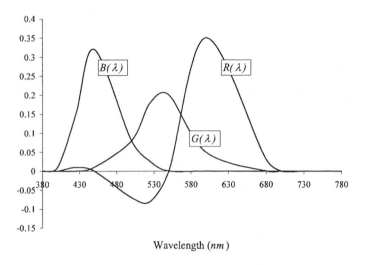

Figure 5.12: Spectral matching functions of the CIE RGB primary system.

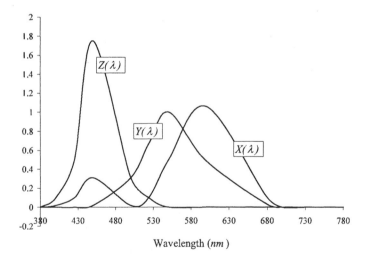

Figure 5.13: Spectral matching functions of the CIE XYZ primary system.

the just noticeable color differences (i.e. colors lying inside the ellipses are perceptually not distinguishable from the center color). Given that the dimensions and the orientations of the ellipses change dramatically from one color region to another, it is evident that color differences computed by the Euclidean distance of the XYZ co-ordinates are not perceptually meaningful, this means that the XYZ color space is not perceptually uniform.

In an attempt to have a perceptually uniform color space CIE has proposed in 1971 the $L^*a^*b^*$ color space. The transformation from XYZ to $L^*a^*b^*$ is not linear and is given by:

$$L^* = 25 \left(\frac{Y}{Y_0}\right)^{\frac{1}{3}} - 16 \qquad\qquad 0.01 \le Y \le 1$$

$$a^* = 500 \left[\left(\frac{X}{X_0}\right)^{\frac{1}{3}} - \left(\frac{Y}{Y_0}\right)^{\frac{1}{3}}\right] \qquad (5.30)$$

$$b^* = 200 \left[\left(\frac{Y}{Y_0}\right)^{\frac{1}{3}} - \left(\frac{Z}{Z_0}\right)^{\frac{1}{3}}\right]$$

where X_0, Y_0, Z_0 are the tristimulus coordinates of the reference white[12]. In this space, L^* approximately represents the lightness (i.e. the perceived luminance), a^* is related to the red-green content, and b^* to the yellow-blue content. In this space the MacAdam ellipses become approximately circular, which means that the Euclidean distance can be considered a good measure of perceptual differences. More recently (1995) CIE has proposed a more flexible way of computing the distance between two colors represented in the CIE $L^*a^*b^*$ space, in such a way to better adapt to experimental conditions.

In practical applications digital images are reproduced through CRT displays, and colors are obtained by mixing three primary sources whose spectral characteristics depend on the three types of phosphors (see footnote 10) used. As an example the color coordinates in the NTSC primary system are called R_N, G_N, B_N and are related to the CIE XYZ coordinates by the following relation:

$$\begin{bmatrix} R_N \\ G_N \\ B_N \end{bmatrix} = \begin{bmatrix} 1.910 & -0.533 & -0.288 \\ -0.985 & 2.000 & -0.028 \\ 0.058 & -0.118 & 0.896 \end{bmatrix} \begin{bmatrix} X \\ Y \\ Z \end{bmatrix} \qquad (5.31)$$

---

[12]It is not needed that this is the same reference white used for defining the CIE XYZ primary system, that is a spectrally constant light source whose tristimulus values would be $(1/3,1/3)$; on the contrary the D65 reference white, which has a spectral power distribution similar to that of sunlight, is usually referred to, whose chromaticity coordinates are $(0.3127,0.3290)$: from the equation it is evident that the reference white to which we refer will have $L^*a^*b^*$ values of $(100,0,0)$.

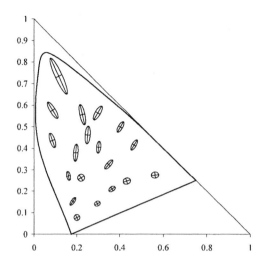

Figure 5.14: Chromaticity diagram for the CIE XYZ color primaries system with the MacAdam ellipses indicating the just noticeable color differences.

and in a similar way a primary system has been defined by ITU-R (standard ITU-R B.709) for contemporary studio monitors, having coordinates

$$\begin{bmatrix} R_{709} \\ G_{709} \\ B_{709} \end{bmatrix} = \begin{bmatrix} 3.240479 & -1.537150 & -0.498535 \\ -0.969256 & 1.875992 & 0.041556 \\ 0.055648 & -0.204043 & 1.057311 \end{bmatrix} \begin{bmatrix} X \\ Y \\ Z \end{bmatrix} \quad (5.32)$$

For transmission purposes these values are then often transformed in a luminance/chrominances system known as YC_rC_b, and defined, according to the ITU-R 601 standard[13], by the equation:

$$\begin{bmatrix} Y \\ C_b \\ C_r \end{bmatrix} = \begin{bmatrix} 16 \\ 128 \\ 128 \end{bmatrix} + \begin{bmatrix} 65.481 & 128.553 & 24.966 \\ -37.797 & -74.203 & 112 \\ 112 & -93.786 & -18.214 \end{bmatrix} \begin{bmatrix} R'_{709} \\ G'_{709} \\ B'_{709} \end{bmatrix}$$
$$(5.33)$$

where Y is related to luminance content, while the C_r and C_b components carry chrominance information, and R'_{709}, G'_{709}, and B'_{709} are the γ pre-corrected values[14] of the ITU-R BT.709 co-ordinates. As we will see better in the following, the HVS is less sensitive to high frequency disturb on

---

[13]This is the most common format for storing digital video. The values of the coordinates span the range of [16, 235] for Y and [16, 240] for C_b and C_r.

[14]This means that the color co-ordinates are passed through a non linear relationship with exponent $1/\gamma$.

chrominance components than on luminance, these two bands are then usually subsampled with respect to the luminance one.

Another useful representation of colors attempt to give a quantitative measure of the perceptual attributes (Lightness, Hue and Saturation) described at the beginning of this section. A simple approximation of these attributes, mainly used in computer graphics applications, can be obtained directly from any RGB color space, which is first transformed (by a simple 3D rotation) into an intermediate space:

$$\begin{bmatrix} V_1 \\ V_2 \\ V_3 \end{bmatrix} = \begin{bmatrix} \frac{1}{\sqrt{3}} & \frac{1}{\sqrt{3}} & \frac{1}{\sqrt{3}} \\ 0 & \frac{1}{\sqrt{2}} & -\frac{1}{\sqrt{2}} \\ \frac{2}{\sqrt{6}} & -\frac{1}{\sqrt{6}} & -\frac{1}{\sqrt{6}} \end{bmatrix} \begin{bmatrix} R \\ G \\ B \end{bmatrix} \qquad (5.34)$$

and then to the HSI color system (Hue, Saturation, Intensity) as:

$$H = \arctan\left(\frac{V_3}{V_2}\right)$$
$$S = \sqrt{V_2{}^2 + V_3{}^2} \qquad (5.35)$$
$$I = V_1$$

The advantage of this color space, which tries to model the attributes of color perception, stands in the fact that it is known that the HVS is much more sensitive to hue than to saturation differences: a higher level of disturb can then be tolerated in the saturation channel. A major drawback is that this space is not perceptually uniform, i.e. the threshold of visibility is not independent on the considered color.

Chromatic contrast sensitivity

The Euclidean distance in the $L^*a^*b^*$ color space is quite suitable and largely used for measuring color differences between uniform patches; for more complex textured images, on the other side, it is not able to adequately model HVS perception. We have already seen, in fact, that perception of achromatic (luminance) differences is much more complicated than a simple Euclidean distance as it would be suggested by the straightforward use of the $L^*a^*b^*$ coordinate system. What it needs to consider, also in this case, thus, is that perception of a color stimulus depends on its frequency (CSF based models) and on the visibility of the colored pattern to which it is superimposed (masking). In fact, it is demonstrated that masking of each color band is influenced by the other bands, and, in particular, that chrominance content can effectively mask luminance disturbs, while, on the contrary, luminance masks make chrominance disturbs easier

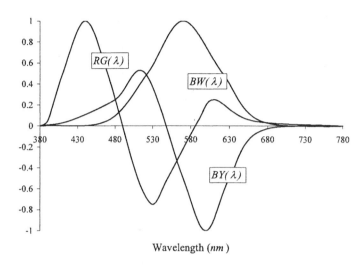

Figure 5.15: Spectral matching functions of the opponent color system BW, RG, BY.

to be perceived over a wide range of masking contrast values. Nevertheless complete and tractable theoretical models of the interactions among the different color channels are not easy to be managed. On the other side, it has also been demonstrated that a good approximated model of color perception can be obtained by considering three independent channels, which are called opponent colors, whose spectral matching functions are shown in figure 5.15 and which practically correspond to the luminance (BW), red-green (RG) and blue-yellow (BY) channels. The coordinates in the opponent system can be obtained from the CIE XYZ values as:

$$
\begin{bmatrix} BW \\ RG \\ BY \end{bmatrix} = \begin{bmatrix} 0.279 & 0.720 & -0.107 \\ -0.499 & 0.29 & -0.077 \\ 0.086 & -0.590 & +0.501 \end{bmatrix} \begin{bmatrix} X \\ Y \\ Z \end{bmatrix} \tag{5.36}
$$

For what regards the CSF of the BW channel in the opponent color space, this corresponds to that already presented for the luminance case (e.g. equation (5.6)) and has a band-pass characteristic with respect to the angular frequencies. On the contrary the contrast sensitivity functions of the two chrominance channels RG and BY have a low pass characteristic. Their mathematical dependence on the angular frequency f can be assumed to be approximately:

$$
S_c(f) = 146e^{-0.45f} \tag{5.37}
$$

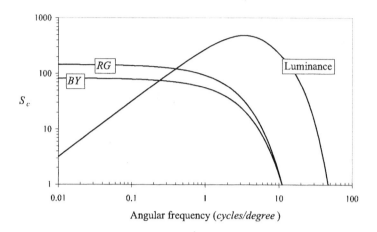

Figure 5.16: Log-log plots of the CSFs with respect to the angular frequency for the opponent colors channels RG and BY. For convenience the CSF of the luminance channel for a background luminance of 30 cd/m^2 is also reported. It is evident the different behaviour of the CSF for the opponent color channels (which have a low pass trend) and for the luminance channel (which have a band pass trend).

for the RG channels and

$$S_c(f) = 83e^{-0.40f} \qquad (5.38)$$

for the BY channels. The dependencies on viewing angle and background luminance are usually not considered. In figure 5.16 the CSFs of the two opponent color bands are plotted in a log-log graph. For convenience the CSF of the luminance channel for a background luminance of 30 cd/m^2 is also reported. The low pass behavior of the CSF for the opponent color channels is evident, in contrast to the band pass behavior of the CSF of the luminance channel. In it also possible to clearly notice the lower spatial resolution exhibited by the chromatic channels (whose CSF has a cut off frequency of around 10 $cycle/degree$) with respect to the luminance channel (whose cut off frequency is about 50 $cycle/degree$). This characteristic of the HVS is largely exploited by video system where chrominance components are usually subsampled (for example by 2 along lines as in 4:2:2 systems, and along lines and columns as in 4:2:0 systems).

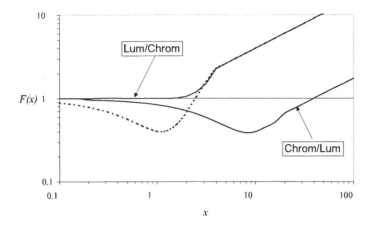

Figure 5.17: Log-log plots of the masking functions (solid lines) for the chrominance disturb over luminance mask and the luminance disturb over chrominance mask cases. The intra band (luminance over luminance and chrominance over chrominance) masking function, is also reported for convenience (dashed line).

Chromatic masking

Let us now pass to consider the masking effects. As it has already been outlined every band has an influence on all other bands. In particular a disturbing stimulus in a band is masked by the stimuli present in the same band and by those in the other bands. For intra-band masking the same rule already described for the luminance channel holds (see figure 5.9). With regard to inter-band masking a quite different behavior is found between the case of a chrominance mask which can reduce visibility of a luminance stimulus and a luminance mask that, on the contrary, increases the visibility of a chrominance disturb (at least for a large range of masking stimulus contrast values). This behavior is evident by observing the shape of the masking functions related to the case of a chrominance disturb superimposed to a luminance mask (curve annotated with Chrom/Lum in figure 5.17) and of a luminance disturb superimposed to a chrominance mask (curve annotated with Lum/Chrom in figure 5.17). These two functions can be well approximated as:

$$F(x) = \begin{cases} 1 - 0.1741x + 0.0154x^2 - 0.0004x^3 & 0 < x < 20 \\ 0.11x^{0.6} & x > 20 \end{cases} \qquad (5.39)$$

and

$$F(x) = \begin{cases} 1 + 0.0447x - 0.0848x^2 + 0.0386x^3 & 0 < x < 4 \\ x^{0.6} & x > 4 \end{cases} \qquad (5.40)$$

respectively. From figure 5.17 appears clearly that while a chrominance masking signal is able to mask both luminance (upper continuous curve) and chromatic disturbs (dashed curve), a luminance mask does not present the same ability with respect to chrominance disturbs (lower continuous curve), on the contrary it increases their visibility over a wide range of masking contrast values extending from around 0.2% to 40% (i.e the pedestal effect is prominent). This effect is anyway compensated by the lower absolute sensitivity of the HVS to chromatic disturbs as it is demonstrated by the previously presented CSFs (figure 5.16).

Opponent color modeling is widely used for estimating perceptual fidelity of color images: usually the inter-band masking effects are neglected, i.e. the three opponent color bands are treated in a completely independent way.

Similarly to what has been told for luminance, also in the case of color images, pixel data are not stored as light power measures, but as color levels: with regard to this issue, the considerations drawn in section 5.1.4 for the problem of mapping grey level values to luminance measures, can be generalized in a straightforward way to the color case.

5.1.6 Perception of time-varying stimuli

What has been told until now was regarding mainly perception of disturbs in static pictures, and is thus applicable to still images. On the other side an important role in the consumer electronic market (but not only) is played by moving pictures, i.e. by video. The goal of this section is just to describe how the perception of disturbs in video can be modeled.

The sensibility of the HVS to spatially and time varying stimuli can be measured by means of experiments in which human observers are asked to choose the just noticeable amplitude of spatially and time varying disturbs, in a way similar to that described in section 5.1.2. For example a stimulus of the form:

$$L(x,t) = L_0 + \Delta L cos(2\pi f x) cos(2\pi f_t t) \qquad (5.41)$$

can be used, where f and f_t are the angular spatial and the temporal frequencies (measured respectively in $cycle/degree$ and Hz), and for simplicity only the x coordinate is considered for spatial frequencies. In this way it is noticed that the temporal frequency value for which the spatiotemporal CSF drop to zero is about 30 Hz (this value is often referred to as Critical Fusion Frequency - CFF).

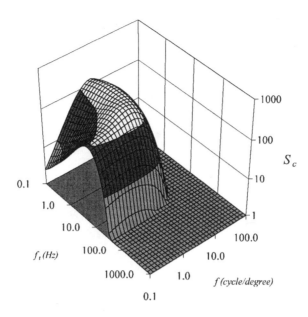

Figure 5.18: Plot of the spatiotemporal CSF with respect to the angular spatial and temporal frequencies obtained by imposing eye fixation and a mean background luminance of 20 cd/m^2.

Anyway this kind of stimulus is not very suitable to model perception in moving pictures. In video, in fact, time varying stimuli are caused by object motion and not by variations of object luminance (as it is modeled by the just described experiment). A more reliable experimental setup considers thus a spatially translating stimulus of the form:

$$L(x,t) = L_0 + \Delta L cos(2\pi f(x - vt)) \tag{5.42}$$

where v is the retinal speed of stimulus translation (measured in $degree/s$), which is easily related to the temporal frequency as $f_t = fv$. By fitting experimental data, a spatiotemporal CSF can be obtained having approximately the form:

$$S_c = \left[6.1 + 7.3 \left|\log_{10}\left(\frac{v}{3}\right)\right|^3\right] v \left(2\pi f\right)^2 e^{-\frac{4\pi f}{45.9}(v+2)} \tag{5.43}$$

and by substituting in this $v = f_t/f$ the plot of figure 5.18 results. The temporal cutoff frequency is confirmed to be around 30 Hz also by this analysis.

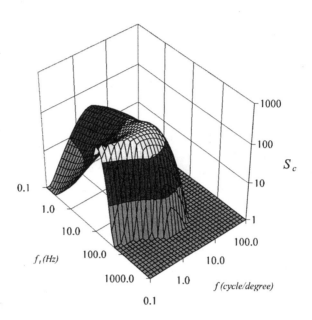

Figure 5.19: Plot of the spatiotemporal CSF with respect to the angular spatial and temporal frequencies obtained by unconstraining eye motion and with a mean background luminance of 20 cd/m^2.

It has to be observed, indeed, that the spatiotemporal CSF as described in (5.43) has been obtained by imposing eye fixation, i.e. constraining the eye not to track the moving stimulus. On the contrary, in normal situations, the eye attempts to track moving objects[15]. A simplified model describing the eye speed during object tracking can be written as:

$$v_{eye} = \max(g \, v_{object}, v_{eye,max}) \qquad (5.44)$$

where the gain g takes into account the fact that perfect tracking is never achieved (i.e. $g < 1$), and $v_{eye,max}$ is due to the fact that tracking speed has a maximum value beyond which the eye is no longer able to properly perceive stimuli. By considering this simple model, the retinal velocity of the projection of the object can thus be assumed to be:

$$v = |v_{object} - v_{eye}| = |v_{object} - \max(g \, v_{object}, v_{eye,max})| \qquad (5.45)$$

---

[15]This kind of eye movement is referred to as "smooth pursuit". At least other two types of eye movements are present: the "drift movement" which is quite slow (around 0.1 $degree/s$) and is always present also when the scene is perfectly static, and the "saccadic movements" which are very fast, and allows the eye to rapidly change its fixation point in the scene (during saccadic movements the HVS is practically blind).

Given that we are interested to model the spatiotemporal CSF with respect to the moving picture angular spatial and temporal frequencies, the relation between these two parameters is strictly related to the object speed, and becomes then $v_{object} = f_t/f$: by substituting this in equation (5.45) and the latter in (5.43) the plot of figure 5.19 is obtained (it has been assumed that $g = 0.82$ and $v_{eye,max} = 80 degree/s$). From this figure it appears with evidence that, thanks to the capability of the eye of tracking moving objects, much larger temporal frequencies can be perceived than it was previously thought. In particular for a spatial frequency of around 1 $cycle/degree$ temporal frequencies up to 180 Hz can be correctly perceived by the HVS.

5.2 The Human Auditory System (HAS)

Although quite complicate as well, the HAS is better known than the HVS and the models describing it are more effective. In general a widely accepted model of the processing that the HAS performs on sound signals is based on a bank of highly overlapping bandpass filters, having asymmetric magnitude and non uniform bandwidth increasing with the filter central frequency.

In the following we will explore the most important characteristics of the HAS perception. As we will see some of the concepts already cited for visual perception (e.g. masking) are valid also for the audio case.

Let us start by defining the metric generally used when dealing with audio perception. Given that audio perception is connected to the pressure that sound waves produce on the eardrum, the natural measure of audio stimuli is basically a pressure measure. Furthermore given the large dynamic range of ear sensibility it is convenient to use a logarithmic formulation. The Sound Pressure Level (SPL) is thus defined as:

$$SPL = 20 \log_{10} \frac{p}{p_0} \tag{5.46}$$

where p is the pressure of the sound stimulus, and p_0 is the standard reference level of 20 $\mu N/m^2$. The standard reference level practically corresponds to the minimum level of pressure that the ear can perceive, and results in an SPL value of 0 dB. Values of SPL larger than 150 dB correspond to the threshold of pain for high intensity stimuli.

By passing to consider the characteristic of the response of the ear to pure tone stimuli we can see that, similarly to the eye, also the ear exhibits a different sensibility to sounds depending on the frequency, and as for luminance this sensibility has a band-pass characteristic. An absolute threshold of hearing function has been standardized for a young listener

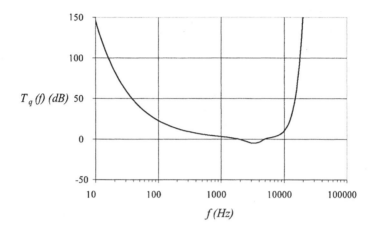

Figure 5.20: Plot of the absolute threshold of hearing expressed as the minimum Sound Pressure Level (SPL) that a young listener with acute hearing can perceive.

with acute hearing in a quiet environment, by measuring the SPL of the minimum perceivable pure tone stimulus in dependence of the frequency. This function (which is measured in dB) is well approximated by:

$$T_q(f) = 3.64 \left(\frac{f}{1000}\right)^{-0.8} - 6.5e^{-0.6\left(\frac{f}{1000}-3.3\right)^2} + 10-3 \left(\frac{f}{1000}\right)^4 \quad (5.47)$$

In figure 5.20 the absolute threshold of hearing is depicted. It appears clearly that the HAS is almost insensitive to sounds having frequency below 10 Hz and above 20 kHz and exhibits a peak of sensitivity at around 3.3 kHz.

5.2.1 The masking effect

The masking effect regards the phenomenon by which a sound stimulus is not perceived by the HAS if it is near in frequency to another higher level stimulus. An useful concept for describing this effect is that of critical bands. It has been observed that if we have a narrow band noise stimulus[16], its intensity perception does not increase if the stimulus bandwidth increases, at least until a certain bandwidth, called critical band, is not

---

[16] Here we mean with noise any sound signal having a spectral response extending over a given range of frequencies, basically noise is everything else than a pure tone sound.

passed. The amplitude of the critical band depends on the frequency (in particular it increases with frequency) as:

$$BE_c(f) = 25 + 75 * \left[1 + 1.4 \left(\frac{f}{1000} \right)^2 \right]^{0.69} \qquad (5.48)$$

In the literature of audio signal processing the distance of one critical band is usually referred to as 1 $Bark$, and the function:

$$z(f) = 13 \arctan(0.00076f) + 3.5 \arctan \left(\frac{f}{7500} \right)^2 \qquad (5.49)$$

can be used to convert from frequency in Hz to the $Bark$ scale[17]. Although critical bands depends continuously on the frequency, it is common use to define a finite number of non-overlapping critical bands (this is for example done in the MPEG Psychoacoustic Model 1).

In order to completely describe masking it is convenient to distinguish three cases: a noise sound masking a pure tone sound (NMT), a pure tone sound masking a noise sound (TMN) and a noise sound masking a noise sound (NMN). Let us consider first the NMT case, and suppose we have a noise masker having a bandwidth of less that 1 $Bark$, a pure tone signal having frequency equal to the central frequency of the band of the noise sound is completely masked if its SPL is at least around 5 dB lower than the SPL of the noise. To be more precise the threshold of perception also slightly depends on the center frequency at hand. The situation for the TMN case is different in that the noise SPL has to be around 21-28 dB lower than the tone SPL: i.e. pure tone sounds are less effective in masking than noise. Finally the case of NMN is the less effective at all, given that the maskee has to have a SPL of around 26 dB lower than the masker. From the already highlighted asymmetry of behavior of the NMT and TMN cases it is evident that for an effective concealment of disturbing signals it is important to identify across the masking signal its pure tone and noise like components.

When the frequencies of the masker and the maskee do not correspond (near frequency masking), the detection threshold is reduced, and this reduction is faster for maskee frequencies lower than that of the masker than for those higher. A good approximation of the amount of reduction (expressed in dB, and often named spreading function) of the threshold is given by:

$$SF(z) = 15.81 + 7.5(z + 0.474) - 17.5\sqrt{1 + (z + 0.474)^2} \qquad (5.50)$$

---

[17]Two frequencies having distance 1 in the Bark scale, are about 1 critical band apart one from the other.

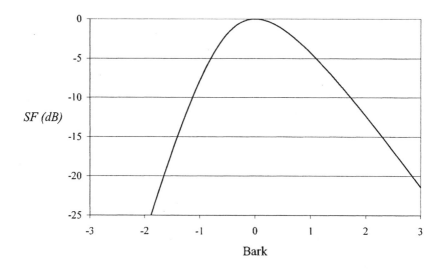

Figure 5.21: Trend of the reduction of detection threshold for near-frequency masking expressed in term of barking units.

where z is the frequency distance expressed in $Bark$. From figure 5.21 it is possible to see that the spreading function is asymmetric (as already mentioned) and have slopes of $+25$ $dB/Bark$ and -10 $dB/Bark$ for maskee frequencies lower and higher of masker frequency respectively.

Masking effects also extend before masker onset and after masker removal. In particular significant backward masking is limited to about a few ms before masker onset, and decays very rapidly, while forward masking can extend up to 100-150 ms, and its decay is slower. Temporal masking (in particular backward masking) are not understood very well yet and related models are difficult to be tuned.

5.3   Concealment through feature selection

Let us now see how the concepts deriving from the analysis of the models of human perception can be exploited for better hiding data into images, audio files and videos.

As a first step some hints can be derived on which features are most suitable to be modified, without dramatically affecting perceptual quality. Secondly (it will be the argument of next section) we will present some techniques based on perceptual considerations for adapting the watermark signal to the local asset perceptual content.

The selection of features is quite difficult to be performed in the asset domain: although some rules can, in fact, be devised for preferring, as embedding areas, some asset regions with respect to others in order to reduce watermark visibility (let us for example think to the three rules of thumb listed at the beginning of the chapter), this approach could make the recovery phase quite unreliable. Let us for example suppose that we have a function $\mathcal{S}$ able of discriminating between the sets of asset samples suitable for embedding and those not suitable, e.g.

$$\mathcal{S}(\mathbf{A}_k) = \begin{cases} 1 & \text{if the set } \mathbf{A}_k \text{ is suitable} \\ 0 & \text{if the set } \mathbf{A}_k \text{ is not suitable} \end{cases} \tag{5.51}$$

and that the embedding procedure only modifies those sets which are suitable. During the recovery phase we should be sure that the sets were embedding was performed are reliably identified, i.e. we need that the function $\mathcal{S}$ classifies in the same way each block of asset samples before and after watermark embedding. Given that the embedder knows the function $\mathcal{S}$ it can perform embedding in such a way to grant this characteristic. As an example, let us suppose that we only want to watermark those asset blocks that have a large degree of texture (according to **Rule 1** of those listed at the beginning of the chapter), a simple function $\mathcal{S}$ can measure the variance of the block and based on this value classifies it as suitable or not suitable for embedding: the embedding procedure could then be designed in such a way to always increase the variance of the blocks. Anyway this is a solution only if we have an application in which we can assume that watermark recovery is performed before any attack, given that the effects of these on the classification produced by the $\mathcal{S}$ function would be really unpredictable.

The situation is different in the case of the transformed domain techniques. In general we have in fact seen that both the HVS and the HAS are less sensitive to disturb having high or very low frequencies it is then common to partition the frequency range into two regions: the high frequencies region, were the watermark signal is embedded, and the low frequencies region, where it is not (due to its very small extension the region of very low frequencies is usually not considered)[18]. Given this partition is fixed, and not estimated time by time during the embedding phase, the problem outlined for the asset domain case does not raise now.

---

[18]Indeed not the whole high frequencies region is used either, because of robustness constraint: in fact, it is not convenient to embed the watermark signal into the highest frequency components that will be the first to disappear due to many common attacks such as compression, low-pass filtering, etc..

Similar considerations are valid for the hybrid techniques. In particular the situation for block-based transforms is identical as for the transform domain case, high frequency coefficient are usually preferred for watermark embedding, in order to reduce watermark visibility. The same objective can be reached in the DWT case by performing embedding in the finest sub-bands.

With regard to the case of color images, we have seen (figure 5.16) that chromatic channels, further to have a smaller cut-off frequency with respect to the luminance channel, also exhibit lower absolute values of sensitivity. This suggests that embedding in the chrominance bands would be more unobtrusive. In particular some watermarking algorithms modify only the pixels of the blue band which is the less sensitive to disturbs. Such an approach can anyway cause problems if the watermark has to survive the conversion of the image to a grey-level format: only a very small amount of watermark energy would survive in this case.

In conclusion we can say that concealment through feature selection is not very easy to be performed, much attention has to be taken in selecting the features, in particular if it desired that watermark recovery has to be achieved also after asset manipulations (attacks), that can make the selected features no longer available or identifiable. The sole exception is when the features are selected on an a-priori fixed basis, and among those which are very likely not to be affected too much by foreseen manipulations (e.g. the case of mid range frequencies). In general, anyway, it is preferable not to select any subset of the feature space, but to modify all its elements with a locally adapted strength. Signal adaptation will be just the subject of next section.

5.4 Concealment through signal adaptation

Given the complexity of human perception it is certainly advantageous to carefully adapt watermark signal to the local asset content for better reducing its perceptibility. In the following we describe two possible approaches for achieving this adaptation.

5.4.1 Concealment through perceptual masks

Let us suppose, as a first step, that a masking function M, having the same cardinality of the asset A, is available, with values, included in the range $[0, 1]$, giving a measure, point by point, of how much insensitive to disturbs is that asset sample. Let us then suppose that we have a copy of the asset A_w watermarked without taking any care about perceptibility issues (e.g. uniformly). A perceptually adapted watermarked asset can be obtained by blending the original and watermarked asset as follows:

$$A'_w = \bar{M} \boxtimes A + M \boxtimes A_w \qquad (5.52)$$

where we have indicated with $\boxtimes$ the sample by sample product, and with $\bar{M}$ the masking function whose elements are the complement to 1 of the elements of M[19]. In practice $\bar{M}$ indicates how sensitive to noise is a given asset sample. Please note that we have indicated the perceptually adapted watermarked asset with A'_w, as we are usually doing for attacked assets, because this kind of adaptation can indeed be considered as a first attack to the originally watermarked data. Indeed this is a peculiar attack, aimed explicitly at increasing watermark robustness given a certain amount of imperceptibility, differently from the major part of attacks that, on the contrary, degrade performances of the watermark recovery procedure. Furthermore this attack is known to the watermark recovery step to have happened: attempts to compensate for it could then, in principle, be made for enhancing recovery performance[20].

For going deeper inside the effects of this masking approach on the watermark let us simply manipulate equation (5.52): by adding and subtracting $M \boxtimes A$ we can easily obtain

$$A'_w = A + M \boxtimes (A_w - A) \qquad (5.53)$$

where $A_w - A$ represent the original watermark signal in the asset domain (independently on the domain where watermark embedding has been performed, and on the embedding rule used this difference always model the signal added to the original asset for carrying the hidden information; of course, in general, this difference will be highly dependent on the original asset). By supposing for example that the embedding has been performed in the asset domain, the difference $A_w - A$ is just the watermarking signal **w**: thus the effect of perceptual masking is to attenuate it in some regions (where disturbs are more perceptible). An analogy immediately raises with the time varying fading channel models which are encountered in mobile communications, in particular the importance of using a diversity approach, i.e. to interleave the to be hidden information in such a way to widely spread it over the whole asset, results with evidence (what has to be avoided, in practice, is that some bits are localized in those regions where only a small amount of watermark can be added because of the perceptual

---

[19]As an example, for an image it means that $\bar{M}(i, k) = 1 - M(i, k)$.

[20]This is usually not done, because all further attacks the asset have undergone are stronger, and thus the advantages of this compensation would be very limited. The situation would be different if, for the data hiding application at hand, further attacks were not expected

constraint). Equation (5.53) also suggests that for additive asset domain techniques, direct weighting of the watermarking signal can be performed before addition to the asset samples, instead of blending. If on the other side embedding is performed in the transform domain, thanks to the linearity property of usual transforms, $A_w - A$ results to be the transform of the watermark signal w: to understand how this is affected, it needs then to understand to which operation the sample by sample multiplication in the asset domain is related in the transform domain. As an example, if the used transform is the DFT, the multiplication in the asset domain corresponds to the convolution in the transform domain: thus blending causes the watermarking signal to be convolved with the transform of the masking function M; one effect of this is that if the watermark signal was originally an uncorrelated process, it looses this property after blending, thus making watermark recovery more complicate (a frequency selective channel is introduced affecting the watermark signal). This suggests that the spectral content of the masking function should be as much concentrated as possible to low frequencies, in order not to degrade to much the watermark (ideally the DFT of the mask should be a delta function, but in this case masking would be ineffective given that the mask would results to be uniform). Similar consideration also hold for the case of DCT, and for block-based transform methods[21].

In general it is interesting to try to understand how the uniformly watermarked asset A_w to be used for blending should be obtained. Given that the effect of the blending mask is only to reduce the watermark strength in the most perceptually sensitive regions, A_w should be obtained by embedding the watermark with a strength that makes it visible in almost all regions of the asset, except for those which are really insensitive to disturbs (in particular just a small increase of watermark strength should make it visible also in these regions). The mask, if properly designed, will then reduce watermark strength on the other asset regions in such a way to make it imperceptible everywhere. This procedure requires obviously a manual tuning of watermark strength and this limit its efficacy when a large amount of assets needs to be watermarked.

As an example, a simple mask for image data hiding can be obtained based on the well known property that the HVS is less sensitive to noise in highly textured areas (see **Rule 1.** at the beginning of section 5.1). For measuring the degree of 'textureness' of a small region, an estimate of the

---

[21]Indeed the masking approach is not very used with block based transform methods, because in this case is easier, and effective as well, to directly deal with visibility constraint in the transform domain, as we will see later in this chapter.

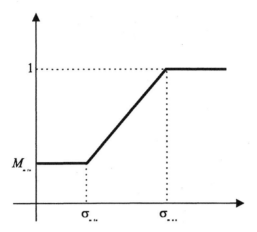

Figure 5.22: Scaling rule for obtaining the masking function M_σ from the local variance values σ_W.

local variance can be employed:

$$\sigma_{\mathcal{N}}^2(i,k) = \frac{1}{|\mathcal{N}|} \sum_{(x,y)\in\mathcal{N}(i,k)} [I(x,y) - \mu_{\mathcal{N}}(i,k)]^2 \tag{5.54}$$

where $\mathcal{N}(i,k)$ is a small square neighborhood centered at the pixel location (i,k), $|\mathcal{N}|$ is the cardinality of the neighborhood, and $\mu_{\mathcal{N}}(i,k)$ is the estimated mean computed on the same window, i.e.:

$$\mu_{\mathcal{N}}(i,k) = \frac{1}{|\mathcal{N}|} \sum_{(x,y)\in\mathcal{N}(i,k)} I(x,y) \tag{5.55}$$

The mask M_σ can then be obtained by scaling the local standard deviation values in the interval $[M_{min}, 1]$ as in figure 5.22 where σ_{max} and σ_{min} are properly chosen limits for the variance. It is in general convenient to choose $M_{min} > 0$ in such a way that at least a small amount of watermark is embedded also in the most sensitive areas. As an alternative the inverse of the local variance can also be used for building the sensitivity mask $\bar{M}$:

$$\bar{M}_\sigma(i,k) = \frac{1}{1 + \sigma_{\mathcal{N}}^2(i,k)} \tag{5.56}$$

thus, instead of the linear relationship depicted in figure 5.22, the following equation results for the mask M_σ:

$$M_\sigma(i,k) = \frac{\sigma_{\mathcal{N}}^2(i,k)}{1 + \sigma_{\mathcal{N}}^2(i,k)} \tag{5.57}$$

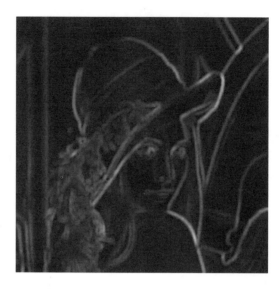

Figure 5.23: Masking function M_σ built based on the computation of the local variance.

The main weakness of the variance based masks is that they do not differentiate between highly textured regions and image edges (on the contrary the latter will produce, often, a higher value of local variance) as can be seen in figure 5.23 (where the mask linearly proportional to the local standard deviation is depicted), while we know that disturbs are more visible around contours than in textured areas (see **Rule 2**). Furthermore this mask does not take into account the dependence of the noise sensitivity level on background brightness. A more sophisticate heuristic mask is presented in section 5.4.3.

Let us now briefly analyze which principles can be used for building masks in the case of video sequences. Let us suppose for simplicity the watermark is independent on the host video signal and does not change sensibly in time. Let us then remember that the eye tends to track moving object and suppose that we want to exploit this characteristic of the HVS. While the eye observe, in the frame, an object that exhibits a large motion, the pupil tracks the object moving texture, which then appears to be almost still in the retina; because of that, a watermark independent from the host data would appear moving in the retina with a speed equal in magnitude and opposite in direction to that of the observed object. For modeling the perception of this watermark we should thus consider the case of perception of moving stimuli with the eye constrained to be still, i.e. exactly the plot

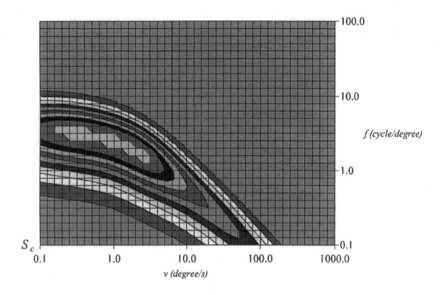

Figure 5.24: Contour levels of the spatiotemporal CSF with respect to the angular spatial frequency and to speed, obtained by imposing eye fixation and a mean background luminance of 20 cd/m^2.

of equation (5.43).

In figure 5.24 the contours levels of the function in (5.43) are depicted, it is thus evident that for large values of object speed, i.e. of watermark apparent speed, the CSF drop to zero for spatial frequencies much lower than for static objects. As an example, for an object moving by 12 $pixel/frame$, which are equivalent, for a 30 Hz frame rate and for a viewing distance of 4 times the height of the frame, to an angular speed of about about 10 $degree/s$, the cut-off frequency of the CSF is about 2.5 $cycle/degree$, while for still object it is about 10 $cycle/degree$. This observation suggests that when building a mask for video watermarking, it should be considered that the watermark signal could be added to lower frequencies in those areas of the frame that move fast than in those that are almost still. Of course this conclusion is valid if the watermarking signal can be considered not to vary too fast from frame to frame. Furthermore this approach, and in general any approach trying to exploit dynamic characteristics of HVS perception, clearly fails if quality requirements are so stringent to impose watermark invisibility also on still frames.

5.4.2 Concealment relying on visibility thresholds

Another possible approach for adapting the watermark to the host signal at hand, is to have a criterion establishing the maximum amount of modification that can be sustained by the host data: this can be defined both in the asset domain (independently on the features used for embedding) or directly in the embedding features domain. Ideally an imperceptibility region around the host data set $\mathcal{R}_p(A)$ should be defined: any modification, caused by watermarking, bringing the host data outside the region will be perceptible, while smaller modifications will not. It is important to stress that it is not needed that the imperceptibility region is defined in the same domain where embedding is performed, it is in fact possible to use optimization techniques for tuning the strength of the watermark signal in the embedding domain in such a way to satisfy the imperceptibility constraint in the domain where this is defined: an example of this approach has been described in section 4.3.2 as *Optimum embedding in block DCT domain*, where the perceptibility constraint is defined in the asset (spatial) domain, while the embedding is performed in the block DCT domain.

Let us for now on concentrate on the case in which the imperceptibility constraint is given in the same domain where the embedding is performed, i.e. is given in terms of the embedding features. The simplest imperceptibility region is defined componentwise, i.e. a maximum possible modification is given for each component of the feature vector:

$$T_l(\mathbf{f}, i) \le f_{w,i} - f_i \le T_u(\mathbf{f}, i) \qquad (5.58)$$

where $T_l(\mathbf{f}, i)$ and $T_u(\mathbf{f}, i)$ are the lower and upper maximum deviation perceptibly tolerable for feature f_i. These two values are, in practice, some perceptibility thresholds, and can depend, in general, on the whole feature set values $\mathbf{f}$, and on the feature index i; usually they are equal[22]. The resulting perceptibility region $\mathcal{R}_p(A)$ is an hyperbox in the feature space. The exploitation of the theoretical results presented in the first two sections of this chapter have brought often to perceptibility constraint of this type, usually defined in the frequency or hybrid domain.

As an example, for a video watermarking system working in the hybrid block DCT domain, the default quantization matrices defined for INTRA[23] coded blocks by the MPEG-2 standard can be used. These are derived by considering the CSF of the HVS, i.e. by considering, for modeling the eye sensitivity to disturbing stimuli, solely the dependence on the frequency.

---

[22]A case in which can be different, is related to the necessity of avoiding under- over-flows in the host data, e.g., for an image, to avoid that watermarked pixel values go beyond their range of validity (which is [0, 255] for usual grey level images).

[23]For INTER blocks the MPEG-2 default quantization matrices are uniform.

In particular the quantization step of each coefficient is usually given by a factor (that is chosen based on the required level of compression, and can vary from block to block) which scales the quantization steps themselves; these quantization steps are given in the form of an 8×8 matrix:

$$
\begin{bmatrix}
8 & 16 & 19 & 22 & 26 & 27 & 29 & 34 \\
16 & 16 & 22 & 24 & 24 & 27 & 29 & 34 \\
19 & 22 & 26 & 27 & 29 & 34 & 34 & 38 \\
22 & 22 & 26 & 27 & 29 & 34 & 37 & 40 \\
22 & 26 & 27 & 29 & 32 & 35 & 40 & 48 \\
26 & 27 & 29 & 32 & 35 & 40 & 48 & 58 \\
26 & 27 & 29 & 34 & 38 & 46 & 56 & 69 \\
27 & 29 & 35 & 38 & 46 & 56 & 69 & 83
\end{bmatrix}
\tag{5.59}
$$

for both luminance and chrominance blocks. For taking into account the lower bandwidth of the CSF on color channels, the matrix suggested[24] by the JPEG standard for chrominance block can be used:

$$
\begin{bmatrix}
17 & 18 & 24 & 47 & 99 & 99 & 99 & 99 \\
18 & 21 & 26 & 66 & 99 & 99 & 99 & 99 \\
24 & 26 & 56 & 99 & 99 & 99 & 99 & 99 \\
47 & 66 & 99 & 99 & 99 & 99 & 99 & 99 \\
99 & 99 & 99 & 99 & 99 & 99 & 99 & 99 \\
99 & 99 & 99 & 99 & 99 & 99 & 99 & 99 \\
99 & 99 & 99 & 99 & 99 & 99 & 99 & 99 \\
99 & 99 & 99 & 99 & 99 & 99 & 99 & 99
\end{bmatrix}
\tag{5.60}
$$

Indeed these matrices do not give an absolute threshold, as defined by equation (5.58), but only establishes how the disturbing noise can be perceptually distributed among the various coefficients. If for video coding systems the tuning (through the setting of the scaling factor) is regulated by the imposed transmission bit-rate, for watermarking applications the human operator tuning is always needed for precisely adapting to the video at hand. This visibility threshold simply depends on the frequency index, and not on the DCT coefficients themselves, i.e., by referring to equation (5.58), the dependence on the feature vector $\mathbf{f}$ is not present. This is due to the fact that the masking effect is neglected (as we have already said only the CSF is considered).

A more refined and quite popular model proposed by Watson et Al. for adapting the quantization matrices of the JPEG still image compression algorithm on a block by block basis, considers also the masking effect (in

---

[24]No default matrices are defined in the JPEG standard, but the presented one is suggested as an example.

particular iso-frequency masking). First of all it is assumed that an absolute visibility threshold $T(u, v, \phi)$ is available for each DCT coefficient, which depends on the frequency (u, v) of the coefficient, on the luminance or chromatic band ϕ that we are considering, and on the viewing conditions. This absolute threshold can for example be obtained by considering the CSFs of the luminance and chrominance bands; alternatively the standard quantization matrices of the JPEG standard can be used[25] In order then to take into account the reduction of HVS sensitivity to local luminance the threshold is adapted to each block mean luminance as:

$$t_a(u, v, \phi, h) = t(u, v, \phi) \left[\frac{C_{Y,h}(0,0)}{\overline{C}_Y(0,0)} \right]^{a_T} \qquad (5.61)$$

where h is the block index, $C_Y(0,0)$ is the DC value of the DCT transform of luminance block h, $\overline{C}_Y(0,0)$ is the mean of the DC coefficients of all luminance blocks in the image, and a_T is a parameter for which a value of 0.649 is proposed, but which can also embed the γ value needed for mapping color levels to color values. It has to be noted that adaptation of all color channel is made with reference to luminance only. Finally the iso-frequency masking effect is considered, yielding to the following perceptually adapted threshold[26]:

$$t_m(u, v, \phi, h) = t_a(u, v, \phi, h) \max \left(1, \left| \frac{C_{\phi,h}(u, v)}{t_a(u, v, \phi, h)} \right|^{0.7} \right) \qquad (5.62)$$

This perceptually adapted threshold is widely used for watermarking applications. The main limit of this approach is that the near-frequency and inter-band masking effects are neglected.

This threshold completely fits the model given by equation (5.58), in that it depends not only on the frequency index, but also on the features values.

A more general approach for defining the perceptibility region is based on the following equation:

$$\Phi \left(\mathbf{f}_w - \mathbf{f} \right) \leq T(\mathbf{f}) \qquad (5.63)$$

In this case the constraint is imposed on a global way, and not component-wise, through a function Φ which combine all components of the difference

---

[25] Indeed Watson proposes to use different matrices, derived from a more accurate model described in [172].

[26] Indeed the masking effect has been defined in terms of contrast values, but given that the local luminance to be used for computing contrast is the same both for the masking and for the maskee contrast, it is easy to verify that equation (5.13) is still valid for modeling visibility thresholds.

between the watermarked and the original feature vector. The perceptibility region can have a very complicate shape, depending on the type of function Φ. In section 5.4.4 a theoretically funded threshold will be derived that fit this last model.

Finally let us make an observation specifically regarding frequency domain embedding. We have seen that the presence of a sinusoid having a given frequency can mask a sinusoidal disturb having the same frequency (iso-frequency masking as described in section 5.1.3). Let us then suppose that we have chosen, as embedding features, the coefficients of the DCT or DFT of the asset: each coefficient represents basically the amplitude of a sinusoid which is present in the image, if we want to have an estimate of the amount of modification that it can sustain, we should refer to equation (5.22). This equation, by supposing that the masking signal (i.e. the coefficient to be modified) is already above the visibility threshold by itself, can be rewritten as:

$$C^m_{Ijn} = C_{Ijn}(I_0) \left[\frac{\Delta I_m / I_0}{C_{Ijn}(I_0)} \right]^w = C_{Ijn}(I_0)^{1-w} \frac{(\Delta I_m)^w}{(I_0)^w} \qquad (5.64)$$

and remembering that $C^m_{Ijn} = \frac{\Delta I^m_{jn}}{I_0}$ and $C_{Ijn} = \frac{\Delta I_{jn}}{I_0}$ we we get that the Just Noticeable modification that can be applied to the coefficient is:

$$\Delta I^m_{jn} = \Delta I^{1-w}_{jn} \Delta I^w_m \qquad (5.65)$$

and given that ΔI_{jn} can be approximated to be almost constant for a large range of grey level values (see figure 5.10), we can write:

$$\Delta I^m_{jn} = \gamma \Delta I^w_m \qquad (5.66)$$

where in general the γ parameter will depend on the considered frequency. This last equation supports the choice of a multiplicative embedding rule for frequency domain watermarking.

5.4.3 Heuristic approaches for still images

In this section two possible heuristic approaches will be detailed for concealing the disturbs caused by watermark embedding on still images. The derivation of both methods is based on the three rules of thumb listed at the beginning of section 5.1: in both cases in fact the goal is to highlight highly textured regions, contours, and dark and bright regions for accurately balancing the watermarking strength among the various image zones.

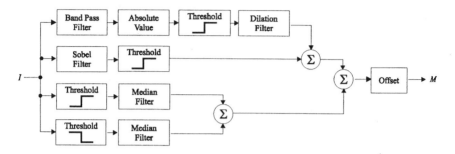

Figure 5.25: Scheme of the procedure to build the heuristic mask based on band-pass filtering, contour extraction and thresholding of the original image.

A locally adaptive mask for transformed domain watermarking

As a first case we describe a masking function M_{bp} originally devised for locally adapting the watermark embedded in the transformed domain by means of a multiplicative rule. Although we have shown that the multiplicative rule has a rationale in that it allows to exploit the iso-frequency masking effect, full frame transform based watermarking techniques dramatically suffers from a lack of spatial localization, i.e. the watermark signal will appear, in the spatial domain, also where the masking component is not really present. Thus the need for spatially adapting the watermark as in equation (5.52).

The procedure for building the mask is based on the already described rules and on the assumption that the watermark will be better hidden in those areas of the image where the watermarked frequencies (which are in this case those belonging to the middle range of the spectrum) are present. The procedure is sketched in figure 5.25. The image is processed through 4 parallel branches. In the first one we have a band-pass filter tuned for extracting just the range of frequencies where the watermark is embedded in; the absolute value of the band-pass filtered image is first thresholded to further highlight these image components; a dilation morphological filter is then applied to the resulting image to fill the holes. The second branch is aimed at enhancing the edges: a Sobel filter followed by a threshold operation is applied. Finally the third and fourth branches are devised for highlighting respectively the brightest and darkest areas of the original image, by means of gray-level thresholding (the median filters in these two branches are useful to regularize the thresholding outputs). The four contributions are then added together, and, finally, an offset value is added to the resulting sum in order to allow at least a minimum amount of watermark to be inserted everywhere (also in uniform areas of medium brightness

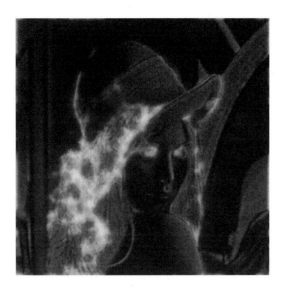

Figure 5.26: Masking image M_{bp} built by the system sketched in Figure 5.25.

a small level of noise can be tolerated). The mask shown in figure 5.26 is thus obtained: it is evident the ability of the method to properly highlight the different areas of the image, by attributing to each the suitable degree of noise sensitivity.

In principle the described heuristic mask can be used with any watermarking system (e.g. for spatial domain systems). In general it is convenient to continue selecting intermediate frequencies for the band of the bandpass filter, to trade-off between watermark invisibility and robustness.

A visibility threshold for DWT based watermarking

We will now analyze how effective exploitation of HVS characteristics for hiding signals can be obtained in the DWT domain. Thanks to its excellent spatiofrequency localization properties, the DWT is very suitable to identify the image areas where a disturb can be more easily hidden. In particular, this property effectively allows to exploit the HVS iso- and near-frequency masking effects: if a DWT coefficient is modified, only the region of the image where the particular frequency corresponding to that coefficient is present will be modified, in contrast to what happens, for example, by using full frame DFT/DCT watermarking.

In this case we suppose the image to be watermarked is first decomposed through DWT in four levels: let us call I_l^θ the sub-band at resolution level

$l = 0, 1, 2, 3$ and with orientation $\theta \in \{0, 1, 2, 3\}$ (see figure 4.7). Given the watermark has to be embedded into the wavelet coefficients we want to evaluate the just noticeable threshold $q_l^\theta(i, k)$ of modification that each coefficient can sustain without degrading the visual quality of the image. This evaluation is performed heuristicaly based mainly on the three rules of thumb presented at the beginning of this chapter. In particular the just noticeable threshold is computed as the weighted product of three terms:

$$q_l^\theta(i, k) = \Theta(l, \theta)\Lambda(l, i, k)\Xi(l, i, k)^{0.2}, \tag{5.67}$$

where the meaning of each term in the above equation is explained below.

Let us start the analysis of $q_l^\theta(i, k)$ by the first term of the right hand side of the expression in (5.67). To take into account how sensitivity to noise changes depending on the band (in particular depending on the orientation and on the level of detail, i.e. on the frequency), we let:

$$\Theta(l, \theta) = \left\{ \begin{array}{ll} \sqrt{2} & \text{if} \quad \theta = 1 \\ 1 & \text{otherwise} \end{array} \right\} \cdot \left\{ \begin{array}{lll} 1.00 & \text{if} & l = 0 \\ 0.32 & \text{if} & l = 1 \\ 0.16 & \text{if} & l = 2 \\ 0.10 & \text{if} & l = 3 \end{array} \right\}. \tag{5.68}$$

The behavior of the HVS CSF (see section 5.1.2) suggests in fact that the eye is less sensitive to disturbs at high frequencies (higher detail bands in the DWT) and for stimuli having diagonal orientation (as those described by the coefficients into the diagonal subbands of the DWT).

The second term takes into account the local brightness, by considering the grey-level values of the low pass version of the image. Based on the consideration that the human eye is less sensitive to changes in very dark and very bright regions (see again section 5.1.2) we define this factor as follows:

$$\Lambda(l, i, k) = \begin{cases} 1 - \Lambda'(l, i, k) & \text{if} \quad \Lambda'(l, i, k) < 0.5 \\ \Lambda'(l, i, k) & \text{otherwise} \end{cases} \tag{5.69}$$

where:

$$\Lambda'(l, i, k) = \frac{1}{256} I_3^3 \left(1 + \left\lfloor \frac{i}{2^{3-l}} \right\rfloor, 1 + \left\lfloor \frac{k}{2^{3-l}} \right\rfloor \right). \tag{5.70}$$

Finally, the third term:

$$\Xi(l, i, k) =$$
$$\sum_{h=0}^{3-l} \frac{1}{16^h} \sum_{\theta=0}^{2} \sum_{x=0}^{1} \sum_{y=0}^{1} \left[I_{h+l}^\theta \left(y + \frac{i}{2^h}, x + \frac{k}{2^h} \right) \right]^2 \cdot$$
$$\mathrm{Var} \left\{ I_3^3 \left(1 + y + \frac{i}{2^{3-l}}, 1 + x + \frac{k}{2^{3-l}} \right) \right\} \begin{array}{l} x = 0, 1 \\ y = 0, 1 \end{array} \tag{5.71}$$

gives a measure of texture activity in the neighborhood of the pixel. In particular, this term is composed by the product of two contributions: the first is the local mean square value of the DWT coefficients in all detail sub-bands, up to the level at hand, while the second is the local variance of the low-pass sub-band. Both these contributions are computed in a small 2×2 neighborhood corresponding to the location (i, k) of the DWT coefficient. Since the first contribution can be considered to represent the distance from the edges, whereas the second one the degree of textureness, we decided to multiply the two terms, according to our consideration that the eye is less sensitive in textured areas, but more sensitive near edges.

The just noticeable amplitude threshold $q_l^\theta(i, j)$ can thus be properly scaled, and used as the maximum amount of modification that each DWT coefficient can undergo to grant imperceptibility.

Main limitation of the heuristic approaches

In general the biggest problem of the presented heuristicaly computed masks is that they only consider the content of the host image, and not the characteristics of the to be embedded watermark: as an example, a highly textured image region having a prominent horizontal orientation seems to be able (according to the heuristic approaches) to mask also vertically oriented component of the watermark, which is clearly false as we have seen in section 5.1.3 with reference to near-frequency masking effects. The objective of next section is just to describe a more theoretically accurate procedure for building masks, which, further to the spectral content of the host signal, considers the spectral content of the watermark as well.

5.4.4 A theoretically funded perceptual threshold for still images

We have seen that the heuristic approaches to watermark concealment, although very simple, present some problems. Some more theoretically funded techniques have also been proposed as those yielding to the JPEG and MPEG quantization matrices: anyway, as we have seen, these also exhibit some limitations. The goal of this section is just to present a theoretically funded method for estimating a visibility threshold, relying on HVS basic concepts, like contrast sensitivity function and contrast masking effect.

A problem is that common masking models only accounts for the presence of a single sinusoidal mask (as we have seen also the sophisticated Watson model consider solely iso-frequency masking). This is not the case in practical applications where the masking signal, namely the host image, is nothing but a sinusoid. Furthermore, the non-sinusoidal nature of the disturbing signal (the watermark) is a further significant deviation from common HVS models, in that such models only permit to foresee the maximum allowable strength of sinusoidal signals.

To solve these problems, the technique described in this section exploits a block-based DCT decomposition of the image. The use of a block-based DCT analysis permits to trade off between spatial and frequency localization of the image features and disturbs. Moreover a combination of both masking and disturbing signal components is taken into account. We will restrict our analysis to the luminance channel, although the extension to color images is straightforward by considering the opponent color space and treating each channel independently from the other[27]. Furthermore the approximation presented in equation (5.22) is used for allowing us to directly work on grey level values.

The host image is divided into blocks $B_l(i,k)$ of size $n_b \times n_b$, to which the DCT is applied; let us rewrite equation (4.12) as:

$$
\begin{aligned}
B_l(i,k) = \frac{1}{n_b} \sum_{u=0}^{n_b-1} \sum_{v=0}^{n_b-1} c(u)c(v)B_l(u,v) \cdot \\
\cdot \left[\cos\left(2\pi\nu_{uv} \left((k+\frac{1}{2})\cos\theta_{uv} + (i+\frac{1}{2})\sin\theta_{uv} \right) \right) + \right. \\
\left. \cos\left(2\pi\nu_{uv} \left((k+\frac{1}{2})\cos\theta_{uv} - (i+\frac{1}{2})\sin\theta_{uv} \right) \right) \right]
\end{aligned}
\tag{5.72}
$$

where $B_l(u,v)$ indicates the DCT coefficients, $\nu_{uv} = \sqrt{u^2+v^2}/2n_b$ is the spatial frequency[28] of the sinusoidal stimulus, and $\theta_{uv} = \arctan(u/v)$ is its orientation. This equation allows us to decompose each image block as the sum of a set of sinusoidal stimuli. In particular for each block l the mean grey level is given by $B_l'(0,0) = B_l(0,0)/2n_b$. Furthermore each coefficient at frequency (u,v) gives birth to two sinusoidal stimuli, having the same amplitude $B_l'(u,v) = B_l(u,v)/n_b$[29], the same spatial frequency, but opposite orientations.

By relying on equation (5.14) (indeed on its correspondent formula depending on grey level), for a DCT coefficient at spatial frequency (u,v) the masking effect of an other coefficient at frequencies (u',v') can be considered. To take into account the non-sinusoidal nature of the masking signal (the host image), for each frequency (u,v) the contributions of all the surrounding frequencies of the block are considered. We introduce a sum of the weighed masking contributes on the whole block. The value of

---

[27] Less trivial would be the case in which we would also to consider inter band masking phenomena.

[28] This can be easily converted to the angular frequency as $f = \pi\nu d/180$ where d is the viewing distance.

[29] When $\theta_{uv} \in \{0,\pi\}$ it results $B_l'(u,v) = B_l(u,v)/\sqrt{2}n_b$

the Just Noticeable Contrast is then obtained from equation (5.14) through the following expression:

$$C_{Ijn}^{m}(u,v,\mathbf{B}_l) = C_{Ijn}(u,v,B_l'(0,0))\cdot$$

$$\cdot \max\left\{1, \left[\sum_{u'=0,v'=0}^{n_b-1,n_b-1}\left|S'(u,u'v,v')\frac{B_l'(u',v')/B_l'(0,0)}{C_{Ijn}(u',v',B_l'(0,0))}\right|\right]^w\right\} \quad (5.73)$$

where $C_{Ijn}^{m}(u,v,\mathbf{B}_l)$ is the Just Noticeable Contrast of coefficient of block l at frequency (u,v), caused by masking of all other coefficients of the block, $C_{Ijn}(u,v,B_l'(0,0))$ is the non masked (absolute) Just Noticeable Contrast for the coefficient at frequency (u,v) as given by equation (5.23), $B_l'(u',v')/B_l'(0,0)$ is the contrast of the masking coefficient, and $S'(u,u'v,v')$ is the weighting function that can be obtained by equation (5.15) as:

$$S'(u,u'v,v') = e^{-\log_2^2\frac{f_{u'v'}}{f_{uv}}/\sigma_f^2} \cdot e^{-\left((\theta_{u'v'}-\theta_{uv})^2+(-\theta_{u'v'}-\theta_{uv})^2\right)/\sigma_\theta^2} \quad (5.74)$$

where the fact that each DCT coefficient gives birth to two sinusoidal components with the same spatial frequencies but opposite orientations, and that the Just Noticeable Contrast has the same value for stimuli having opposite orientations, has been considered. Expression (5.73) has been derived by considering that if we have two sinusoidal stimuli at very similar frequency, these should contribute to the masking effect mostly like a unique sinusoidal stimulus having the sum of the single amplitudes, this motivated the summing operation to be performed before the w nonlinearity is applied. A similar consideration yields also to make the sum, before the maximum is selected.

In order to guarantee the invisibility of a sinusoidal disturb in a given block, the contrast of the component of the disturb at a given frequency (u,v) must be smaller than the value of the C_{Ijn}^{m} obtained by (5.73). Anyway the non-sinusoidal nature of the inserted watermark has to be taken into account. Let us suppose that to each coefficient we add a watermarking component of amplitude $w_l(u,v)$[30], we have to consider that nearby watermarking coefficients will enforce each other, i.e. the visibility of a couple of neighboring coefficients can not be considered the same as if they were alone. Thus we write an equivalent contrast of disturb at coefficient (u,v) in block l as:

$$C_D^{eq}(u,v,\mathbf{w}_l) = \sum_{u'=0,v'=0}^{n_b-1,n_b-1}\left|S'(u,u'v,v')\frac{c(u')c(v')}{n_b}\frac{w_l(u',v)'}{B_l'(0,0)}\right| \quad (5.75)$$

---

[30]This does not imply the watermark to be additive, this term $w(u,v)$ is simply the difference between the watermarked coefficient and the original one $w_l(u,v) = B_{l,w}(u,v) - B_l(u,v)$; what we want to get is a threshold on the maximum allowable modification that the coefficient can sustain.

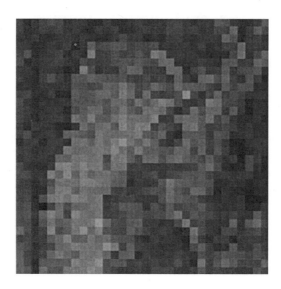

Figure 5.27: Mask obtained for the Lenna image by means of the block-DCT based perceptual model.

where $\frac{c(u)c(v)}{n_b} \frac{w_l(u,v)}{B_l'(0,0)}$ is the contrast of the added signal, and we have assumed that the same weighting function can be used for modeling the enforcing effect of neighboring disturbs.

The invisibility constraint results then to be:

$$C_D^{eq}(u, v, \mathbf{w}_l) \leq C_{Ijn}^m(u, v, \mathbf{B}_l) \tag{5.76}$$

which has clearly the general form given by equation (5.63), i.e. the constraint is imposed in a global way.

Based on this approach it is also possible to build a masking function for spatially shaping any kind of watermark. By referring to equation (5.53) let us suppose that the mask M is block-wise constant, and let us indicate with M_l the value asumed by the mask in block l; let us also indicate with $w_h(u, v)$ the DCT coefficients of block h of the difference between the original and the uniformly watermarked image; it is the possible to compute $C_D^{eq}(u, v, l)$. By exploiting the linearity property of the DCT transform, it is easy to verify that, for satisfying the invisibility constraint, we must have

$$M_l \cdot C_D^{eq}(u, v, \mathbf{w}_l) \leq C_{Ijn}^m(u, v, \mathbf{B}_l) \quad \forall(u, v) \tag{5.77}$$

and we can thus set:

$$M_l = \min_{u,v} \frac{C_{Ijn}^m(u, v, \mathbf{B}_l)}{C_D^{eq}(u, v, \mathbf{w}_l)} \tag{5.78}$$

In figure 5.27 the resulting masking function is shown for the Lenna image. The mask computed with the approach described produces reliable results, especially on textured areas. This is mainly due to the fact that the disturbing signal frequency content is also considered for building the mask. Moreover this method allows the maximum amount of watermarking energy that each image can tolerate to be automatically obtained.

5.4.5 MPEG-based concealment for audio

In this section we overview the salient points of the procedure for building the Psychoacoustic Model 1 of the MPEG-1 standard. The aim of this model is to allow to estimate an audibility threshold adapted to a short segment of audio signal, and taking into account the absolute hearing sensitivity of the HAS (figure 5.20) and the masking effects (section 5.2.1).

As it as already been mentioned, to effectively exploit masking characteristics, it needs to identify the pure tone and noise like components of the audio segment. For this goal a 512^{31} points FFT is first computed on a segment of audio signal samples. Before FFT the audio samples $S(i)$ are normalized by scaling them by 512 (the FFT length) and by $2^{N_b} - 1$ where N_b is the number of bits per sample, a Hanning window, which is basically a raised cosine, is then used for reducing the cropping effect in the computation of the FFT. An estimate of the power spectral density of the segment can thus be obtained as

$$P(u) = 90.302 + 10 \log_{10} |F(u)|^2, \quad 0 \le u \le 511 \qquad (5.79)$$

where $F(u)$ are the FFT samples, and 90.302 is a normalizing value granting that a full scale sinusoid will bring a $P(u)$ value of around 80 dB, while a very low amplitude (ideally ± 1) input signal at the maximum acuity frequency of around 4 kHz will produce an SPL around 0 dB.

Tonal components are then identified as the local maxima among the sample PSD, in particular, the tonal set S_T is defined as:

$$S_T = \{P(u)|P(u) > P(u \pm 1) \text{ and } P(u) > P(u \pm \Delta_u) + 7\} \qquad (5.80)$$

where

$$\Delta_u \in \begin{cases} 2 & 2 < u < 63 \ \ (0.17 - 5.5kHz) \\ \{2,3\} & 63 \le u < 127 \ \ (5.5 - 11kHz) \\ \{2,3,4,5,6\} & 127 \le u \le 256 \ \ (11 - 20kHz) \end{cases} \qquad (5.81)$$

---

[31]The number of FFT points depends on the used sample frequency.

Table 5.1: Idealized critical bands. For each of the 25 critical bands the band number (No.) the central frequency (Freq.) and the bandwidth (BW) are reported.

No.	Freq.	BW	No.	Freq.	BW	No.	Freq.	BW
1	50	100	10	1175	190	19	4800	900
2	150	100	11	1370	210	20	5800	1100
3	250	100	12	1600	240	21	7000	1300
4	350	100	13	1850	280	22	8500	1800
5	450	110	14	2150	320	23	10500	2500
6	570	120	15	2500	380	24	13500	3500
7	700	140	16	2900	450	25	19500	6500
8	840	150	17	3400	550			
9	1000	160	18	4000	700			

Tonal maskers $P_{TM}(k)$ are then computed for each element of the set S_T as:

$$P_{TM}(u) = 10 \log_{10} \sum_{h=-1}^{1} 10^{0.1P(u+h)} \qquad (5.82)$$

i.e. the sum of each maximum and of its two neighbors is considered as the energy of the tone.

A finite number of non-overlapping critical bands is then considered as from Table 5.1 and for each band a single noise masker $P_{NM}(\bar{u})$ is computed by combining all spectral lines that are not within a $\pm\Delta_u$ neighborhood of a tonal masker, i.e.

$$P_{NM}(\bar{u}) = 10 \log_{10} \sum 10^{0.1P(u)} \qquad (5.83)$$

where the sum is extended to all the spectral lines within the critical band, that do not lie inside a $\pm\Delta_u$ interval of a tonal masker, and the representative frequency is the geometric mean of all the critical band frequencies. In practice all other spectral lines not being identified at tonal maskers, and far enough from a tonal masker, are considered to contribute to the noise maskers.

Once selected, the sets of tone and noise maskers are simplified by trying to reduce not relevant information, for example the maskers that are below the absolute threshold of hearing are eliminated. Based on the remaining maskers, the masking thresholds are computed. In particular the masking contribution at the frequency v of the masking tone at frequency u is given by:

$$T_{TM}(v, u) = P_{TM}(u) - 0.275z(u) + SF(P_T(u), v, u) - 6.025 \qquad (5.84)$$

for the tone maskers and by:

$$T_{NM}(v, u) = P_{NM}(u) - 0.175z(u) + SF(P_N(u), v, u) - 2.025 \qquad (5.85)$$

for the noise maskers, where $z(u)$ is the Bark frequency corresponding to u and the spreading function $SF(P, v, u)$ is modeled as:

$$SF(P, v, u) = \begin{cases} 17\Delta_z - 0.4P + 11 & -3 \le \Delta_z < -1 \\ (0.4P + 6)\Delta_z & -1 \le \Delta_z < 0 \\ -17\Delta_z & 0 \le \Delta_z < 1 \\ (0.15P - 17)\Delta_z - 0.15P & 1 \le \Delta_z < 8 \end{cases} \qquad (5.86)$$

where $\Delta_z = z(v) - z(u)$.

All masking contribution are finally put together, to obtain a global masking threshold, by assuming that masking effects are additive, i.e.:

$$T_g(v) = 10 \log_{10} \left(10^{0.1T_q(v)} + \sum_{u=1}^{N_T} 10^{0.1T_{TM}(v,u)} + \sum_{u=1}^{N_N} 10^{0.1T_{NM}(v,u)} \right) \qquad (5.87)$$

where N_T and N_N are the number of tone and noise maskers respectively.

Such a threshold represents the maximum amount of disturbing stimulus that can be added to each frequency v before being perceived. It can thus be used to limit the amount of watermarking signal directly added in the frequency domain. Anyway it can also be used for shaping the spectral content of a time domain watermark: to this aim the watermark is first transformed through the DFT, it is then weighted by the global masking threshold, then transformed back and added to the audio signal.

5.5 Application oriented concealment

What we have described until now is how to conceal the watermark in such a way to make it imperceptible for a human observer/listener. Nevertheless there are some applications where the primary asset final user is not an human, but (typically) a computer program that has to extract some information from the data. In this cases the concealment problem can be quite different, and it is likely that the unobtrusiveness[32] constraint has to be formulated in a different way from what we have seen until now in this chapter.

In the following we give some examples of applications where the unobtrusiveness constraint is discussed just having in mind the final asset user.

---

[32]Let us here talk about unobtrusiveness, given that the concept of imperceptibility is tightly related to human perception.

As we will see, the approach we will follow for this analysis is not very different from that one we used for the imperceptibility case: first a model of the observer is built, then from the characteristics of the model some hints are drawn on the best strategy to hide the data inside the asset.

5.5.1   Video surveillance systems

Video surveillance (VS) systems main goal is to control an area of an environment by means of a video acquisition, transmission and storage system. Usually the acquired video is also automatically analyzed by some Automatic Visual Inspection (AVI) system. One of the requirements of VS systems is that they should be able to give an authentication proof of the stored or transmitted video material. For satisfying this requirement, watermarking based authentication can be useful; let us then analyze which unobtrusiveness constraint it needs to satisfy in this case.

In a VS case the watermark visibility issue is not crucial as it is for other video materials, since VS data do not exhibit a quality comparable to that of visual data used in arts/media. In fact, VS data are usually acquired by inexpensive, low quality devices and in highly uncontrolled environment, thus being of low visual quality. In addition, the visual analysis that would possibly be carried out in a law court will focus on the semantic content of the images, rather than on their visual quality. It is thus better to concentrate our investigation on what unobtrusiveness means from AVI point of view.

An AVI system usually works at two levels: a low level, where the images acquired by the camera are processed to detect moving objects, and a higher level where the behavior of moving objects is analyzed, in order to understand if an alarm situation is currently present. Let us consider first how low level processing could be affected by the authentication process. As a first consideration, it could be said that the allowable modification level is related to the noise level tolerable by the AVI system. Given that AVI systems have to operate in uncontrolled environments, they usually exhibit a high degree of robustness with respect to image perturbations. These considerations permit to conclude that the invisibility requirement is more relaxed in the VS framework.

To move the analysis one step further, the design of a watermarking algorithm for VS data authentication could follow two different guidelines.

A first possibility derives from the observation that if the embedded signal (i.e. the embedded noise pattern) is the same for all video frames, the presence of the watermark will not disturb the correct functioning of the system in any way, given that AVI techniques aim at detecting differences between frames. Of course, this would imply that the watermark is neither frame-dependent, nor time-varying (which means that it will not be able to help in maintaining the correct sequential frame order). Furthermore a major drawback of this approach is that a frame-independent watermark can be easily found by a comparative analysis of all image sequence frames, and then could be easily added again to fake frames.

As a second possibility, it has to be considered that low level image processing techniques usually employed in AVI systems, largely exploit the time coherence of natural object motion; this means that a certain degree of correlation is assumed to be present between subsequent frames, and that uncorrelated frame differences are filtered out as noise. Thus, it is important that watermarks embedded in successive frames are highly uncorrelated, in order to avoid producing misinterpreted moving objects, thus originating false alarms. Let us, for example, consider a spatial-domain watermarking technique, which embeds a binary code in an image by using a spread-spectrum amplitude modulation, and let us assume that the sequential number of each frame has to be embedded within the frame. The technique is by itself image-independent, but the watermark becomes frame-dependent because it contains the frame number. Given that frame numbers of subsequent frames are highly correlated, it is important that the spreading sequence is changed from frame to frame, to avoid confusing the AVI system.

The considerations regarding the opportunity of using watermarks which are highly uncorrelated among frames, are still valid when the watermark effect on high level processing stages is considered. High level reasoning procedures usually examine the moving objects detected by low level analysis, and try to estimate their trajectories in subsequent frames. A strong assumption underlying such procedures is related to motion coherence: moving objects should follow a naturally smooth trajectory. This assumption allows a drastic reduction of noise effects. With high probability, possible fake moving objects due to the watermark will not exhibit a natural motion, and thus, will not be interpreted by high level analysis as true moving objects.

However, the most important consequence of the particular meaning of the transparency in VS, is that stronger (and thus more robust) watermarking procedures can be applied in this case than in classical authentication applications. This could surely boost their acceptance as legal proofs. As an example, the invisibility constraint can be relaxed, thus allowing to obtain a more robust authentication scheme which resists moderate JPEG compression. Of course the strength of the watermark should not seriously degrade the image, thus reducing its value as a legal proof.

5.5.2   Remote sensing images

The application of data hiding technologies to remote sensing images has recently emerged, the goal of this section is just to investigate the watermark unobtrusiveness issue with regard to this type of images.

Due to the very large number of possible applications of remote sensing data, the approach of defining sufficiently specific watermarking requirements, albeit of a general relevance, is a hardly viable one. In general, we can view the production of a remote sensing image as a measurement process, where the value of each pixel represents a measure of a physical property of the terrain (and its interaction with the atmosphere). This is particularly true for the case of multispectral or hyperspectral images, that are constituted of a large set of bands, each one recording the amount of electromagnetic energy emitted by each ground pixel area, in a small range of frequencies. Since any watermarking algorithm ultimately modifies pixel values, the introduction of the watermark can be seen as an additional source of noise in the measurement process. For such a noise to be tolerable, it must be lower than (or at least comparable to) the noise normally introduced by the measurement process, whose level is usually expressed by means of an upper bound on the maximum measurement error.

These considerations naturally lead to the introduction of the near-lossless watermarking concept, where a watermark is said to be near-lossless if its insertion introduces an error whose maximum value can be exactly specified by the user[33]. Thus the unobtrusiveness constraint result to be very simple in this case. To be more precise, multispectral and hyperspectral images are mainly used for automatic classification purposes: multidimensional pixel values are then fed to classification tools that assign each pixel to a particular class. What could be granted, for example, is that the percentage of pixels classified differently before and after watermark embedding, results to be not much different from the percentage of pixels classified differently by different classification tools (there is always a minimum amount of pixels that have an inherent degree of uncertainty). The near-lossless constraint try just to achieve this objective.

Of course, remote sensing images can also be subject to visual analysis, especially in the panchromatic case. Such an analysis can be carried out in a multimedia-like framework (e.g. preparation of advertising material), or in a scientific context, e.g. for the extraction of some ground features (e.g. roads, buildings, etc.). In the former case the watermarking unobtrusiveness requirements can be expressed again in terms of transparency to the

---

[33]According to the above definition, the near-lossless watermarking concept closely resembles the concept of near-lossless image coding usually encountered in the remote sensing literature.

human visual system. Conversely, in the latter case it is desirable that the watermark does not negatively impact on the visual image analysis task; this also holds in case that automatic analysis algorithms are operated on the images. This remark can be recast into the requirement that the watermark should be spatially uncorrelated, thus not raising fake structures in the image data.

5.6   Further reading

The studies on human perception go back to the 19th century, in particular E. E. Weber [230] firstly proposed the law that brings his name for describing the relation between the sensation that an human being gets corresponding to a given external stimulus, by highlighting that the perceptual sensibility decreases when the stimulus increases. He studied in particular the sensory response to weight, temperature, and pressure: his work was extended to visual stimuli by his former student G. T. Fechner [76].

An excellent source of details about the photometric unity of measures used for quantifying visual stimuli is the book by G. Wyszecki and W. S. Stiles [235] where a very comprehensive overview of color science theory is presented.

Many analytical expression have been proposed in the literature for the visual Contrast Sensitivity Function (CSF), we reported in this chapter the one due to P. G. J. Barten [21] because it is the most complete in modeling the dependency of this parameter on all observation parameters (background luminance, angular frequency, observer viewing angle). This, anyway, does not include the dependency on the stimulus orientation, we were thus referring to the work by S. Comes and B. Macq [49] for finally getting the expression given by equation (5.6). Other expressions for the CSF can be found in [191] and [190] where, however, there is not explicit dependence on background mean brightness and angle of observation.

Readers interested to get more details about the luminance contrast masking phenomenon and its modeling can refer to the classical paper by C. F. Stromeyer and B. Julesz [207] and to those by G. E. Legge [132], and mainly [133].

With regard to the standard color space definitions it is possible to directly refer to the CIE document: in particular to [98] for the RGB and CIE XYZ color spaces, and to [99] for CIE $L^*a^*b^*$. The details of the new perceptual distance defined by CIE for colors, and mentioned on page 5.1.5 are given in [100]. On the other side the definition of NTSC color primaries can be found in [112], while the ITU-R color primaries are defined in [110] for the B.709 analogue standard and in [109] for the digital video standard.

The human perception oriented HSI color space is described in computer graphic books, as for example [89].

The literature about opponent color modeling of chromatic vision is very wide, anyway a good starting point on this topic is constituted by the paper by A. B. Poirson and B. A. Wandell [182]. The transformation to pass from XYZ to opponent color coordinates is given in [236], while the CSFs for the opponent color channels are presented in [160]. The masking phenomena among color channels and their modeling are described in [212]. Based on opponent color theory many tools have been developed for evaluating the perceptual quality of color images (see for example [231]), however interband masking effects are usually neglected, that is the single opponent channels are treated independently.

With regard to the perception of time varying stimuli, we have seen that quite different results are obtained if the eye is constrained to stay fixed or is left free to move: anyway by modeling the eye movement it is possible to make the different results appear in agreement. More details about the spatio-temporal stabilized (i.e. obtained by fixing eye gaze) CSF we presented in section 5.1.6 can be found in [119], while the velocity compensated version is proposed in [63].

In the description of the characteristics of the Human Auditory System we referred to the classical paper of T. Painter and A. Spanias, where an excellent overview of acoustic perception phenomena is presented.

More details about the watermarking technique that embeds the information into the blue channel, and mentioned in section 5.3, can be found in [129].

The definition of MPEG-2 and default JPEG quantization matrices can be found in the official standard document (i.e. respectively in [105] and in [107]); interested readers can also refer, for a good introduction to compression standards, to the book by A. M. Tekalp [214]. On the other side the more advanced quantization matrices proposed by A. B. Watson are presented in [228], while in [229] their extension for managing color images is described.

Some general methods for exploiting HVS characteristics in watermarking applications have been presented in [233, 181]

The heuristic mask whose computational scheme is briefly sketched in figure 5.25, was originally presented in [22], interested readers can refer to that document for a deeper analysis of its performance, and a comparison with other methods. The method for estimating the just noticeable amount of modification tolerable by DWT coefficients, and described in section 5.4.3, is based on work by A. S. Lewis and G. Knowles [134], aimed at perceptually optimizing the quantization process of DWT coefficients for still image compression, more details about the adaptation of this method to image watermarking are given in [18]. On the other side a comprehensive analysis of the theoretically funded perceptual threshold described in section 5.4.4 is given in [64].

With regard to audio watermarking, more details about the MPEG-1 Psychoacoustic Model 1 can be found in the standard [102], or in the already cited paper by T. Painter and A. Spanias [165], while its use for shaping a time domain watermark is presented in [211].

Two examples of application oriented unobtrusiveness requirements are given in [23] regarding video surveillance systems, and in [15] for remote sensing images.

6

Data recovery

The definition of a reliable procedure to retrieve the information hidden within the host signal is of fundamental importance for the proper development of any data hiding system. This is not an easy task, because of the many modifications the host asset may undergo after embedding. As a matter of fact, modelling the watermark channel in the presence of attacks is a very complicated problem, due to the wide variety of possible attacks to be taken into account. Another factor which somewhat complicates the recovery of the hidden information is the unavailability of the original asset at the detector/decoder. As we will discuss in chapter 9, in line of principle this is not a problem, however, in practice, blind systems are much more complicated than non-blind ones.

Due to the wide variety of attacks and to the difficulties of developing an accurate statistical model of host features, the structure of the detector/decoder is, usually, derived by considering a simplified channel model. The performance of the system in the presence of more complicated channels, then, is evaluated either theoretically (by assuming that the detector/decoder structure is known) or experimentally. In this chapter we follow a similar approach: we derive the detector/decoder structure in some simple cases, dealing with over-simplified channel models, where attacks are either absent or modelled as noise addition. Then we evaluate the error probability of the system for the simplified channel, being aware that a more accurate, experimental, analysis is needed to assess the performance of the systems in more realistic situations.

We first consider the detection problem, in which the detector is only asked to decide whether the asset at hand contains a given watermark or not. Then we pass to the decoding problem, in which the decoder has to take a decision on which message, among those possible, is actually embedded within the asset under analysis. In this second case, it is usually assumed that the host asset is surely marked, nevertheless this is not necessarily the case. For this reason, we also touch the problem of watermark presence assessment in readable watermarking systems.

6.1   Watermark detection

We start by considering the detection problem, i.e. given a digital asset A and a watermark code b, decide whether A contains b or not. Of course, the problem depends on the particular embedding rule adopted by the system. When embedding passes through the injection of a watermarking signal w into the host feature set, the problem can be easily formulated as one of signal detection in a noisy environment, where noise accounts for both the unknown host signal and the possible presence of attacks. In informed embedding systems, the situation is rather different. In line of principle, the detector structure, i.e. the detection regions associated to each watermark message, could be defined without making any reference to the embedding process, for example, by adopting random coding arguments. In this case embedding would reduce to mapping the host asset into the detection region subject to the invisibility constraint. Moreover, the performance of the system would largely depend on the adopted embedding strategy and the asset at hand, since it would depend on the position of the marked asset within the embedding region.

In most cases, though a more tortuous path is followed. Detection regions are computed (often optimally) by assuming that a blind embedding strategy is adopted (e.g. by applying additive or multiplicative spread spectrum watermarking), then the informed embedding paradigm is applied to actually watermark the host asset[1]. Additionally, (informed coding) the same watermark may be associated to several detection regions. If the above approach is followed, informed embedding does not have any impact on the structure of the detector, however it contributes to diminish the error probability (or improve imperceptibility, or increase the payload). The actual computation of the error probability in the presence of informed embedding is usually a cumbersome task, since it intimately depends on the host asset. The only noticeable exception is when a constant robustness strategy is adopted (see "MD" approach in section 4.3.1), since in this case the same error probability is obtained regardless of the asset at hand.

With the above considerations in mind, while deriving the structure of

---

[1]This way of designing the embedding and detection part of a watermarking system is not necessarily optimum, since, in general, the joint optimization of the detection and embedding processes would be preferable, however no such system has been developed yet for detectable watermarking.

the various detectors proposed so far, we will not take explicitly into account whether blind or informed embedding was used, since we will always assume that detection amounts to the extraction of a signal **w** immersed within noise.

In the attempt to be as general as possible, we first reformulate the detection problem as a classical hypothesis testing problem. Then we consider more specific situations in which simple statistical models are used to characterize the host feature set and attack noise. We develop most of our analysis by considering single channel systems (e.g. the watermarking of grey level images), where features assume scalar values. Only at the end of the section, we will give some hints on how the analysis can be extended to multichannel cases.

6.1.1 A hypothesis testing problem

In order to formalize the detection problem, let us assume we want to verify whether an asset A' contains the watermark code b^* or not. The host asset is indicated by A' instead of A or $A_{\mathbf{w}}$ to make it explicit that A' may coincide neither with the original asset nor with the marked asset $A_{\mathbf{w}}$. Dealing with blind embedding systems, we can assume that looking for the presence of b^* within A' amounts to looking for the presence of a certain watermark signal $\mathbf{w}^*$ [2]. The decision must be taken on the basis of a set of observed variables coinciding with the set of features $\mathbf{f}'$ extracted from A'. In the framework of statistical detection theory, the above problem corresponds to deciding in which of a finite number of states the observed system, namely the possibly marked asset, resides. More specifically, let us consider the following alternative hypothesis:

H_0: A' does not contain $\mathbf{w}^*$;

H_1: A' contains $\mathbf{w}^*$.

where H_0 is a composite hypothesis accounting for the following two situations:

Case a_0: A' is not watermarked;

Case b_0: A' contains a watermark other than $\mathbf{w}^*$.

Watermark detection amounts to defining a test of the simple hypothesis H_1 versus the composite alternative H_0 that is optimum with respect to a certain criterion.

---

[2]In some cases, e.g. with certain informed embedded systems, the detector has to look for the presence of one out many signals associated to the same message b^*.

Likelihood ratio

In Bayes theory of hypothesis testing, the criterion is minimization of risk. Bayes risk is defined as the average of a loss function L_{ij}, where L_{01} is the loss sustained when hypothesis H_0 is in force but H_1 is chosen, and L_{10} is the loss sustained when hypothesis H_1 is in force and H_0 is chosen. By remembering that in our case observation variables correspond to the vector $\mathbf{f}'$, the decision criterion can be given the form of a decision rule Φ mapping each $\mathbf{f}'$ into 1 or 0, corresponding to H_1 and H_0:

$$\Phi(\mathbf{f}') = \begin{cases} 1, & \mathbf{f}' \in R_1 \quad (H_1 \text{ is in force}) \\ 0, & \mathbf{f}' \in R_0 \quad (H_0 \text{ is in force}) \end{cases} \tag{6.1}$$

where R_1 and R_0 are acceptance and rejection regions for hypothesis H_1. Minimization of Bayes risk leads to a decision criterion which is based on the, so called, likelihood ratio $\ell(\mathbf{f}')$:

$$\ell(\mathbf{f}') = \frac{p(\mathbf{f}'|H_1)}{p(\mathbf{f}'|H_0)}, \tag{6.2}$$

where $p(\mathbf{f}'|H_i)$ is the pdf of vector $\mathbf{f}'$ conditioned to hypothesis H_i. More specifically, minimum Bayes risk is achieved by letting:

$$R_1 = \{\mathbf{f}' : \ell(\mathbf{f}') > p_0 L_{01}/p_1 L_{10}\}, \tag{6.3}$$

or, equivalently:

$$\Phi(\mathbf{f}') = \begin{cases} 1, & \ell(\mathbf{f}') > p_0 L_{01}/p_1 L_{10} \\ 0, & \text{otherwise} \end{cases} \tag{6.4}$$

where p_0 and p_1 are the a priori probabilities of H_0 and H_1.

The exact specification of Φ requires that the watermark embedding rule is specified and that both the host features and the attack noise are characterized statistically, which will be the goal of next sections.

Threshold selection

By analyzing the decision rule defined by equation (6.4), we can see that the detector operates by comparing the likelihood ratio $\ell(\mathbf{f}')$ against a detection threshold λ, where:

$$\lambda = \frac{p_0 L_{01}}{p_1 L_{10}}. \tag{6.5}$$

A common approach to set λ consists in trying to minimize the overall error probability P_e. By letting P_f be the probability of revealing the presence of $\mathbf{w}^*$ when $\mathbf{w}^*$ is not actually present (false alarm probability), and $P_m = 1 - P_d$ the probability of missing the watermark presence (P_d denotes the probability of correctly revealing the watermark), we have:

$$P_e = p_0 P_f + p_1 (1 - P_d). \tag{6.6}$$

From decision theory it is known that to minimize P_e we must set $\lambda = 1$, which corresponds to the common situation in which $L_{01} = L_{10}$ and $p_0 = p_1$. Note also that in this case we have $P_f = P_m$, that is the minimum error probability is obtained by letting the probability of missing the watermark and that of falsely revealing its presence equal.

A problem with the above, minimum error, detector is that usually the model used to derive the detection rule only accounts for very simple attacks, e.g. addition of white Gaussian noise. When facing different kinds of attacks, however, it leads to a probability of missing the watermark which is considerably higher than the probability of falsely revealing the watermark presence, which is not a desirable behavior in many cases (the reason for such a behavior will be clear after the analysis in the next sections). In addition, in many applications, false detection probability can not fall below a certain level, regardless of the probability of missing the watermark. In these cases, it is preferable to minimize the probability of missing the watermark subject to a constraint on the maximum false detection probability. This is the aim of the Neyman-Pearson detection criterion,False alarm probability!Neyman Pearson criterion according to which the probability of correctly detecting the watermark is maximized subject to a prescribed limit on P_f.

As for the Bayes criterion, detection relies on the comparison of the likelihood ratio against a threshold λ:

$$\Phi(\mathbf{f}') = \begin{cases} 1, & \ell(\mathbf{f}') > \lambda \\ 0, & \text{otherwise} \end{cases} \tag{6.7}$$

what changes here is how the threshold is computed. More specifically, λ is calculated so that the desired false detection probability is achieved, i.e. we must have:

$$P\{\ell(\mathbf{f}') > \lambda | H_0\} = \bar{P}_f; \tag{6.8}$$

where $\bar{P}_f$ is the target false detection probability. By letting $p(\ell | H_0)$ be the pdf of ℓ under hypothesis H_0, we can rewrite the condition in (6.8) as:

$$\int_{\lambda}^{+\infty} p(\ell | H_0) d\ell = \bar{P}_f. \tag{6.9}$$

By solving the above equation for λ, we obtain the threshold to be adopted in the Neyman-Pearson criterion. Once λ has been fixed (thus determining

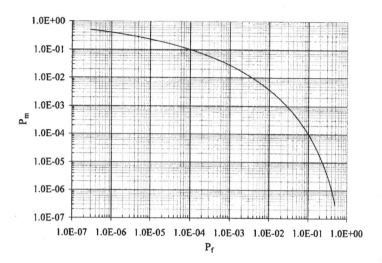

Figure 6.1: Example of a ROC curve defining the characteristic of a watermark detector. Each point of the curve corresponds to a different detection threshold, for which the false alarm and missed detection probabilities given in the figure are obtained.

P_f), the probability of missing the watermark can be calculated as:

$$P_m = \int_{-\infty}^{\lambda} p(\ell|H_1)d\ell. \tag{6.10}$$

The performance of a detector based on the Neyman-Pearson criterion are usually expressed by ROC (Receiver Operating Characteristic) curves, in which P_m is plotted against P_f, as exemplified in figure 6.1[3].

It is often convenient to replace the likelihood ratio with a log-likelihood ratio, defined as:

$$\mathcal{L}(\mathbf{f}') = \ln \ell(\mathbf{f}'), \tag{6.11}$$

leading to a decision rule having the form:

$$\mathcal{L}(\mathbf{f}') \lessgtr \ln \lambda = \Lambda, \tag{6.12}$$

where Λ can be derived by exploiting equation (6.9), or calculated directly by letting:

$$P\{\mathcal{L}(\mathbf{f}') > \Lambda|H_0\} = \bar{P}_f; \tag{6.13}$$

---

[3]Instead of P_m, sometimes P_d is plot against P_f, however the meaning of ROC curves does not change

yielding:

$$\int_{\Lambda}^{+\infty} p(\mathcal{L}|H_0)d\mathcal{L} = \bar{P}_f. \tag{6.14}$$

The actual implementation of a watermark detector based on the Neyman-Pearson criterion, requires that the exact relationship between the watermark signal, the host features and attack noise is defined. This in turn, requires that the watermark embedding rule is specified and that proper statistical models are available to describe the watermark, the host features and the noise introduced by attacks. With reference to noise modelling, it has to be noted that in some cases, the detector structure is derived in the absence of attacks, the only source of uncertainty at the detector being the host features. In such a case, the optimality of the detector is clearly lost, when attacks are taken into account, and the robustness of the watermark in the presence of attacks must be verified experimentally.

In the sequel we apply the above statistical framework to some practical cases, thus deriving the structure of some of the most popular watermark detectors. We start by considering a very simple case following the basic AWGN (Additive White Guassian Noise) assumptions, to finish with more complicated situations dealing with multiplicative watermarks hosted by non-Gaussian features.

6.1.2 AWGN channel

The simplest channel model to deal with is the Additive White Gaussian channel model, in which both host features and attacks are modelled as uncorrelated, Gaussian noise added to the watermark signal. In this case we can write:

$$f_{w,i} = f_i + \gamma w_i + n_i, \tag{6.15}$$

where n_i is a white Gaussian noise accounting for attacks. Actually, the degradation introduced by attacks is much more complicated than pure white noise addition, however by modelling attacks as additive white noise, the problem is considerably simplified, thus allowing the derivation of the detector structure in closed form. Additionally, under certain assumptions, AWGN channel may be considered as a worst-case attack with the corresponding analysis giving an upper bound on the achievable performance.

As to host features, we are making two important assumptions. The first one is that f_i's follow a Gaussian distribution. This is only rarely the case, hence the results we derive below are clearly suboptimal, thus calling for an experimental validation of detector performance. The second assumption is that host features, as well as noise, form an uncorrelated sequence. This is approximately true for techniques operating in the frequency domain, since both DCT and DFT coefficients can be assumed to be uncorrelated. However, this assumption is clearly invalid in other domains, such as space or time domain, since both audio signals and images exhibit a strong correlation between adjacent samples.

Particular attention must paid to the simultaneous presence in equation (6.15) of two sources of uncertainty, i.e., f_i and n_i. This is actually the case in blind embedding systems, where the host signal must be treated as disturbing noise. If the informed embedding approach is used, though, the presence of f_i can be totally or partially compensated for by the embedder. Ideally, the embedder could let $w_i' = w_i - f_i$ so that after addition we have $f_{w,i} = w_i$. Practically, the total rejection of the host signal may lead to unacceptable distortion, thus only a partial rejection is possible. In general, we can say that watermarked features assume the form:

$$f_{w,i} = \gamma w_i + a f_i + n_i, \tag{6.16}$$

where a is a rejection factor taking value in the $[0, 1]$ interval. For the sake of simplicity, we will always assume that $a = 1$, i.e. that no host rejection is performed by the embedder. In other words, the detector structure is derived by assuming that a blind embedder is used. Once the detector structure, and hence, the detection region, has been defined, the embedder may still exploit knowledge of f_i to improve system performance, e.g. by outputting a marked asset which is more inside the detection region. If both f_i and n_i are assumed to be normally distributed, equation (6.16) can be considerably simplified, since a new signal $x_i = a f_i + n_i$ can be introduced which is still white and normally distributed, though with different variance and mean, leading to

$$f_{w,i} = x_i + w_i, \tag{6.17}$$

where only one source of noise is considered.

Let us observe that we also assumed the channel to be stationary, i.e. f_i's and n_i's are modelled as identically distributed random variables. It is important to keep in mind, though, that this is not necessarily true, since different host features may follow different pdf's, or attacks may affect different features in a different way. This is the case, for example, of frequency domain watermarking, where coefficients corresponding to different frequencies usually have different energies and are modified in a different way by attacks such as low pass filtering, or lossy compression.

Detector structure

We start the analysis by considering the meaning of hypotheses H_0 and H_1 in the AWGN case. We clearly have[4]:

$$H_0 : \begin{cases} \textbf{Case } a_0 : & f_i' = x_i \\ \textbf{Case } b_0 : & f_i' = x_i + \gamma v_i \end{cases}$$
$$H_1 : f_i' = x_i + \gamma w_i, \tag{6.18}$$

where cases a_0 and b_0 correspond, respectively, to a non marked host asset and to a host asset containing a watermark $\mathbf{v} \neq \mathbf{w}$. Note that unlike $\mathbf{w}$, the watermark signal $\mathbf{v}$ is not known since it may be any signal in $\mathbf{W}$. Note also that cases a_0 and b_0 can be treated together if $\mathbf{v}$ is allowed to coincide with the null sequence.

Computation of $\ell(\mathbf{f}')$ is complicated by the fact that H_0 is a composite hypothesis, since the possibility that $\mathbf{f}'$ is marked with any watermark other than $\mathbf{w}$ must be taken into account. For this reason, the likelihood ratio takes the following general form:

$$\ell(\mathbf{f}') = \frac{p(\mathbf{f}'|\mathbf{w})}{\int_{\mathbb{R}^n} p(\mathbf{f}'|\mathbf{v}) p(\mathbf{v}) d\mathbf{v}}. \tag{6.19}$$

Indeed, the integral at the denominator should be computed on $\mathbb{R}^n - \mathbf{w}$, however it is known from the theory of measure that the integrals of a function over two integration domains differing by a set of measure zero are the same (as it is our case, given that $\mathbf{w}$ is a single point in $\mathbb{R}^n$) [5].

The numerator of $\ell(\mathbf{f}')$ can be calculated by remembering that $\mathbf{x}$ is a stationary, white, normally distributed sequence. We have:

$$p(\mathbf{f}'|\mathbf{w}) = \prod_{i=1}^{n} \frac{1}{\sqrt{2\pi\sigma_x^2}} \exp\left(-\frac{(f_i' - \mu_x - \gamma w_i)^2}{2\sigma_x^2}\right), \tag{6.20}$$

where μ_x and σ_x^2 indicate the mean and variance of the samples of $\mathbf{x}$, respectively. For the same reason, the denominator of $\ell(\mathbf{f}')$ can be written

---

[4]When H_0 holds, we can say that $f_i' = x_i$ only because we assumed that $a = 1$, otherwise it would have been $f_i' = f_i + n_i \neq x_i = af_i + n_i$.

[5]Actually, given that the number of possible watermarks is finite, it would be more exact to let

$$\ell(\mathbf{f}') = \frac{p(\mathbf{f}'|\mathbf{w})}{\sum_{\mathbf{v} \in \{\mathbf{W}-\mathbf{w}\}} p(\mathbf{f}'|\mathbf{v}) p(\mathbf{v})},$$

however, due to the large number of elements in $\mathbf{W}$ and to the pseudo-random nature of the watermarks, it is reasonable, and convenient, to use the continuous formulation adopted in (6.19).

as:

$$p(\mathbf{f}'|H_0) = \prod_{i=1}^{n} \int_{\mathbb{R}} \frac{1}{\sqrt{2\pi\sigma_x^2}} \exp\left(-\frac{(f_i' - \mu_x - \gamma v_i)^2}{2\sigma_x^2}\right) p(v_i) dv_i. \quad (6.21)$$

The evaluation of (6.21) is very cumbersome, then it is necessary to simplify it. A common simplification which is usually done is to let

$$p(\mathbf{f}'|H_0) = p(\mathbf{f}'|\mathbf{0}), \quad (6.22)$$

where by $\mathbf{0}$ the null watermark is meant. In other words, we assume that H_0 only consists of case a_0, thus neglecting the possibility that the host asset is marked with a watermark different than $\mathbf{w}$. Though intuitively appealing, neglecting the presence of case b_0 in H_0 is not always possible, then assumption (6.22) must be given a more rigorous justification. To do so, let us assume that the standard deviation σ_x is much larger than γv_i. This is surely the case if x corresponds, at least in part, to the host features, since $\sigma_x \gg \gamma v_i$ derives from the imperceptibility constraint (on the contrary, for $a = 0$, when x only accounts for attack noise, such an assumption may not hold)[6]. In this case we can use Taylor's series to expand $p(\mathbf{f}'|\mathbf{v})$ up to the first term, leading to:

$$\frac{1}{\sqrt{2\pi\sigma_x^2}} \exp\left(-\frac{(f_i' - \mu_x - \gamma v_i)^2}{2\sigma_x^2}\right) \approx \frac{1}{\sqrt{2\pi\sigma_x^2}} \exp\left(-\frac{(f_i' - \mu_x)^2}{2\sigma_x^2}\right) + cv_i. \quad (6.23)$$

By inserting the above equation into (6.21), and by noting that for virtually all the watermarking schemes $E[v_i] = \mu_v = 0$, we have:

$$p(\mathbf{f}'|H_0) = \prod_{i=1}^{n} \frac{1}{\sqrt{2\pi\sigma_x^2}} \exp\left(-\frac{(f_i' - \mu_x)^2}{2\sigma_x^2}\right) = p(\mathbf{f}'|\mathbf{0}). \quad (6.24)$$

When we can not assume that $\sigma_x \gg \gamma v_i$, the validity of (6.22) can still be maintained (at least heuristically) if we assume that the watermark follows a Gaussian pdf. In this case, in fact, x_i's are always normally distributed regardless of whether case a_0 or b_0 holds. By also noting that μ_x and σ_x^2 are usually estimated a posteriori on the to-be-examined asset, we can argue that an asset which is marked with a watermark $\mathbf{v} \neq \mathbf{w}$, can be regarded just as any other non-marked asset. In other words there is no mean, and no

---

[6]As a matter of fact, we are integrating the conditional pdf on the whole real axis, hence, if the support of $p(v)$ is not limited, we can not maintain that $\sigma_x \gg \gamma v_i$ over the whole integration domain. However, in practical systems v_i can not take large values due to the imperceptibility constraint, hence allowing us to calculated the integral in 6.21 on a limited domain.

reason, to distinguish between a non-marked asset and an asset containing a watermark signal $\mathbf{v} \neq \mathbf{w}$.

Having simplified the form of $\ell(\mathbf{f}')$, we can go on with the derivation of the detector structure. To this aim, we rewrite $\ell(\mathbf{f}')$ as follows:

$$\ell(\mathbf{f}') = \frac{p(\mathbf{f}'|\mathbf{w})}{p(\mathbf{f}'|\mathbf{0})} = \frac{\prod_{i=1}^{n} \exp\left(-\frac{(f_i' - \mu_x - \gamma w_i)^2}{2\sigma_x^2}\right)}{\prod_{i=1}^{n} \exp\left(-\frac{(f_i' - \mu_x)^2}{2\sigma_x^2}\right)}, \tag{6.25}$$

which, by passing to a logarithmic formulation, yields:

$$\mathcal{L}(\mathbf{f}') = \sum_{i=1}^{n} \frac{1}{2\sigma_x^2} \left[(f_i' - \mu_x)^2 - (f_i' - \mu_x - \gamma w_i)^2 \right]$$

$$= \frac{1}{2\sigma_x^2} \left[\sum_{i=1}^{n} 2\gamma f_i' w_i - \sum_{i=1}^{n} 2\gamma \mu_x w_i - \sum_{i=1}^{n} \gamma^2 w_i^2 \right]. \tag{6.26}$$

By noting that the last two terms in square brackets do not depend on $\mathbf{f}'$, we conclude that the linear correlation between $\mathbf{f}'$ and $\mathbf{w}$, i.e.:

$$\rho = \sum_{i=1}^{n} f_i' w_i = \mathbf{f}' \cdot \mathbf{w}, \tag{6.27}$$

is a sufficient statistic for watermark detection. Stated in another way, in order to decide whether a given watermark is present in A' or not, the detector needs only look at the correlation between the to-be-searched watermark and the host feature vector extracted from A', and compare it against a detection threshold. Instead of using ρ as expressed in (6.27), in the sequel we will find useful to adopt the following modified definition of correlation

$$\rho = \frac{1}{n} \sum_{i=1}^{n} f_i' w_i, \tag{6.28}$$

since while not changing the meaning of ρ, division by n will simplify notation later.

The next step requires that the detection threshold T_ρ is defined (we prefer using the notation T_ρ instead of Λ to make it clear that we are adopting a correlation-based detector). To do so, we apply the Neyman-Pearson criterion, i.e. we set T_ρ by equalling the false detection probability to a target value $\bar{P}_f$. By specializing equation (6.14) to the case at hand, we obtain:

$$\int_{T_\rho}^{\infty} p(\rho|H_0) d\rho = \bar{P}_f, \tag{6.29}$$

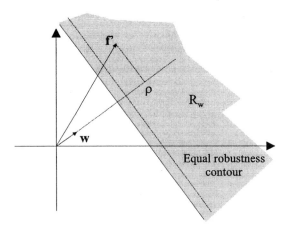

Figure 6.2: 2D representation of the detection region R_W for correlation-based detection. The shape of an equal robustness contour is shown as well.

where by $p(\rho|H_0)$ the pdf of ρ when it is assumed that A' is not marked, is meant.

Equation (6.27), along with (6.7) and (6.29), completely specifies the detection region R_w. By observing that ρ is nothing but a scaled projection of $\mathbf{f}'$ on $\mathbf{w}$, R_w can be given the simple 2D representation shown in figure 6.2. Under the same assumption, equal-robustness contours are easily seen to be straight lines parallel to the detection region border.

Interestingly, it can be seen that when a detector based on correlation is used, informed embedding does not provide any advantage with respect to blind embedding, since the best embedding strategy always consists in adding a scaled version of $\mathbf{w}$ to the host asset[7]. No need say that, while informed embedding may be of no help, informed coding as described in section 4.3.1 may still be conveniently applied.

The exact specification of T_ρ through equation (6.29), requires that the statistic of ρ is derived. This is the goal of the next section, where an analytic expression for T_ρ is given, along with the error probability characterizing the detector.

Error probability

The computation of the probability of falsely detecting the watermark presence:

$$P_f = P\{\rho > T_\rho|H_0\}, \tag{6.30}$$

---

[7]This is only true as long as the regions of acceptable distortion have a circular shape.

and the probability of missing it:

$$P_m = P\{\rho < T_\rho | H_1\}, \tag{6.31}$$

passes through the evaluation of the statistical behavior of

$$\rho = \frac{1}{n} \sum_{1=1}^{n} f_i' w_i. \tag{6.32}$$

Preliminarily, we have to decide whether to average the error probabilities over the host asset samples, the watermark samples, or both. To this aim, let us note that when computing the false detection probability for setting T_ρ at the detector, the watermark signal is known. In this case, then, we only have to average over all possible host assets. In contrast, when evaluating the performance of the whole watermarking system, we have to average over all possible watermarks as well. Some authors also proposed to fix the host asset and averaging over all possible watermarks (see section 6.1.3), in the attempt to derive the overall performance of the system when marking a given host asset. When used to derive the detection threshold, though, this approach is sub-optimum, since the detector is constrained to always use the same threshold without adapting it to the characteristics of the watermark it is looking for. For this reason, we decided to use the latter approach only when the others solutions are not analytically viable.

Before going on with the statistical analysis of ρ, let us recall briefly the assumptions our analysis will rely on. We assume that watermark samples are zero mean i.i.d. random variables[8]. We also assume that attack noise and host features are i.i.d. normal variables, however we will not make any assumption on their means.

We start by noting that according to our model f_i''s are independent Gaussian random variables and w_i's are fixed parameters which are known to the detector, hence we conclude that ρ follows a normal distribution. To completely characterize it, it is sufficient to estimate its mean and variance. Let us assume that H_0 holds. In this case we have $f_i' = x_i$, hence:

$$\mu_{\rho|H_0} = E[\rho|H_0] = \frac{1}{n} E\left[\sum_{i=1}^{n} x_i w_i\right] = \frac{1}{n} \sum_{i=1}^{n} E[x_i] w_i = \mu_x \overline{w}, \tag{6.33}$$

where we let $\overline{w} = \sum w_i / n$ denote the sample average of the sequence **w**. As to the variance of ρ under hypothesis H_0, we can write:

$$\sigma^2_{\rho|H_0} = \text{var}\left(\frac{1}{n} \sum_{i=1}^{n} x_i w_i\right) = \frac{1}{n^2} \sigma_x^2 \sum_{i=1}^{n} w_i^2 = \frac{1}{n} \sigma_x^2 \overline{w^2}, \tag{6.34}$$

---

[8]This is only required to replace $p(f'|H_0)$ with $p(f'|0)$, but does not affect the analysis given in the rest of the chapter.

with

$$\overline{w^2} = \frac{1}{n}\sum_{i=1}^{n}w_i^2 = \frac{\|\mathbf{w}\|^2}{n}, \tag{6.35}$$

denoting the sample mean square value of $\mathbf{w}$. We can now calculate the false detection probability, obtaining[9]:

$$P_f = \int_{T_\rho}^{\infty} p(\rho|H_0)d\rho = \frac{1}{2}\,\text{erfc}\left(\sqrt{\frac{(T_\rho - \mu_{\rho|H_0})^2}{2\sigma_{\rho|H_0}^2}}\right). \tag{6.36}$$

This expression of P_f can be inverted to calculate the detection threshold, yielding:

$$T_\rho = \sqrt{2}\sigma_{\rho|H_0}\text{erfc}^{-1}(2P_f) + \mu_{\rho|H_0}. \tag{6.37}$$

By inserting equations (6.33) and (6.34) in the above expression, we finally have:

$$T_\rho = \sqrt{\frac{2\sigma_x^2\overline{w^2}}{n}}\text{erfc}^{-1}(2P_f) + \mu_x\overline{w}. \tag{6.38}$$

In order to evaluate the probability of missing the watermark presence, we now turn the attention to hypothesis H_1. In this case we have, $f_i' = x_i + \gamma w_i$, thus yielding:

$$\mu_{\rho|H_1} = \frac{1}{n}E\left[\sum_{i=1}^{n}(x_i + \gamma w_i)w_i\right] = \mu_x\overline{w} + \gamma\overline{w^2}, \tag{6.39}$$

$$\sigma_{\rho|H_1}^2 = \text{var}\left(\frac{1}{n}\sum_{i=1}^{n}(x_i + \gamma w_i)w_i\right) = \frac{1}{n}\sigma_x^2\overline{w^2} = \sigma_{\rho|H_0}^2 = \sigma_\rho^2, \tag{6.40}$$

where we have introduced the new symbol σ_ρ^2 to indicate the variance of ρ under both hypotheses H_0 and H_1. Having derived the statistics of ρ for a watermarked asset, we can express the missed detection probability as a function of T_ρ[10]:

$$P_m = \frac{1}{2}\,\text{erfc}\left(\sqrt{\frac{(\mu_{\rho|H_1} - T_\rho)^2}{2\sigma_\rho^2}}\right). \tag{6.41}$$

By using the result in (6.38), it is easy to obtain the following expression for P_m, in which the missed detection probability is expressed as a function

---

[9]Equation (6.36) holds only if $T_\rho > \mu_{\rho|H_0}$, as it is usually the case.

[10]Equation 6.41 holds only if $\mu_{\rho|H_1} > T_\rho$, which is always the case for small missed detection probabilities.

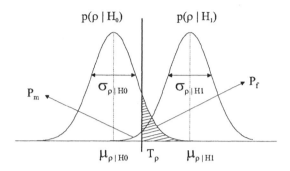

Figure 6.3: Statistics of ρ under hypothesis H_0 and H_1.

of P_f:

$$P_m = \frac{1}{2} \operatorname{erfc}\left(\frac{\gamma\sqrt{\overline{nw^2}} - \sqrt{2}\,\operatorname{erfc}^{-1}(2P_f)\sigma_x}{\sqrt{2}\sigma_x}\right). \tag{6.42}$$

Equation (6.42) completely characterizes the detector performance. Such performance are usually summarized through ROC curves where the missed detection probability is plotted against P_f. The statistics of ρ under hypothesis H_0 and H_1 are summarized in figure 6.3. In the same figure it is also shown how setting T_ρ according to the Neyman-Pearson amounts to letting the grey area under the tail of $p(\rho|H_0)$ be equal to $\bar{P}_f$. In addition, fixing T_ρ results in a missed detection probability which is equal to the area under the left tail of $p(\rho|H_1)$ [11].

So far we have analyzed the system behavior by fixing the watermark signal **w** the detector looks for. As we noted earlier, though, in order to evaluate the overall system performance, we must average the error probabilities derived so far over all possible watermarks. With regard to P_f, nothing changes, since, according to the Neyman-Pearson criterion, the detection threshold is expressly chosen to achieve the same false detection probability regardless of **w**. On the contrary, P_m may depend on the particular realization of **w**. In order to derive a global figure of merit for the system, we must average the expression in (6.42) over **w**, i.e.:

$$\overline{P}_m = E\left[\frac{1}{2} \operatorname{erfc}\left(\frac{\gamma\sqrt{\overline{nw^2}} - \sqrt{2}\,\operatorname{erfc}^{-1}(2P_f)\sigma_x}{\sqrt{2}\sigma_x}\right)\right]. \tag{6.43}$$

Though analytically cumbersome, computation of $\overline{P}_m$ may still be performed numerically, thus providing an overall measure of system perfor-

---

[11] If the minimum error probability criterion was used, the threshold should have been set midway between $\mu_{\rho|H_0}$ and $\mu_{\rho|H_1}$.

mance. Alternatively we can observe that if $\mu_w = 0$, then for large values of n we have:

$$\overline{w^2} \approx E[w^2] = \sigma_w^2, \tag{6.44}$$

regardless of the particular realization of $\mathbf{w}$. Hence we can write:

$$\overline{P}_m \approx P_m \approx \frac{1}{2} \operatorname{erfc}\left(\frac{\gamma \sigma_w \sqrt{n} - \sqrt{2} \operatorname{erfc}^{-1}(2P_f)\sigma_x}{\sqrt{2}\sigma_x}\right). \tag{6.45}$$

By observing equation (6.45), it is readily seen that system performance ultimately depends only on the ratio

$$n \cdot \mathrm{SNR} = n \, \frac{\gamma^2 \sigma_w^2}{\sigma_x^2}, \tag{6.46}$$

where attention has to be paid to meaning of the SNR parameter, since it may not coincide with the true watermark to host asset power ratio, both because the mean value of $\mathbf{f}$ is neglected, and because the host asset power may not be equal to the power of the host feature set (e.g. because we marked only a subset of the host feature set).

Among the parameters contributing to $n \cdot$ SNR, n and γ are design parameters that can be adjusted to obtain a desired performance level. Accordingly, two instruments are available to designers to control system performance: to increase the watermark strength γ or to augment n. In the first case the distance between $\mu_{\rho|H_0}$ and $\mu_{\rho|H_1}$ increases, whereas in the second case the variance σ_ρ^2 diminishes. Interestingly, detector performance does not depend on the pdf of $\mathbf{w}$.

Equation (6.45) can be used to draw the overall ROC curve of the system. For example, in figure 6.4 the ROC curves which are obtained for different values of SNR and for $n = 1000$ and $n = 10000$ are shown. As it can be seen, with $n = 1000$ an SNR at least equal to -10dB is needed to achieve acceptable performance, whereas for $n = 10000$ an SNR of -20dB is already enough to provide very low error probabilities.

Finally, let us observe that, while for deriving the detector structure we exploited the assumption that x_i's are normally distributed, such an assumption is not needed to derive the error probability. In order to demonstrate normality of ρ, in fact, one can still resort to the central limit theorem[12]. When either the attack or the host features do not follow a normal distribution, then, error probabilities can still be computed through equations (6.36) and (6.45), though the correlation detector is no longer optimum.

---

[12] A general discussion on the applicability of the central limit theorem for the computation of P_f is given at the end of section 6.1.6

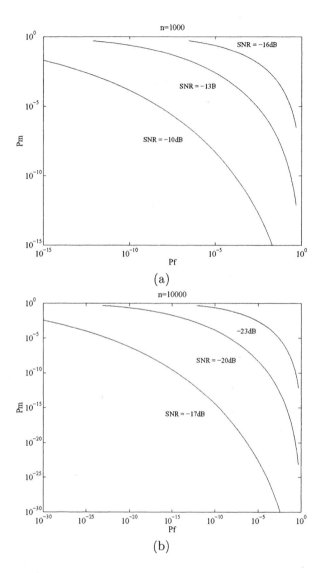

Figure 6.4: ROC curves for additive watermarking in additive white Gaussian noise, for different values of SNR.

Error probability in the presence of a different watermark

So far we calculated the false detection probability by assuming that the host asset is not marked, however, we must remember that H_0 is a composite hypothesis accounting also for the case in which the asset contains a watermark $\mathbf{v} \neq \mathbf{w}$. It was only because of the properties of the watermark, noticeably we exploited the fact that $\mu_v = 0$, that we could replace $p(\mathbf{f'}|H_0)$ with $p(\mathbf{f'}|0)$. However, when computing the error probabilities we must go back to the true meaning of H_0 and calculate P_f by assuming that a watermark other than $\mathbf{w}$ is present in A. Under this assumption we have:

$$f_i' = f_i + n_i + \gamma v_i, \tag{6.47}$$

$$\rho = \frac{1}{n} \sum_{i=1}^{n} (f_i + n_i + \gamma v_i) w_i, \tag{6.48}$$

easily leading to[13]:

$$p(\rho|H_0) = \mathcal{N}\left(0, \frac{\sigma_{x'}^2 \overline{w^2}}{n}\right), \tag{6.49}$$

with $\sigma_{x'}^2 = \sigma_f^2 + \sigma_n^2 + \gamma^2 \sigma_v^2$. The only difference with respect to the P_f computed in the previous section is that now the term $\gamma^2 \sigma_v^2$ contributes to increase the variance of ρ. If, as it is usually the case, $\gamma \ll 1$, such a term can be neglected and subcase b_0 of H_0 treated as subcase a_0.

Practical implementation of the detector

From the analysis in the previous section we concluded that:

$$T_\rho = \sqrt{\frac{2\sigma_x^2 \overline{w^2}}{n}} \, \text{erfc}^{-1}(2P_f) + \mu_x \overline{w}, \tag{6.50}$$

from which it immediately comes out that the detector does not need to know γ, i.e. the strength used to embed the watermark. This is a very important consequence of basing the detection on the Neyman-Pearson criterion. The probability of false alarm, in fact, is solely determined by the $p(\rho|H_0)$ curve which, of course, does not depend on γ. The most important consequence of adopting a detector which does not depend on γ, is that the embedder can adjust the watermark strength to the asset at hand, e.g. to trade-off between imperceptibility and robustness, without informing the detector of its choice.

Another important consequence of equation (6.50) is that for the implementation of the detector it is required that μ_x and σ_x^2 are known. This poses some practical problems, since due to the wide variability of host assets, the mean and variance of host features are very difficult to estimate.

---

[13]Note that if the pdf of v_i's is not Gaussian, the central limit theorem must be invoked to justify normality of ρ.

To get around the problem, μ_x and σ_x^2 are usually estimated on the to-be-inspected asset. To analyze the impact of such a strategy on the detector structure let us assume that $\mu_x = 0$ and that σ_x^2 is estimated as[14]:

$$\sigma_x^2 \approx \frac{1}{n} \sum_{i=1}^{n} x_i^2 = \frac{1}{n} \sum_{i=1}^{n} (f_i')^2 = \frac{1}{n} \|\mathbf{f}'\|^2, \qquad (6.51)$$

where the second equality follows from the fact that, while fixing the false detection probability, we are assuming that A' is not marked, i.e. $f_i' = x_i$. Note that by estimating σ_x^2 on A', the possible presence of attack noise is automatically taken into account, since $x_i = f_i + n_i$. When x_i's can not be assumed to be identically distributed, e.g. because they have a different variance, the host feature sequence is split into subsequences, within which the stationarity assumption holds, then the above analysis is applied separately to each subsequence.

Though equation (6.51) has been introduced as an approximation of σ_x^2, its adoption at the detector considerably modifies the detector structure itself. By remembering that we assumed $\mu_x = 0$, and by exploiting equations (6.51) and (6.50), the decision rule can be rewritten as:

$$\Phi(\mathbf{f}') = \begin{cases} 1, & \frac{\mathbf{f}' \cdot \mathbf{w}}{\|\mathbf{w}\| \|\mathbf{f}'\|} > T_{\rho_n} \\ 0, & otherwise \end{cases} \qquad (6.52)$$

with

$$T_{\rho_n} = \sqrt{\frac{2}{n}} \ \mathrm{erfc}^{-1}(2P_f). \qquad (6.53)$$

The quantity

$$\rho_n = \frac{\mathbf{f}' \cdot \mathbf{w}}{\|\mathbf{w}\| \|\mathbf{f}'\|}, \qquad (6.54)$$

representing the observation variable the modified detector relies on, is called normalized correlation and depends on the observed features $\mathbf{f}'$ in a more complicated manner than ρ, due to the presence of $\|\mathbf{f}'\|$ at the denominator. As it will be detailed in section 6.1.4, ρ_n also comes out when considering the possible dependence of attacks on the host signal. The performance of the approximated detector introduced in this section, then, is closely related to that of the detector described in section 6.1.4, and hence, we will postpone its analysis to such a section.

When considering the practical applicability of the correlation-based detector described so far, we must note that many of the assumptions we made are not matched in practical scenarios. Noticeably, host features

---

[14]Similar considerations hold if $\mu_x \neq 0$.

and attacks may not be normally distributed, host features and noise samples may not be uncorrelated, noise may depend on the host signal. In all of these cases, correlation-based detection is far from being the optimum choice. Nevertheless, correlation detection is a very popular solution, both for its simplicity and because in many cases an accurate statistical model of the host features and the attacks is not available, thus precluding the possibility of deriving the optimum detector for the case at hand. Another reason to opt for correlation-based detection, is that in this way the exhaustive search of the watermark at several different positions in the feature space can be implemented efficiently by resorting to FFT-based correlation computation. The most common situations correlation-based detection is used in, include additive watermarking in the asset domain and additive watermarking in the wavelet domain.

6.1.3 Additive / Generalized Gaussian channel

Image watermarking in the DCT domain, be it full-frame or block DCT watermarking, is a situation in which the assumption that the host features follow a Gaussian distribution is invalid. It is known, in fact, that the pdf of DCT coefficients deviates significantly from normality. In this case, the correlation-detector derived in the previous section is clearly suboptimum. A pdf which is commonly adopted to characterize DCT coefficients is the zero-mean Generalized Gaussian (GG) density, given by:

$$p(x) = A \exp(-|\beta x|^c), \tag{6.55}$$

where the parameters A and β can be expressed as a function of the shape parameter c and the standard deviation of x:

$$\beta = \frac{1}{\sigma_x} \sqrt{\frac{\Gamma(3/c)}{\Gamma(1/c)}}, \tag{6.56}$$

$$A = \frac{\beta c}{2\Gamma(1/c)}, \tag{6.57}$$

with $\Gamma(x)$ denoting the standard gamma function

$$\Gamma(x) = \int_0^\infty t^{x-1} e^{-t} dt. \tag{6.58}$$

The parameter c controls the shape of $p(x)$. For example, it can be seen that the Gaussian and the Laplacian distributions are special cases of the generalized Gaussian pdf, given by $c = 2$ and $c = 1$, respectively. In figure 6.5, the shape of the generalized Gaussian for different values of c is shown.

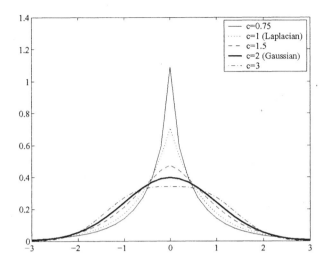

Figure 6.5: Shape of the Generalized Gaussian pdf for different values of c.

Detector structure

When the host features f_i's follow a GG pdf, the optimum detector structure in the presence of noise can not be derived analytically. This is due to the simultaneous presence in equations (6.15) and (6.16) of two sources of noise, namely f_i' and n_i, having different pdf, and to the consequent difficulty in deriving the pdf of $f_i' + n_i$. In order to get around the problem, the optimum detector structure is derived in the absence of n_i. The actual performance of the detector when attacks are present, then, has to be evaluated experimentally.

Let us start the analysis by rewriting equation (6.25) in the GG case (the usage of (6.25) instead of (6.19) can be justified as for the AWGN case). We have

$$\ell(\mathbf{f}') = \frac{p(\mathbf{f}'|\mathbf{w})}{p(\mathbf{f}'|\mathbf{0})} = \frac{\prod_{i=1}^{n} \exp\left(-|\beta(f_i' - \gamma w_i)|^c\right)}{\prod_{i=1}^{n} \exp\left(-|\beta f_i'|^c\right)}, \tag{6.59}$$

where, as usual, we have assumed that the host features are i.i.d. random variables. By passing to a logarithmic formulation, we have:

$$\mathcal{L}(\mathbf{f}') = \sum_{i=1}^{n} \beta^c (|f_i'|^c - |f_i' - \gamma w_i|^c), \tag{6.60}$$

which represents the sufficient statistic for detection in the GG case. Complete specification of the detection rule, requires that the detection threshold Λ is fixed. To do so, the Neyman-Pearson criterion is applied, thus permitting Λ to be calculated as in (6.14).

Error probability

To go on with the computation of Λ and the evaluation of the error probability, the statistic of $\mathcal{L}$ under hypothesis H_0 and H_1 must be derived. Before proceeding, we observe that in the GG case averaging over f_i does not lead to a closed-form expression of the error probability, thus it is customary to fix **f** and average over **w** by further assuming that w_i's follow a discrete distribution with equiprobable values $\{-1, +1\}$. This is a suboptimum approach, since the detector is not allowed to adapt itself to the particular mark it is looking for. In spite of this, it has been shown that, when DCT coefficients are chosen as the host features, a significant performance improvement can be obtained by adopting the detector derived under the GG assumption instead of the classical correlation-based detector stemming from AWGN analysis.

With the above assumptions in mind, let us start by assuming that H_0 is in force. By noting that $\mathcal{L}(\mathbf{f}')$ is a sum of statistically independent terms, we can approximate $p(\mathcal{L})$ by a Gaussian distribution[15]. The mean of $\mathcal{L}(\mathbf{f}')$ under H_0 is easily derived, by noting that in this case $f_i' = f_i$ since we assumed attacks are not present, we have:

$$\mathcal{L}(\mathbf{f}') = \sum_{i=1}^{n} \beta^c (|f_i|^c - |f_i - \gamma w_i|^c), \tag{6.61}$$

yielding

$$E[\mathcal{L}|H_0] = \mu_{\mathcal{L}|H_0} = \sum_{i=1}^{n} \beta^c |f_i|^c - \frac{1}{2} \sum_{i=1}^{n} \beta^c (|f_i - \gamma|^c + |f_i + \gamma|^c). \tag{6.62}$$

To calculate the variance of $\mathcal{L}(\mathbf{f}')$ under H_0, we note that such a variance is the sum of the variances of $\beta^c (|f_i|^c - |f_i - \gamma w_i|^c)$. By remembering that now f_i is a fixed value, we obtain:

$$\sigma_{\mathcal{L}|H_0}^2 = \frac{1}{4} \sum_{i=1}^{n} \beta^{2c} (|f_i + \gamma|^c - |f_i - \gamma|^c)^2. \tag{6.63}$$

---

[15]Though the central limit theorem truly applies, the speed of approach to normality may be rather slow. An accurate design procedure, then, should be based on a more precise characterization of the statistics behind $\mathcal{L}(\mathbf{f}')$. In this section, we will not proceed along this path to come back on this topic later on in the text, when dealing with multiplicative watermarking schemes.

When H_1 is in force, we have $f_i' = f_i + \gamma w_i$, hence:

$$\mathcal{L}(\mathbf{f}') = \sum_{i=1}^{n} \beta^c(|f_i + \gamma w_i|^c - |f_i|^c), \qquad (6.64)$$

easily yielding:

$$\mu_{\mathcal{L}|H_1} = -\mu_{\mathcal{L}|H_0} \triangleq \mu, \qquad (6.65)$$

$$\sigma^2_{\mathcal{L}|H_1} = \sigma^2_{\mathcal{L}|H_0} \triangleq \sigma^2. \qquad (6.66)$$

By remembering that $p(\mathcal{L})$ is approximately Gaussian, we have:

$$P_f = \frac{1}{2} \operatorname{erfc}\left(\frac{\Lambda + \mu}{\sqrt{2}\sigma}\right), \qquad (6.67)$$

which can be used to fix Λ, and

$$P_m = \frac{1}{2} \operatorname{erfc}\left(\frac{\mu - \Lambda}{\sqrt{2}\sigma}\right). \qquad (6.68)$$

Equations (6.67) and (6.68) can be given a more pleasant form, in which P_m is expressed as a function of P_f and the ratio between μ and σ (a sort of signal to noise ratio). More specifically, it is easy to show that:

$$P_m = \frac{1}{2} \operatorname{erfc}\left(\frac{\sqrt{2}\mu}{\sigma} - \operatorname{erfc}^{-1}(2P_f)\right). \qquad (6.69)$$

The performance of the detector, then, are completely specified by the ratio μ/σ.

As a last note we remind that when a watermark $\mathbf{v} \neq \mathbf{w}$ is present, an analysis similar to that carried out for the AWGN case should be carried out to get a more precise expression for P_f. Nevertheless, if we assume that the watermark strength is sufficiently low, the analysis developed here is accurate enough.

Practical implementation of the detector

Optimum detection based on the GG assumption has been successfully applied to the detection of an additive image watermark embedded within the block-DCT coefficients of the host image. Block-DCT coefficients, in fact, fit rather accurately the GG model. For example, it is known that in most cases, the low frequencies DCT coefficients interested by the watermark are well approximated by the generalized Gaussian with $c = 1/2$.

A first problem with the application of the optimum GG detector to watermarking schemes operating in the block-DCT domain, is that DCT coefficients at different frequencies usually do not have the same energy. Specifically, whereas we can assume that all the coefficients follow a GG pdf with the same shape parameter c, the variance of coefficients (and hence β) varies with frequency. Accordingly, equations defining the GG detector must be rewritten by letting β vary with i, i.e. the position of the coefficient within the 8×8 block. As to the exact values of β_i's, they can be obtained by estimating σ_i's on the image at hand, trusting that watermark presence does not alter them significantly.

The problem of obtaining a good value of c which fits well all the DCT coefficients regardless of frequency, is a more complicated one. A possible choice consists in letting $c = 0.8$, since such a value has been reported to be the best choice for the DCT frequencies usually hosting the watermark. Alternatively, the sample mean absolute value and variance of DCT coefficients can be calculated and matched to those of the GG distribution. Maximum likelihood estimation is another reportedly good solution. Finally, the value of c resulting in the highest μ/σ ratio may be used, thus attempting to maximize the detector response.

Once c and β_i's have been estimated, the performance of the detector for a given host image can be obtained by calculating the ratio μ/σ and subsequently applying equation (6.69).

A further improvement of the detector can be obtained by observing that, sometimes, block DCT coefficients exhibit very high values which are not accurately modelled by the Generalized Gaussian pdf. In order to avoid this mis-matching between true host features and the theoretical model, we can let the detector work only on those features whose magnitude is below a certain threshold. This technique is usually referred to as detection with point elimination. The error analysis of a detector adopting point elimination is slightly more complicated, however the resulting detector may show a considerable performance improvement.

6.1.4 Signal dependent noise with host rejection at the embedder

In the previous sections we assumed that attack noise does not depend on the host signal. This is not a realistic assumption for many common attacks and signal processing tools. Examples of signal-dependent attacks include, but are not limited to, low-pass filtering, signal amplification, lossy coding, quantization, sharpening, digital-to-analog and analog-to-digital conversion. In such cases, the detectors developed in the previous sections are no longer optimum, since the assumptions they rely on do not accurately model reality.

In the attempt to take noise-signal correlation into account we now introduce a new noise model. To be specific, let $\mathbf{f}_{nf}$ be the noise-free, possibly marked feature sequence. Note that we have introduced the new symbol $\mathbf{f}_{nf}$ to explicitly signify that we are looking at the possibly marked host features, prior to noise addition. If a watermark $\mathbf{w}$ is actually present in $\mathbf{f}_{nf}$, we have $\mathbf{f}_{nf} = \mathbf{f} + \mathbf{w}$, otherwise $\mathbf{f}_{nf} = \mathbf{f}$. In order to take the correlation between $\mathbf{n}$ and $\mathbf{f}_{nf}$ into account, we assume that $\mathbf{n}$ is a zero mean Gaussian random vector with covariance:

$$\Sigma_{\mathbf{n}} = \sigma_n^2[r\mathbf{f}_{nf}^t\mathbf{f}_{nf} + (1-r)\mathbf{I}], \qquad (6.70)$$

where the superscript t indicates transposition of the row vector $\mathbf{f}_{nf}$, σ_n^2 is the overall noise strength, $\mathbf{I}$ is the identity matrix, and r accounts for the correlation between $\mathbf{f}_{nf}$ and $\mathbf{n}$.

Detector structure

Derivation of the optimum detector structure under the noise model expressed by equation (6.70) is not viable, then some simplifying assumptions must be made. First, we assume that a perfect informed embedder is available, thus making it possible the total rejection of the host features ($a = 0$ in equation (6.16)). Under such hypothesis we have $\mathbf{f}_{nf} = \mathbf{w}$, and $\mathbf{f}' = \mathbf{n} + \mathbf{w}$ yielding:

$$p(\mathbf{f}'|H_1) = \frac{1}{(2\pi)^{n/2}\,|\Sigma_{\mathbf{w}}|^{1/2}}\exp(-(\mathbf{f}'-\mathbf{w})\Sigma_{\mathbf{w}}^{-1}(\mathbf{f}'-\mathbf{w})^t/2), \qquad (6.71)$$

where $\Sigma_{\mathbf{w}}$ is obtained by letting $\mathbf{f}_{nf} = \mathbf{w}$ in (6.70). Note that the above equation completely characterizes the statistics of $\mathbf{f}'$ under hypothesis H_1, since $\mathbf{w}$ is known.

When hypothesis H_0 is in force, equation (6.71) should be rewritten by replacing $\mathbf{w}$ with $\mathbf{f}$, the non-marked feature sequence. However, the resulting pdf should be averaged over $\mathbf{f}$, since $\mathbf{f}$ is not known at the detector. In order to simplify the analysis, we assume that in this case $\mathbf{f}' = \mathbf{f} + \mathbf{n}$ follows a Gaussian distribution with covariance matrix given by $(\sigma_f^2 + \sigma_n^2)\mathbf{I}$. With this simplification, we can write:

$$p(\mathbf{f}'|H_0) = \frac{1}{(2\pi)^{n/2}(\sigma_f^2 + \sigma_n^2)^{n/2}}\exp\left(-\frac{\sum_{i=1}^n (f_i')^2}{2(\sigma_f^2 + \sigma_n^2)}\right). \qquad (6.72)$$

Having defined the channel models for H_0 and H_1, statistical decision theory can be applied, leading to a detector based on the following decision rule, if

$$\frac{(\mathbf{f}' \cdot \mathbf{w})^2}{\|\mathbf{f}'\|^2\|\mathbf{w}\|^2} > \frac{k}{\|\mathbf{f}'\|^2} + \frac{\sigma_f^2 + r\sigma_n^2}{\sigma_f^2 + \sigma_n^2}, \qquad (6.73)$$

then decide for watermark presence, otherwise for watermark absence. Parameter k is chosen so that the desired false detection probability is achieved. Note also that equation (6.73) holds for large values of n and by assuming that $\|\mathbf{w}\|^2$ grows linearly with n[16]. It has been demonstrated that detection regions based on (6.73) have a hyperbolic form.

Implementation of a detector relying on equation (6.73) is a difficult piece of work, because of the complicated form of the sufficient statistic behind it. A considerable simplification can be obtained by disregarding the term $k/\|\mathbf{f}'\|^2$, leading to a detector deciding for watermark presence if and only if:

$$\rho_n = \frac{\mathbf{f}' \cdot \mathbf{w}}{\|\mathbf{f}'\|\|\mathbf{w}\|} > T_{\rho_n}, \tag{6.74}$$

where T_{ρ_n} is set according to the Neyman-Pearson criterion, and where we can recognize ρ_n to be the normalized correlation between $\mathbf{f}'$ and $\mathbf{w}$.

Before proceeding with the analysis of the form of detection regions, and with the computation of error probabilities, it is worth remembering that ρ_n-based detection was already introduced in section 6.1.2. The adoption of normalized correlation as the observed variable, then, can be motivated in two different ways: as an approximation of the optimum detector derived under AWGN assumptions, or as an approximation of the optimum detector when correlation between host signal and noise is modelled through (6.70).

By observing that ρ_n is nothing but the cosine of the angle between $\mathbf{f}'$ and $\mathbf{w}$, the form of the detection region for a given watermark $\mathbf{w}$ is readily seen to be a hypercone with the axis pointing in the direction of $\mathbf{w}$.

Error probability

We now have to calculate the detection threshold so that a target false detection probability is obtained. To this aim it is necessary that the pdf of ρ_n under hypothesis H_0 is derived. In order to simplify the analysis, we will assume that f_i's have a zero mean. A first possibility consists in approximating $p(\rho_n|H_0)$ with a normal distribution, having zero mean and standard deviation equal to $n^{-1/2}$, i.e.

$$p(\rho_n|H_0) = \mathcal{N}(0, 1/\sqrt{n}). \tag{6.75}$$

In this case, the false detection probability can be easily calculated since we have:

$$P_f = \frac{1}{2}\text{erfc}\left(\sqrt{\frac{nT_{\rho_n}^2}{2}}\right). \tag{6.76}$$

---

[16]More details on the derivation of equation 6.73 may be found in [57].

When T_{ρ_n} is much smaller than 1, the above equation provides a sufficiently accurate approximation of the true error probability. As T_{ρ_n} increases, though, the false detection probability is significantly overestimated. The reason for such a lack of accuracy can be readily understood by observing that whereas the normalized correlation is always less than one, the tails of the Gaussian pdf never reach zero, thus resulting in a non-null false detection probability even for $T_{\rho_n} > 1$. This can also be explained by noting that normalized correlation can not be written as a sum of independent random variables, hence the central limit theorem does not apply.

A better approximation can be obtained by exploiting a well known result of statistic theory regarding the correlation coefficient. Let r_n be defined as follows:

$$r_n = \frac{(\mathbf{f}' - \overline{\mathbf{f}'}) \cdot (\mathbf{w} - \overline{\mathbf{w}})}{\|\mathbf{f}' - \overline{\mathbf{f}'}\| \, \|\mathbf{w} - \overline{\mathbf{w}}\|}, \qquad (6.77)$$

with $\overline{\mathbf{f}'}$ and $\overline{\mathbf{w}}$ denoting the sample mean of $\mathbf{f}'$ and $\mathbf{w}$ respectively, and let Fisher Z-statistic be defined as:

$$Z = \frac{1}{2} \ln \frac{1 + r_n}{1 - r_n}. \qquad (6.78)$$

If the elements of $\mathbf{f}'$ are i.i.d. samples drawn from a unitary Normal distribution, then the values of Z roughly follows a Gaussian pdf with zero mean and $\sigma = 1/\sqrt{n-3}$. In our case, the correlation coefficient has to be replaced by normalized correlation. Nevertheless, as it is argued in [152], when r_n is replaced by ρ_n, Z still follows a normal pdf with a standard deviation which is equal to $1/\sqrt{n-2}$.

The detection threshold T_{ρ_n} can now be calculated by reasoning on Z instead of ρ_n. More specifically, by observing that Z is an increasing function of r_n, we can let:

$$P_f = P(Z > T_Z | H_0) = \frac{1}{2} \mathrm{erfc} \left(\sqrt{\frac{T_z^2(n-2)}{2}} \right), \qquad (6.79)$$

and calculate the detection threshold on ρ_n as:

$$T_Z = \frac{1}{2} \ln \frac{1 + T_{\rho_n}}{1 - T_{\rho_n}} \quad \Rightarrow \quad T_{\rho_n} = \tanh T_Z. \qquad (6.80)$$

The approximation calculated through Z-statistic is much more accurate than that obtained through the Gaussian approximation, nevertheless, when T_{ρ_n} approaches one, it starts underestimating P_f.

In order to calculate the exact error probability let us recall that ρ_n is nothing but the cosine of the angle between $\mathbf{f}'$ and $\mathbf{w}$. By remembering that

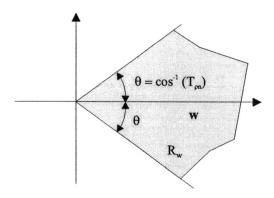

Figure 6.6: Detection region for correlation coefficient watermark recovery.

f_i'''s are independent normally distributed random variables, we note that all the directions of $\mathbf{f}'$ have the same probability. Thresholding ρ_n, hence, amounts to verifying whether $\mathbf{f}'$ lies in a hypercone subtending an angle θ equal to $\cos^{-1} T_{\rho_n}$, pointing in the direction of $\mathbf{w}$. Figure 6.6 illustrates such an hypercone for $n = 2$. The false detection probability, now, can be calculated as the ratio between the angle subtended by the hypercone and half n-round angle. By resorting to n-dimensional geometry, it can be shown that such a ratio amounts to:

$$P_f = \frac{I_{n-2}(\theta)}{2I_{n-2}(\pi/2)}, \tag{6.81}$$

$$I_d(\theta) = \int_0^\theta \sin^d(u)\,du, \tag{6.82}$$

with $\theta = \cos^{-1} T_{\rho_n}$ and where $I_d(\theta)$ can be calculated through the following recursive formula:

$$\begin{aligned} I_0(\theta) &= \theta, \\ I_1(\theta) &= 1 - \cos(\theta), \\ &\quad\ldots \\ I_d(\theta) &= \frac{d-1}{d} I_{d-2}(\theta) - \frac{\cos(\theta)\sin^{d-1}(\theta)}{d}. \end{aligned} \tag{6.83}$$

Though more cumbersome, the above equations provide an exact expression for P_f and hence should be preferred to the approximated solutions given previously. Such solutions, though, are still valid design tools, when a first rough idea of the error probability is needed.

An important consequence of basing detection on normalized correlation is that in so doing the need to know the variance of $\mathbf{f}'$ is avoided. This is a very important result, since such a variance is usually very difficult to estimate thus raising serious implementation problems [17].

At this point we should derive the pdf of ρ_n under hypothesis H_1 and calculate the probability of missing the watermark. Such a computation, though, is rather complicated due to the difficulty of deriving the pdf of ρ_n which now assumes the form:

$$\rho_{n|H_1} = \frac{(\mathbf{f} + \mathbf{w}) \cdot (\mathbf{w})}{\|\mathbf{f} + \mathbf{w}\| \|\mathbf{w}\|}. \tag{6.84}$$

For this reason we will not go further with the analysis of the detector based on normalized correlation, we only observe that if n is sufficiently large, and the watermark sufficiently weak, we may see the detector derived in this section as an approximation of that based on linear correlation. Hence, we can use the analysis carried out in such a case to roughly estimate the missed detection probability of the detector.

As a last comment, we observe that, in this section, normalized correlation detection was initially derived by assuming that informed embedding principles are used to completely reject the host feature samples. In this case, the computation of missed detection probability goes through a completely different path, being it dependent, in addition to attack noise, on the extent of the intersection between the admissible distortion region and the hyperconic detection region described before.

Informed embedding and normalized correlation detection

In section 6.1.2, we noted that when detection is based on linear correlation between the host features and the watermark, no advantage has to be expected from the exploitation of informed embedding principles. On the contrary, if a detector based on the normalized correlation is used, a significant performance improvement can be obtained by exploiting host asset knowledge at the embedder. This can be seen by reasoning as in section 4.3, where the case of a hyperconic detection region was taken as a typical example in which informed embedding has the potential to bring advantage with respect to blind methods (figures 4.17 through 4.19).

More specifically, in figure 4.19, the advantages brought by three possible informed embedding strategies are shown as opposed to blind embedding. Whereas the strategy aiming at maximizing the detector response (MNC), i.e. normalized correlation, does not need any further explanation,

---

[17]Interestingly, we showed in section 6.1.2 that if we try to estimate σ_f^2 in the context of linear correlation detection, we still come down to normalized correlation detection.

the MR (maximum robustness) and MD (minimum distortion) approaches need that a measure of watermark robustness is specified. For example, in [57] and [153] the following robustness function is used:

$$R(\mathbf{f}') = \frac{(\mathbf{f}' \cdot \mathbf{w})^2}{\|\mathbf{w}\|^2 T_{\rho_n}^2} - \|\mathbf{f}'\|^2. \qquad (6.85)$$

6.1.5 Taking perceptual masking into account

Another deviation of real systems from the theoretical models described so far, is that these models do not take perceptual masking into account. In other words, the presence of the term m_i in equation (4.37) is not taken into account. In order to get around the problem, two solutions are possible: i) treat m_i as a random unknown and redesign the detector so to account for its presence, or ii) try to estimate m_i and then look for the composite watermark:

$$\mathbf{w}' = \{w_1', w_2' \ldots w_n'\} = \{m_1 w_1, m_2 w_2 \ldots m_n w_n\}. \qquad (6.86)$$

The latter solution is by far the most common, since it only needs that the same procedure used by the embedder to calculate the sequence $\mathbf{m}$ is applied by the detector. Of course, such an approach is effective only if the watermark presence does not affect significantly the computation of $\mathbf{m}$, which is usually the case due to the imperceptibility constraint, and if the same can be said of attacks. This second requirement is a more critical one, since distortion due to attacks may be heavier than that introduced by the watermark, in addition attacks could be designed just to make the estimation of $\mathbf{m}$ difficult, so to fool the detector.

As a last remark, it has to be noted that the presence of m_i may change the statistics of the watermark signal, thus invalidating some of the assumptions the optimum detectors rely on. In spite of this, even if no theoretical analysis is available, experimental results tend to indicate that some gain can be obtained by estimating m_i and then designing the detector by relying on $\mathbf{w}'$ instead of $\mathbf{w}$.

6.1.6 Multiplicative Gaussian channel

So far we have only considered the case of additive watermarking. In addition to being the most popular embedding strategy, additive watermarking also has the advantage that theoretical analysis is relatively simple, especially when both host features and attacks are modelled as white Gaussian processes. In some cases, though, the choice of a multiplicative watermark may be preferable. This is the case, for example, of algorithms operating in the frequency domain, where the adoption of a multiplicative embedding rule permits to exploit the masking characteristics of the Human Visual and Auditory Systems (see end of section 5.4.2).

In this section, we derive the structure of the optimum detector for a multiplicative watermark hosted by uncorrelated, normally distributed features, in the presence of additive Gaussian noise. We also make the simplifying assumption that both noise samples and host features have a zero mean[18]. While such a model may be rarely encountered in practice, its analysis provides some useful insight into the performance of multiplicative systems. It will also be useful to compare theoretically the performance of multiplicative and additive watermarking methods. Such a theoretical comparison, in fact, is only viable in this simple case, and results are extremely instructive.

A more realistic, more complicated, case, will be considered in the next section (6.1.7), however the analysis will not be as complete as the one we are going to carry out in the Gaussian case.

Under the assumption of multiplicative watermarking and additive attack noise, observed features assume the form[19]:

$$f_i' = f_i + \gamma w_i f_i + n_i = f_i(1 + \gamma w_i) + n_i. \tag{6.87}$$

Note that in this case we do not take into account the possibility that due to informed embedding the host signal is partially or totally rejected, since host feature rejection can not be modelled as simply as in the additive case. On the contrary, we will use the model in (6.87) to derive the optimum detector structure. No need saying that an informed embedder may always exploit the knowledge of detector structure and original host features, to improve the effectiveness of the system, e.g. by letting the watermarked asset lie as inside as possible within detection region.

Detector structure

Once again, the analysis starts by writing the likelihood function. As in section 6.1.2, we will use the simplified form of $\ell(\mathbf{f}')$, where the composite hypothesis H_0 is replaced by the hypothesis that $\mathbf{w} = \mathbf{0}$. By assuming that the watermark strength is much lower than host signal energy, this approximation can be justified as follows:

$$\int_{\mathbb{R}} p(f'|v)p(v)dv \approx \int_{\mathbb{R}} (p(f'|0) + kv)p(v)dv = p(f'|0), \tag{6.88}$$

---

[18]If this is not the case, the detector will have to subtract the sample mean of $\mathbf{f}'$ from the host features prior to detection.

[19]We neglect the possible presence of a spatial masking function $\mathbf{m}$.

where we have exploited the assumption that $\mu_v = 0$. With this approximation in mind, we can write

$$\ell(\mathbf{f}') = \frac{p(\mathbf{f}'|\mathbf{w})}{p(\mathbf{f}'|0)} = \prod_{i=1}^{n} \frac{p(f_i'|w_i)}{p(f_i'|0)}. \tag{6.89}$$

When H_0 holds, $f_i' = f_i + n_i$ and each f_i' follows a normal pdf with zero mean and variance $\sigma_n^2 + \sigma_f^2$, i.e.

$$p(f_i'|H_0) = \mathcal{N}(0, \sigma_n^2 + \sigma_f^2) = \mathcal{N}(0, \sigma_0^2), \tag{6.90}$$

where we let $\sigma_n^2 + \sigma_f^2 = \sigma_0^2$ to simplify notation. Conversely, if H_1 is in force we have $f_i' = f_i(1 + \gamma w_i) + n_i$, hence:

$$p(f_i'|H_1) = \mathcal{N}(0, \sigma_n^2 + \sigma_f^2(1 + \gamma w_i)^2) = \mathcal{N}(0, \sigma_{1,i}^2), \tag{6.91}$$

where the dependence of $\sigma_{1,i}^2$ upon the watermark sample w_i is explicitly indicated through the subscript i. Substituting $p(f_i'|H_0)$ and $p(f_i'|H_1)$ in equation (6.89) yields:

$$\ell(\mathbf{f}') = \prod_{i=1}^{n} \frac{\sigma_0}{\sigma_{1,i}} \exp\left(-\frac{(f_i')^2}{2} \cdot \left(\frac{1}{\sigma_{1,i}^2} - \frac{1}{\sigma_0^2}\right)\right), \tag{6.92}$$

which, by passing to a logarithmic formulation, gives

$$\mathcal{L}(\mathbf{f}') = \sum_{1=1}^{n} \left[\frac{(f_i')^2}{2}\left(\frac{1}{\sigma_0^2} - \frac{1}{\sigma_{1,i}^2}\right) + \ln\frac{\sigma_0}{\sigma_{1,i}}\right]. \tag{6.93}$$

By retaining only the terms depending on $\mathbf{f}'$, we derive the following sufficient statistic for watermark detection:

$$\frac{1}{n}\sum_{i=1}^{n}(f_i')^2\left(\frac{\sigma_{1,i}^2 - \sigma_0^2}{\sigma_{1,i}^2}\right), \tag{6.94}$$

where division by n has been introduced for normalization purpose. The optimum decision rule, then, can be expressed as follows:

$$\Phi(\mathbf{f}') = \begin{cases} 1, & \frac{1}{n}\sum_{i=1}^{n} k_i (f_i')^2 > \Lambda \\ 0, & \text{otherwise} \end{cases} \tag{6.95}$$

where Λ is set by fixing the false detection probability, and where:

$$k_i = \frac{\sigma_{1,i}^2 - \sigma_0^2}{\sigma_{1,i}^2}. \tag{6.96}$$

The analysis of the form of detection regions is very complicated, since it depends in a complicated manner on the values assumed by w_i's, noticeably on their sign. We only note that advantages are likely to be got by applying informed embedding principles to this case, even if, as to date, neither a theoretical nor an experimental analysis is available.

Error probability

Setting Λ according to the Neyman Pearson criterion, requires that the false detection probability is computed. This, in turn, requires that the statistics of

$$z = \frac{1}{n} \sum_{i=1}^{n} k_i (f_i')^2 \tag{6.97}$$

are derived. Let us start by assuming that H_0 holds. We observe that z is the sum of n independent random variables, each of them being the squared value of a Gaussian random variable, namely f_i'. At this point, the central limit theorem is usually invoked to argue that z tends to the Gaussian distribution. While the applicability of the central limit theorem can be easily proved, the speed of convergence raises some questions about the accuracy of the analysis under the normality assumption. This is especially true since we will use this assumption to compute very small error probabilities, which can be severely affected by the inaccuracy of the normal model. In spite of these concerns, we first go on with the approximated analysis, and discuss the limits and usefulness of the approximation in a subsequent paragraph.

In order to compute the mean and variance of z, let us note that, by letting σ_0^2 be the variance of f_i' under H_0, we have:

$$E[(f_i')^2] = \sigma_0^2, \tag{6.98}$$

and

$$\text{var}((f_i')^2) = 2\sigma_0^4. \tag{6.99}$$

By summing over i, we have:

$$\mu_{z|H_0} = \frac{1}{n} \sum_{i=1}^{n} k_i \sigma_0^2 = \overline{k} \sigma_0^2, \tag{6.100}$$

$$\sigma_{z|H_0}^2 = \frac{1}{n^2} \sum_{i=1}^{n} 2k_i^2 \sigma_0^4 = \frac{2}{n} \, \overline{k^2} \sigma_0^4, \tag{6.101}$$

where by $\overline{k}$ and $\overline{k^2}$ the sample averages of k_i and k_i^2 are meant. In other words, we have:

$$p(z|H_0) = \mathcal{N}\left(\overline{k}\sigma_0^2, \frac{2}{n}\,\overline{k^2}\sigma_0^4\right). \tag{6.102}$$

The above equation can be used to calculate the false detection probability, thus permitting to set the detection threshold Λ. It is, in fact, easy to see that:

$$P_f = \frac{1}{2}\,\mathrm{erfc}\left(\sqrt{\frac{n(\Lambda - \overline{k}\sigma_0^2)^2}{4\overline{k^2}\sigma_0^4}}\right), \tag{6.103}$$

from which it follows:

$$\Lambda = \sqrt{\frac{4\,\overline{k^2}}{n}}\,\sigma_0^2\,\mathrm{erfc}^{-1}(2P_f) + \overline{k}\sigma_0^2. \tag{6.104}$$

We now calculate the probability of missing the watermark. To this aim, let us consider the statistic of z under hypothesis H_1. As before, we assume that for large values of n, z follows a Gaussian distribution. We now have:

$$E[(f_i')^2] = \sigma_{1,i}^2, \tag{6.105}$$

and

$$\mathrm{var}((f_i')^2) = 2\sigma_{1,i}^4, \tag{6.106}$$

with $\sigma_{1,i}^2$ given by equation (6.91). By reasoning as in the H_0 case, we obtain:

$$\mu_{z|H_1} = \frac{1}{n}\sum_{i=1}^{n} k_i\sigma_{1,i}^2 = \overline{k}\sigma_1^2, \tag{6.107}$$

$$\sigma_{z|H_1}^2 = \frac{1}{n^2}\sum_{i=1}^{n} 2k_i^2\sigma_{1,i}^4 = \frac{2}{n}\,\overline{k^2}\sigma_1^4, \tag{6.108}$$

$$p(z|H_1) = \mathcal{N}\left(\overline{k}\sigma_1^2, \frac{2}{n}\,\overline{k^2}\sigma_1^4\right). \tag{6.109}$$

We can now express the missed detection probability as a function of Λ. More specifically we have:

$$P_m = \frac{1}{2}\,\mathrm{erfc}\left(\sqrt{\frac{n(\overline{k}\sigma_1^2 - \Lambda)^2}{4\overline{k^2}\sigma_1^4}}\right). \tag{6.110}$$

By inserting equation (6.104) into the above expression, we obtain the relationship relating P_f to P_m, which completely characterizes the performance of the detector:

$$P_m = \frac{1}{2} \, \text{erfc} \left(\sqrt{\frac{(\sqrt{n}(\overline{k\sigma_1^2} - \overline{k\sigma_0^2}) - \sigma_0^2 \sqrt{\overline{4k^2}} \, \text{erfc}^{-1}(2P_f))^2}{4\overline{k^2}\sigma_1^4}} \right). \quad (6.111)$$

Equation (6.111) may be used to plot the ROC curves of the detector for a given $\mathbf{w}$. If the overall system performance has to be calculated, the above expression has to be averaged overall all possible marks. This is a very cumbersome piece of work. However, as in the AWGN case a simplified expression may be obtained by substituting $\overline{k}$, $\overline{k^2}$, $\overline{k\sigma_1^2}$ and $\overline{k^2\sigma_1^4}$ with the expected values of k_i, k_i^2, $k_i\sigma_{1,i}^2$, $k_i^2\sigma_{1,i}^4$ (note that expected values must be calculated by averaging over $\mathbf{w}$), an approximation which is valid for sufficiently large values of n. With regard to the computation of P_f when a watermark other than $\mathbf{w}$ is present, an analysis similar to that carried out for the AWGN case holds.

In figure 6.7, an example of ROC curves obtained from equation (6.111) is given[20]. The curves have been obtained in the absence of noise ($\sigma_n^2 = 0$) by letting $\sigma_w^2 = 1$, $\sigma_f^2 = 0.25$, $n = 1000$ (part (a) of the figure) and $n = 10000$ (part (b)). In figure 6.8 the same curves are reported when noise is added ($\sigma_n^2 = 0.1$) so to evaluate the impact of noise on detector accuracy. Note that the amount of noise we added is rather high since its power is comparable to that of host features.

Limits of the normality assumption

In this paragraph we discuss the use of the central limit theorem to claim that z is normally distributed. Our aim is to evaluate the limits of such an assumption and investigate possible ways to improve it.

As we already noted, each term in the sum defining z is the square value of a Gaussian random variable, as such each term turns out to follow a χ^2 distribution with one degree of freedom. We remember that a χ^2 distribution with n degrees of freedom is obtained by summing the square values of n independent Gaussian random variables, all with the same mean and variance. The expression defining the χ^2 pdf with n degrees of freedom is:

$$p_{\chi_n^2}(z) = \frac{1}{2^{n/2}\Gamma(n/2)} z^{n/2-1} e^{-z/2} u(z). \quad (6.112)$$

---

[20]The values we used are typical of a still image watermarking system operating in the DCT mid-frequency domain.

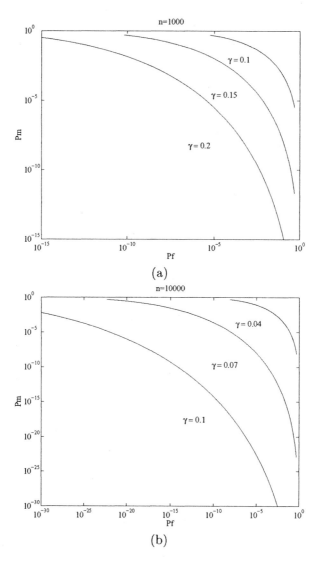

(a)

(b)

Figure 6.7: ROC curves for multiplicative watermarking in the absence of noise. The curves have been obtained by letting $\sigma_w^2 = 1$ and $\sigma_f^2 = 0.25$.

When calculating the false detection probability, we are interested in the statistic of z under hypothesis H_0. When H_0 holds, $(f_i')^2$'s are identically distributed random variables, nevertheless, due to the presence of coefficients k_i's, we can not conclude that z follows a χ_n^2 pdf. In fact, k_i's

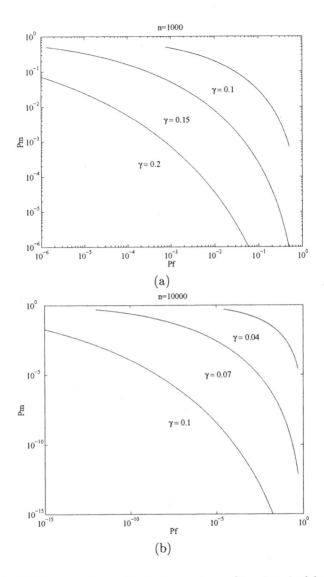

Figure 6.8: ROC curves for multiplicative watermarking impaired by additive white Gaussian noise. The curves have been obtained by letting $\sigma_w^2 = 1$, $\sigma_f^2 = 0.25$, and $\sigma_n^2 = 0.1$.

depend on w_i's which can not be assumed to be constant over i. Each term in the summation defining z, then, has a different variance, whereas for the χ_n^2 model it is required that the Gaussian variables in the sum are equally

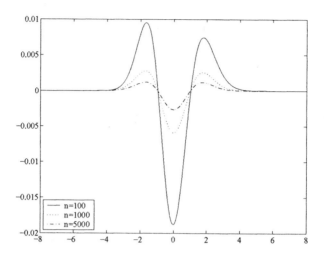

Figure 6.9: Difference between normal and χ_n^2 cumulative density functions. In both cases we let $\mu = n$ and $\sigma^2 = 2n$. The error is plotted against normalized variables, i.e. $(z_n - \mu_{z_n})/\sigma_{z_n}$.

distributed.

In spite of the above observation, in order to get some insight into the validity of the normal approximation, we make the simplifying hypothesis that all the terms in z has the same variance (i.e. we neglect the dependence of k_i on i). Our problem, then, is to verify whether the χ_n^2 pdf actually tends to a normal pdf for $n \to \infty$, and to evaluate the convergence speed.

Let us consider first the applicability of the central limit theorem. Even without considering the set of sufficient and necessary conditions for the theorem to hold, we can note that a sufficient condition ensuring the applicability of the theorem, is that the variables in the sum are independent random variables, whose pdf has a finite third order moment. This is surely the case for the χ^2 distribution, hence we can conclude that $p(z)$ tends to a normal pdf for $n \to \infty$. It is known, however, that for the sum of χ^2 variables the convergence speed is rather slow.

To get more insight into the error we make when replacing the χ_n^2 pdf with a normal pdf, let us consider the following normalized variables: $z_n = \sum_{i=1}^n (f_i')^2$ following a χ_n^2 distribution with mean $\mu_{z_n} = n$ and variance $\sigma_{z_n}^2 = 2n$, and a variable $\tilde{z}_n$ following a normal pdf with the same mean and variance of z_n[21]. Figures 6.9 and 6.10 illustrate the behavior of the difference between the normal and the χ_n^2 cdf's. The figures refer to

---

[21]In practice we are assuming that each f_i in $\sum_i (f_i')^2$ has zero mean and unitary variance.

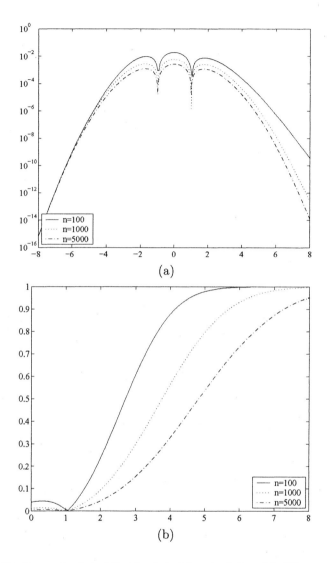

Figure 6.10: Log-scale plot of the absolute (a) and relative (b) difference between normal and χ_n^2 cumulative density functions. Relative difference is computed with respect to the right percentile of pdf's.

cumulative density functions, thus giving an immediate idea of the impact of the normal approximation on error probabilities. The error is plotted against normalized variables, i.e. $(z_n - \mu_{z_n})/\sigma_{z_n}$. In figure 6.9 the absolute error is given. It is interesting to look at the sign of the error. When considering the false detection probability, we are interested in the area under the right tail of $p(z_n|H_0)$. As it can be seen from the figure, for large values of z_n (as those involved in the computation of P_f) the error is always positive, thus allowing us to conclude that the normal approximation tends to underestimate the true false detection probability[22]. The converse is true for the computation of P_m. In this case, in fact, we are interested in the left tail of $p(z_n|H_1)$. If z_n is sufficiently far apart from μ_{z_n}, as it is always the case when dealing with small missed detection probabilities, the error is still positive, thus ensuring that the normal approximation results in an overestimate of the true P_m.

In part (a) of figure 6.10 the absolute error is plotted along a log scale, thus allowing us to better appreciate its magnitude for a large deviation regime. Part (b) of the figure reports the relative error on the right percentile, i.e. the error on the cdf divided by 1 minus the true value. As it can be seen for large value of z_n the relative error may become large[23], even if for large values of n such an error is clearly reduced.

In order to evaluate the accuracy of the normal approximation, when k_i's are not constant, it is possible to resort to Montecarlo simulations. In figure 6.11, the comparison between the simulation results and those obtained via the normal approximation are given. As it can be seen, the approximation error is sufficiently small (smaller than that obtained for constant k_i's), even if it has to be considered that such an error is likely to increase for more effective watermarks yielding lower error probabilities[24].

Having concluded that the analysis based on the central limit theorem is an imprecise one, we must ask whether and how we should replace it, since an estimation of the false detection probability is needed in order to set the detection threshold. A possibility consists in using the exact analysis we developed by assuming that k_i's are constant (even if this is surely not the case since k_i's depend on the sequence of w_i, which can not be assumed to be constant). Alternatively, the false detection probability given by Montecarlo simulations may be used. The problem with Montecarlo simulations is that they are too computationally expensive to be used within practical detectors. The analysis based on the central limit theorem, hence, can still be of great importance, since it permits to determine an approximate value of the detection threshold. No need saying that, in this case, the detector must be designed so to compensate for the

---

[22]Being interested in calculating the area under the right tail $p(z_n|H_i)$ a larger value of the cdf results in a lower error probability.

[23]Similar considerations hold for the left percentile.

[24]The reason why we did not draw the curves for lower error probabilities is a practical one, since simulations get more and more computationally expensive as the error probabilities under analysis decrease.

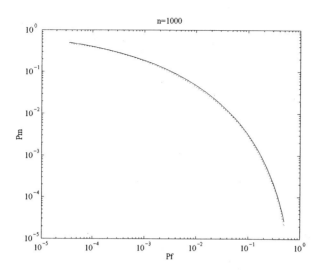

Figure 6.11: Comparison between the ROC curve derived under the normal approximation (solid line) and that resulting from Montecarlo simulations (dotted line). Results refer to an antipodal watermark in the absence of noise, with $\sigma_f^2 = 0.25$, $n = 1000$ and $\gamma = 0.09$.

approximation error introduced by the normality assumption. To this aim, the comparison between the approximated false detection probability and that obtained through Montecarlo simulations may be of great help.

Implementation issues

A possible difficulty with the practical implementation of the optimum detector derived above is that the separate knowledge of σ_n^2 and σ_f^2 is required. Unlike the additive case, where it is only requested that the sum $\sigma_n^2 + \sigma_f^2$ is known, computation of $\sigma_{1,i}^2$ requires that both the variance of the host features and that of noise are known. For a practical implementation of the detector, then, it is required that such variances are estimated separately by relying on the to-be-inspected asset.

6.1.7 Multiplicative Weibull channel

Multiplicative watermarking is a rather popular choice for image watermarking systems operating in the frequency domain. The main reason for such a choice has been detailed in section 5.4.2, where we showed how visual masking can be taken into account through multiplicative watermarking.

For frequency-domain systems, the Gaussian assumption we made in the previous section is not valid. For systems operating in the DCT domain, for instance, the more fitting Generalized Gaussian model should be used. As an alternative to DCT-based systems, watermarking in the DFT domain is often considered. This because by inserting the watermark into the magnitude of DFT coefficients, invariance to temporal/spatial shifts is automatically achieved. It is evident, that when the host features coincide with the magnitude of DFT coefficients, the assumption that f_i's are normally distributed is completely wrong, all the more that DFT magnitudes only assume positive values, whereas for any Gaussian pdf the probability of getting a negative value is never zero.

In this section we derive the optimum detector structure for a multiplicative watermarking system operating in the magnitude-of-DFT domain. For readers interested in DCT domain systems some references are given at the end of the chapter.

As opposed to the Gaussian case, derivation of the optimum detector structure in the presence of noise is not viable, since we can not explicitly calculate the pdf of noisy marked coefficients. Then, we will derive the detector structure in the absence of attacks, i.e. we will let

$$f_i' = f_i(1 + \gamma w_i), \tag{6.113}$$

where f_i's are the magnitudes of host DFT coefficients. In order to preserve the magnitude nature of host features, we assume that $|\gamma w_i|$ is always smaller than unit, so that f_i' is always greater than 0. Such a condition is surely met if the pdf of w_i's is zero outside the $[-1/\gamma, 1/\gamma]$ interval, thus preventing the adoption of a normally distributed watermark. Of course when attacks are present the detector is no longer optimum, hence its validity must be verified experimentally. We also assume that embedding is carried out blindly, without attempting to compensate for the host feature presence. No need saying that informed embedding may still be conveniently applied, to improve the performance of the system.

Detector Structure

As a first step we must identify a suitable statistical model to describe the magnitude of DFT coefficients. To do so, we note that a parametric pdf is needed which is nonzero on the positive real axis only, and which is both flexible and easy to handle from a mathematical point of view. The solution we adopt here consists in assuming that the magnitude of DFT coefficients follows a Weibull pdf, for which:

$$p_W(x) = \frac{\beta}{\alpha} \left(\frac{x}{\alpha}\right)^{\beta-1} \exp\left[-\left(\frac{x}{\alpha}\right)^{\beta}\right], \tag{6.114}$$

where $\alpha > 0$ and $\beta > 0$ are real-valued positive constants determining the mean and variance of the pdf. More specifically, we have:

$$\mu_x = \alpha \Gamma \left(1 + \frac{1}{\beta} \right), \qquad (6.115)$$

$$\sigma_x^2 = \alpha^2 \Gamma \left(1 + \frac{2}{\beta} \right) - \mu_x^2. \qquad (6.116)$$

Limit cases are obtained for $\beta = 1$, yielding an exponential pdf, and $\beta = 2$, resulting in a Rayleigh distribution. The dependence of Weibull's shape upon α and β is exemplified in figure 6.12.

Having identified a suitable model for the host features, we must write the likelihood ratio. As usual we start by letting:

$$\ell(\mathbf{f}') = \frac{p(\mathbf{f}'|\mathbf{w})}{p(\mathbf{f}'|\mathbf{0})} = \prod_{i=1}^{n} \frac{p(f_i'|w_i)}{p(f_i'|0)}. \qquad (6.117)$$

When H_0 holds, $f_i' = f_i$ and each f_i' follows a Weibull pdf with parameters equal to those of the original coefficients:

$$p(f_i'|H_0) = \frac{\beta}{\alpha} \left(\frac{f_i'}{\alpha} \right)^{\beta-1} \exp \left[-\left(\frac{f_i'}{\alpha} \right)^{\beta} \right], \qquad (6.118)$$

where, for sake of simplicity, host features have been assumed to be identically distributed. If H_1 is in force, then we have $f_i' = f_i(1 + \gamma w_i)$, hence:

$$p(f_i'|H_1) = \frac{\beta}{\alpha(1 + \gamma w_i)} \left(\frac{f_i'}{\alpha(1 + \gamma w_i)} \right)^{\beta-1} \exp \left[-\left(\frac{f_i'}{\alpha(1 + \gamma w_i)} \right)^{\beta} \right]. \qquad (6.119)$$

Substituting $p(f_i'|H_0)$ and $p(f_i'|H_1)$ in equation (6.117) yields:

$$\ell(\mathbf{f}') = \prod_{i=1}^{n} \left(\frac{1}{1 + \gamma w_i} \right)^{\beta} \exp \left[-\left(\frac{f_i'}{\alpha(1 + \gamma w_i)} \right)^{\beta} + \left(\frac{f_i'}{\alpha} \right)^{\beta} \right], \qquad (6.120)$$

which, by passing to a logarithmic formulation, gives

$$\mathcal{L}(\mathbf{f}') = \sum_{1=1}^{n} [-\beta \ln(1 + \gamma w_i)] + \sum_{1=1}^{n} \left[-\left(\frac{f_i'}{\alpha(1 + \gamma w_i)} \right)^{\beta} + \left(\frac{f_i'}{\alpha} \right)^{\beta} \right]. \qquad (6.121)$$

By applying some algebra and by retaining only the terms depending on $\mathbf{f}'$, we derive the following sufficient statistic for watermark detection:

$$z = \sum_{i=1}^{n} v_i (f_i')^{\beta}, \qquad (6.122)$$

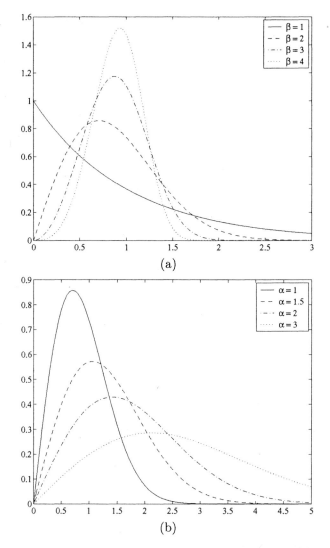

Figure 6.12: Shape of the Weibull pdf for different values of β (a) and α (b).

with

$$v_i = \frac{(1 + \gamma w_i)^{\beta} - 1}{\alpha^{\beta}(1 + \gamma w_i)^{\beta}}. \tag{6.123}$$

The optimum decision rule, then, can be expressed as follows:

$$\Phi(\mathbf{f'}) = \begin{cases} 1, & \sum_{i=1}^{n} v_i (f_i')^\beta > \Lambda \\ 0, & \text{otherwise} \end{cases} \tag{6.124}$$

where Λ is set by fixing the false detection probability. As for the multiplicative Gaussian case, the analysis of the form of detection regions is very complicated, hence we will not carry it out. We only note that, once again, advantages are likely to be got by applying the informed embedding principle.

Error probability

By following the path of the previous sections, we now fix Λ by means of the Neyman-Pearson criterion. To do so, it is necessary that the false detection probability is computed. This requires that the statistic of

$$z = \sum_{i=1}^{n} v_i (f_i')^\beta \tag{6.125}$$

is derived. We first focus on $(f_i')^\beta$. By remembering that under hypothesis H_0, $f_i' = f_i$ follows a Weibull pdf with parameters α and β, it can be easily demonstrated that the pdf of $(f_i')^\beta$ is an exponential one, with mean α^β and variance $\alpha^{2\beta}$. By also taking into account the presence of v_i, we conclude that each term of the sum in (6.125) follows an exponential pdf. More specifically, by letting $z_i = v_i (f_i')^\beta$, we have that:

$$p(z_i | H_0) = \lambda_i \exp(-\lambda_i z_i) \, u(z_i), \tag{6.126}$$

where

$$\lambda_i = \frac{(1 + \gamma w_i)^\beta}{(1 + \gamma w_i)^\beta - 1}. \tag{6.127}$$

Equation (6.126) only holds if $v_i > 0$, i.e. for $w_i > 0$. In fact, it is easy to show that when $w_i < 0$, λ_i is a negative quantity and the following pdf is obtained:

$$p(z_i | H_0) = |\lambda_i| \exp(-\lambda_i z_i) \, u(-z_i), \tag{6.128}$$

To go on, the pdf of the sum of n exponentially distributed random variables has to be evaluated. A first possibility consists in assuming that z is normally distributed. As we outlined previously, in some cases this may introduce a significant approximation error, thus leaving to the analysis derived under the normality assumption only an indicative value. A second possibility, consists in trying to derive the exact pdf of z. Though very cumbersome, this second solution is a viable one due to the particularly simple form of the pdf of z_i's. For the sake of clarity, we will first derive the error probabilities under the normality assumption, then, in the next paragraph, we will describe the more precise solution based on the exact computation of $p(z)$.

Let us assume, then, that the central limit theorem may be invoked to maintain that, for the values of n encountered in practical applications, z is normally distributed. Under this assumption, to calculate P_f we only need to compute $\mu_{z|H_0}$ and $\sigma^2_{z|H_0}$. This is an easy task, since due to the independence of z_i's, we simply have:

$$\mu_{z|H_0} = \sum_{i=1}^{n} \frac{1}{\lambda_i} = \sum_{i=1}^{n} \frac{(1+\gamma w_i)^\beta - 1}{(1+\gamma w_i)^\beta}, \tag{6.129}$$

$$\sigma^2_{z|H_0} = \sum_{i=1}^{n} \frac{1}{\lambda_i^2} = \sum_{i=1}^{n} \left[\frac{(1+\gamma w_i)^\beta - 1}{(1+\gamma w_i)^\beta} \right]^2, \tag{6.130}$$

whose insertion in:

$$P_f = \frac{1}{2} \operatorname{erfc} \left(\sqrt{\frac{(\Lambda - \mu_{z|H_0})^2}{2\sigma^2_{z|H_0}}} \right), \tag{6.131}$$

gives the false detection probability. Inversion of equation (6.131) gives the detection threshold as a function of P_f:

$$\Lambda = \sqrt{2}\sigma_{z|H_0} \operatorname{erfc}^{-1}(2P_f) + \mu_{z|H_0}. \tag{6.132}$$

To calculate the probability of missing the watermark, we must derive the statistic of z by assuming the H_1 holds. In this case, $f_i' = f_i(1+\gamma w_i)$, hence marked host coefficients are still distributed according to a Weibull pdf with parameters $\beta' = \beta$ and $\alpha' = \alpha(1+\gamma w_i)$:

$$p(f_i'|H_1) = \left(\frac{\beta}{\alpha(1+\gamma w_i)} \right) \cdot \left(\frac{f_i'}{\alpha(1+\gamma w_i)} \right)^{\beta-1} \exp \left[-\left(\frac{f_i'}{\alpha(1+\gamma w_i)} \right)^\beta \right], \tag{6.133}$$

leading to

$$p(z_i|H_1) = \lambda_i' \exp\left(-\lambda_i' z_i\right) u(z_i), \tag{6.134}$$

with

$$\lambda_i' = \frac{\lambda_i}{(1+\gamma w_i)^\beta} = \frac{1}{(1+\gamma w_i)^\beta - 1}, \tag{6.135}$$

or

$$p(z_i|H_1) = |\lambda_i'| \exp\left(-\lambda_i' z_i\right) u(-z_i), \tag{6.136}$$

if $v_i < 0$. As before we will assume that for large values of n, z follows a Gaussian distribution, whose mean and variance turn out to be:

$$\mu_{z|H_1} = \sum_{i=1}^{n} \frac{1}{\lambda_i} = \sum_{i=1}^{n} [(1 + \gamma w_i)^\beta - 1], \qquad (6.137)$$

$$\sigma_{z|H_1}^2 = \sum_{i=1}^{n} \frac{1}{\lambda_i^2} = \sum_{i=1}^{n} [(1 + \gamma w_i)^\beta - 1]^2. \qquad (6.138)$$

In figure 6.13, an example of the two Gaussian pdf's obtained under hypotheses H_0 and H_1 is given. Results of part (a) refer to a watermark consisting of $n = 1000$ samples, embedded in a host feature sequence for which $\beta = 1.8$ (the value of α has no effect on $p(z)$). The watermark strength γ was set to 0.1. In part (b) of the figure the same plot is given for $n = 10000$. As it was expected, increasing n has a beneficial effect, in that the two pdf's move far apart. Note also that, since we have not normalized the detection statistic by dividing z by n, increasing n also results in a shift of the pdf's and in a larger standard deviation. The larger standard deviation, however is over-compensated by the larger distance between the pdf's (distance tends to increase proportionally to n, whereas standard deviation only increases as $\sqrt{n}$).

To conclude the analysis, we now express the missed detection probability as a function of Λ. We have:

$$P_m = \frac{1}{2} \operatorname{erfc} \left(\sqrt{\frac{(\mu_{z|H_1} - \Lambda)^2}{2\sigma_{z|H_1}^2}} \right), \qquad (6.139)$$

leading to:

$$P_m = \frac{1}{2} \operatorname{erfc} \left(\frac{\mu_{z|H_1} - \mu_{z|H_0} - \sqrt{2}\sigma_{z|H_0} \operatorname{erfc}^{-1}(2P_f)}{\sqrt{2}\sigma_{z|H_1}} \right). \qquad (6.140)$$

As usual, the above equation completely characterizes the performance of the detector[25]. For example, it may be used to plot the ROC curves of the detector for a given $\mathbf{w}$. If the overall system performance has to be calculated, the above expression must be averaged overall all possible marks. In figure 6.14, an example of a ROC curve obtained from equation (6.140) is given. The curve has been obtained by letting $n = 5000$, $\gamma = 0.08$, and $\beta = 1.8$.

---

[25]With regard to the computation of P_f when a watermark other than $\mathbf{w}$ is present, an analysis similar to that carried out for the AWGN case has to be developed.

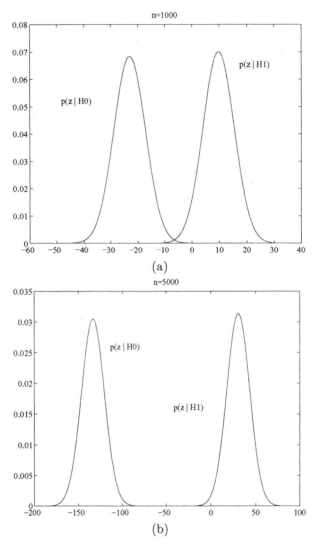

Figure 6.13: Probability density functions of z conditioned to H_0 and H_1, for $\beta = 1.8$, $\gamma = 0.1$, $n = 1000$ (a) and $n = 10000$ (b).

Removing the normality assumption

We already noticed, when treating the multiplicative Gaussian channel, that assuming the sufficient statistic z to be normally distributed leads to an approximation error which in some cases may be a significant one.

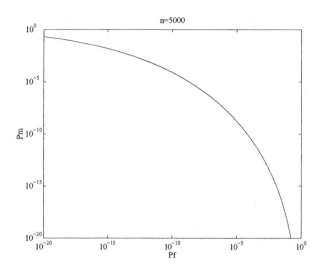

Figure 6.14: ROC curve for multiplicative watermarking of Weibull-distributed host features. The curve has been obtained under normal approximation of $p(z)$, by letting $n = 5000$, $\beta = 1.8$ and $\gamma = 0.08$.

The same observation can be made now, since while the sum in equation (6.122) truly tends to a normal pdf, the convergence rate can be rather slow, and introduce an error affecting the probabilities computed in the previous section, especially when they are very small (large deviation regime). In the Gaussian case, though, it was not possible to develop an exact analysis of the detector behavior, hence we limited our investigation to an estimate of the error in a simplified case, i.e. when all the terms in (6.97) are equally distributed, and to Montecarlo simulations.

With reference to the case treated in this section, namely multiplicative watermarking of Weibull distributed features, an exact analysis is possible, though at the expense of a considerable complication of the analysis.

Let us start by recalling our goal, i.e. deriving the statistic of

$$z = \sum_{i=1}^{n} v_i (f_i')^{\beta} = \sum_{i=1}^{n} z_i, \qquad (6.141)$$

and use it to set the detection threshold and calculate the false and missed detection probabilities. We already observed that under hypothesis H_0, z_i's follow an exponential pdf with exponent parameters λ_i's (equations (6.126) through (6.128)). In order to go on, we split the sum in two parts, a first part with all z_i's for which $\lambda_i > 0$ (let us indicate such terms as z_i^+), and a second one with the z_i's for which $\lambda_i < 0$ (we indicate these terms by z_i^-). We have:

$$z = \sum_{i|v_i>0} v_i(f_i')^\beta + \sum_{i|v_i<0} v_i(f_i')^\beta = \sum_{i=1}^{n^+} z_i^+ + \sum_{i=1}^{n^-} z_i^- = z^+ + z^-, \quad (6.142)$$

where by n^+ and n^- we indicated the number of terms with positive and negative λ_i respectively. Clearly, the pdf of z under hypothesis H_0, is the convolution between the pdf's of z^+ and z^-. By noting that all z_i^+'s follow an exponential pdf with positive λ_i, it can be proven that[26]:

$$p(z^+|H_0) = \sum_{i=1}^{n^+} \left[\frac{\prod_{k=1}^{n^+} \lambda_k^+}{\prod_{k \neq i}(\lambda_k^+ - \lambda_i^+)} \exp(-\lambda_i^+ z^+) \right] u(z^+), \quad (6.143)$$

where the terms λ_i^+ have been introduced to indicate explicitly that we are dealing with positive λ_i's. The above equation is only valid if $\lambda_i^+ \neq \lambda_k^+$, for each $k \neq i$. This implies that $w_i \neq w_k$, for $k \neq i$. This is a reasonable hypothesis as long as the watermark coefficients follow a continuous pdf, e.g. they are uniformly distributed in $[-1, +1]$. If this is not the case, e.g. if w_i's only take values $+1$ or -1, the analysis must be slightly modified, however the main results still hold. With reference to z^- a similar formula can be obtained, since we have:

$$p(z^-|H_0) = \sum_{j=1}^{n^-} \left[\frac{\prod_{k=1}^{n^-} |\lambda_k^-|}{\prod_{k \neq j}(\lambda_j^- - \lambda_k^-)} \exp(-\lambda_i^- z^-) \right] u(-z^-), \quad (6.144)$$

where the terms λ_i^- are used to indicate negative λ_i's. In order to calculate the pdf of z, the convolution between $p(z^+|H_0)$ and $(z^-|H_0)$ must be computed. This is an easy, though tedious, task leading to:

$$p(z|H_0) = \prod_{k=1}^{n^+} \lambda_k^+ \prod_{k=1}^{n^-} |\lambda_k^-| \sum_{i=1}^{n^+} \sum_{j=1}^{n^-} \left[\frac{h(\lambda_i^+, \lambda_j^-, z)}{\prod_{k \neq i}(\lambda_k^+ - \lambda_i^+) \prod_{k \neq j}(\lambda_j^- - \lambda_k^-)} \right],$$
$$(6.145)$$

where $h(\lambda_i^+, \lambda_j^-, z)$ is defined as follows:

$$h(\lambda_i^+, \lambda_j^-, z) = \begin{cases} \dfrac{\exp(-\lambda_i^+ z)}{\lambda_i^+ - \lambda_j^-} & z > 0 \\[2ex] \dfrac{\exp(-\lambda_j^- z)}{\lambda_i^+ - \lambda_j^-} & z < 0 \end{cases} \quad (6.146)$$

---

[26]Proof of equation (6.143) can be achieved either by induction or by exploiting the properties of characteristic functions.

In order to calculate the false detection probability, and hence set the detection threshold Λ, the cdf (cumulative density function) of z must be calculated. This requires that the pdf given in equation (6.145) is integrated between $-\infty$ and z, yielding:

$$\int_{-\infty}^{z} p(x|H_0)dx = \text{Err}_0(z) =$$

$$\prod_{k=1}^{n+} \lambda_k^+ \prod_{k=1}^{n^-} |\lambda_k^-| \sum_{i=1}^{n^+} \sum_{j=1}^{n^-} \frac{-\exp(-\lambda_j^- z)}{\lambda_j^- (\lambda_i^+ - \lambda_j^-) \prod_{k \neq i}(\lambda_k^+ - \lambda_i^+) \prod_{k \neq j}(\lambda_j^- - \lambda_k^-)},$$

$$(6.147)$$

for $z < 0$, and

$$\text{Err}_0(z) =$$

$$\prod_{k=1}^{n+} \lambda_k^+ \prod_{k=1}^{n^-} |\lambda_k^-| \sum_{i=1}^{n^+} \sum_{j=1}^{n^-} \frac{\lambda_j^- [1 - \exp(-\lambda_i^+ z)] - \lambda_i^+}{\lambda_j^- \lambda_i^+ (\lambda_i^+ - \lambda_j^-) \prod_{k \neq i}(\lambda_k^+ - \lambda_i^+) \prod_{k \neq j}(\lambda_j^- - \lambda_k^-)},$$

$$(6.148)$$

for $z > 0$, where we introduced the function $\text{Err}_0(z)$ to indicate the cdf of z under hypothesis H_0. The detection threshold Λ can now be calculated. We have:

$$P_f = p(z > \Lambda|H_0) = 1 - \text{Err}_0(\Lambda), \qquad (6.149)$$

or

$$\Lambda = \text{Err}_0^{-1}(1 - P_f). \qquad (6.150)$$

To complete the analysis of the optimum detector, we must calculate the missed detection probability as a function of P_f and plot the corresponding ROC curves. To do so, we must derive the statistics of z under hypothesis H_1. In the previous section we already noticed that under hypothesis H_1, z_i's still follow an exponential distribution, with parameter:

$$\lambda_i' = \frac{\lambda_i}{(1 + \gamma w_i)^\beta}. \qquad (6.151)$$

Equations (6.145) through (6.148), then, still hold: it only needs to replace λ_i's with λ_i''s. By letting $\text{Err}_1(z)$ indicate the cdf of z under hypothesis H_1, we obtain:

$$P_m = \text{Err}_1 \left[\text{Err}_0^{-1}(1 - P_f) \right], \qquad (6.152)$$

which can be used to plot the ROC curves of the detector.

A major problem with the exact analysis given above is that numerical implementation of the Err_0 and Err_1 functions requires a very high accuracy, due to the necessity of dealing with the huge numbers resulting from the products contained in the formula. In most cases, it is necessary to use ad hoc routines providing a computational accuracy which is much higher than that available on normal computers where double precision floating point numbers are represented by means of 64-bit words. Another possibility consists in resorting to Montecarlo simulations, even if, as we noted earlier, the computational complexity rapidly becomes unaffordable when very low error probabilities are involved. Finally, the results provided by the approximated analysis may be used, by paying attention to set the detection threshold in such a way to compensate for the approximation errors introduced by the normal assumption.

Practical implementation of the detector

The actual implementation of the detector on the basis of equations (6.129) through (6.131) requires that the parameters α and β are known. Though such parameters refer to the pdf of non-marked coefficients, their actual value can be estimated a-posteriori on the watermarked image supposed that the presence of the watermark does not alter them significantly, that is the watermark strength is sufficiently small ($\gamma << 1$). As we noted earlier, this is often the case, since γ is limited by the necessity of ensuring the invisibility of the watermark.

A second problem with the theoretical analysis carried out so far is that we assumed α and β to be constant over the whole host feature set. This may not be the case, thus, in order to make the analysis more realistic, the DFT region hosting the watermark is usually split into a number of subregions within which α and β can be assumed to be constant. For example, in figure 6.15, the region with host DFT coefficients is split into 16 subregions. Under the hypothesis that the coefficients within each subregion follow the same pdf, the parameters α and β can be evaluated through Maximum Likelihood (ML) estimation. With this approximation, the structure of the detector slightly changes, since the sufficient statistic is now given by:

$$z = \sum_{k=0}^{S-1} \sum_{i=1}^{n_s} \frac{(1 + \gamma w_{k \cdot n_s + i})^{\beta_k} - 1}{\alpha^{\beta_k}(1 + \gamma w_{k \cdot n_s + i})^{\beta_k}} (f_i')^{\beta_k}, \qquad (6.153)$$

where we assumed that the host feature sequence is split into S subregions, each with n_s coefficients. Detection threshold is still given by (6.132), where

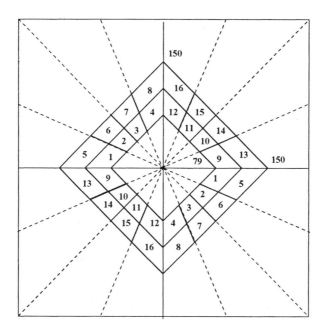

Figure 6.15: In order to estimate the parameters α's and β's, the host DFT region is split into 16 subregions inside which α and β are assumed to be constant. In the figure the marked region goes from the 79-*th* to the 150-*th* diagonal.

$\mu_{z|H_0}$ and $\sigma^2_{z|H_0}$ are now given by:

$$\mu_{z|H_0} = \sum_{k=0}^{S-1} \sum_{i=1}^{n_s} \frac{(1 + \gamma w_{k \cdot n_s + i})^{\beta_k} - 1}{(1 + \gamma w_{k \cdot n_s + i})^{\beta_k}}, \tag{6.154}$$

$$\sigma^2_{z|H_0} = \sum_{k=0}^{S-1} \sum_{i=1}^{n_s} \left[\frac{(1 + \gamma w_{k \cdot n_s + i})^{\beta_k} - 1}{(1 + \gamma w_{k \cdot n_s + i})^{\beta_k}} \right]^2. \tag{6.155}$$

Finally let us observe that watermark robustness is usually improved by means of spatial masking. The optimum detector structure, though, has been derived without taking spatial masking into account. The actual performance of the detector in presence of masking, then, must be validated experimentally.

6.1.8 Multichannel detection

So far we treated only the case in which host features are scalar values. When this is not the case, we must deal with a vector watermarking channel. Watermarking of color images is the most common case in which multichannel watermarking comes into play. The derivation of an optimum multichannel detector requires that a multichannel model of the host features is available.

The main problem with multichannel detection is that a multivariate model which fits well the features used in practical applications is very difficult to find. The most straightforward model would be the multivariate Gaussian model in which the correlation between feature vector components, e.g. color image bands, is modelled through a multivariate gaussian. When such a model applies, the derivation of the optimum watermark detector goes the same path followed in section 6.1.2 for the scalar case.

If the multivariate Gaussian model does not apply, e.g. because the watermark is inserted within the magnitude of DTF coefficients of image color bands, a suitable multivariate model is very difficult to find. In order to retain optimality, then, one may desire to decorrelate host feature components, e.g. by applying the Karhunen-Loeve Transform as indicated in chapter 4. If this approach is adopted, then multichannel detection reduces to the scalar cases treated previously.

Finally, a suboptimum approach can be followed. If we accept to loose optimality, then a solution could consist in adopting a correlation-based detector, as the one described in section 6.1.2. The only difference with respect to the scalar case comes into play when calculating the false and missed detection probabilities, since the correlation between feature components has to be taken into account. We will not describe such an analysis here, however, interested readers can refer to the further reading section at the end of the chapter for an indication of some detailed papers dealing with multichannel watermark detection.

6.2 Decoding

In this paragraph we treat multi-bit, or readable, watermarking. In this case recovery of hidden information does not consists in deciding whether the host asset contains a given message or not. On the contrary, the hidden message must be retrieved from the data without knowing it in advance. As the problem of watermark detection was solved by resorting to statistical detection theory, watermark decoding is naturally modelled as a digital communication problem, where the receiver must decide which message/signal was transmitted among the set of possible messages/signals.

The derivation of the optimum decoder structure depends on many factors, including the particular strategy used to encode the message, the embedding rule, the choice of the host feature set, the adoption of an informed or a blind embedding strategy. A thorough discussion of all possible combinations of the above choices would require more space than is available in this book, hence we only treat the most important cases, trusting that this will give the reader sufficient insight to deal with all the situations which are not covered explicitly here.

We will focus the analysis on independent bit signalling schemes, where each bit of the to-be-hidden message is treated independently of the others. We will consider the following channel models: additive watermarking in Gaussian noise, additive watermarking of Generalized Gaussian host features, multiplicative watermarking in Gaussian noise, multiplicative watermarking of Weibull-distributed features and QIM watermarking. In all cases we will assume that each bit is repeatedly hidden in r, independent, host features. The possible usage of a spread spectrum sequence to improve secrecy will be considered as well.

Note that we will not explicitly investigate the difference between systems exploiting informed embedding/coding principles to hide the watermark and blind ones, nevertheless, the reader should keep in mind that some of the schemes we will treat were expressly designed to exploit such principles. This is the case, for example, of QIM systems, where complete rejection of the host signal is obtained by jointly applying informed embedding and informed coding principles.

6.2.1 General problem for binary signalling

Let us recall the problem we are trying to solve. We are given a sequence of n observed host features $\mathbf{f}'$. By assuming that $\mathbf{f}'$ contains a k-bit long hidden message belonging to the set of possible messages $\mathbf{B}$, we must define a decoding rule $\phi(\mathbf{f}')$ mapping each $\mathbf{f}'$ into an element of $\mathbf{B}$. If we assume that channel coding is not used, $\mathbf{B}$ contains all the 2^k possible k-bit-long sequences. Definition of the decoding rule requires that the feature space is split into 2^k non overlapping regions $\mathcal{R}_j$ such that $\phi(\mathbf{f}') = \mathbf{b}_j$ if $\mathbf{f}' \in \mathcal{R}_j$. Regions $\mathcal{R}_j$'s, called decision, or decoding, regions, completely define the decoder.

We assume that each bit in $\mathbf{b}$ is tied to a different subset of $\mathbf{f}'$, $\{\mathcal{S}_m\}$, where the subscript m indicates that $\{\mathcal{S}_m\}$ hosts the m-th bit of $\mathbf{b}$. For sake of simplicity, we assume that $\{\mathcal{S}_m\}$'s are obtained by splitting $\mathbf{f}'$ into k consecutive chunks each containing $r = n/k$ samples (in order not to complicate the analysis we assume n/k is an integer). In other words, we let $\mathcal{S}_j = \{f'_i\}_{i=(j-1)r+1}^{jr}$. Before embedding, the sequence $\mathbf{b}$ is transformed into an antipodal sequence $\mathbf{t}$, however, not to introduce a new symbol, we will assume that b_i's take values ± 1. Finally, prior to embedding, the sequence $\mathbf{b}$ may be multiplied by a spreading sequence $\mathbf{w}$, thus producing the watermark signal $\mathbf{w}'$. Note that if sequence spreading is not used we simply have $\mathbf{b} = \mathbf{w}'$.

Given the framework described above, we look for the optimum decoding rule ϕ, where optimality refers to the maximization of the probability of a correct decision or, equivalently, to the minimization the probability of error. By assuming that all the possible 2^k messages are equally probable, the probability of outputting a wrong sequence is:

$$
\begin{aligned}
P_e &= \frac{1}{2^k} \sum_{m=1}^{2^k} P(e|\mathbf{b}_m) = \frac{1}{2^k} \sum_{m=1}^{2^k} P(\mathbf{f}' \notin \mathcal{R}_m | \mathbf{b}_m) = \\
&= \frac{1}{2^k} \sum_{m=1}^{2^k} [1 - P(\mathbf{f}' \in \mathcal{R}_m | \mathbf{b}_m)] = 1 - \frac{1}{2^k} \sum_{m=1}^{2^k} P(\mathbf{f}' \in \mathcal{R}_m | \mathbf{b}_m),
\end{aligned}
\tag{6.156}
$$

where $P(e|\mathbf{b}_m)$ denotes the error probability conditioned to $\mathbf{b}_m$. The probability $P(\mathbf{f}' \in \mathcal{R}_m | \mathbf{b}_m)$ can be expressed as:

$$
P(\mathbf{f}' \in \mathcal{R}_m | \mathbf{b}_m) = \int_{\mathcal{R}_m} p(\mathbf{f}' | \mathbf{w}, \mathbf{b}_m) d\mathbf{f}',
\tag{6.157}
$$

where conditioning to $\mathbf{w}$ is a consequence of the fact that we assumed that the spreading sequence, if any, is known by the decoder. In order to minimize P_e we must let, for each m:

$$
\mathcal{R}_m = \{ \mathbf{f}' \mid p(\mathbf{f}' | \mathbf{w}, \mathbf{b}_m) \geq p(\mathbf{f}' | \mathbf{w}, \mathbf{b}_l), \ \forall l \neq m \}.
\tag{6.158}
$$

Stated in another way, the decoded sequence is obtained by looking for the sequence that maximizes the pdf $p(\mathbf{f}' | \mathbf{w}, \mathbf{b})$, thus leading to a maximum-likelihood (ML) optimum criterion:

$$
\hat{\mathbf{b}} = \arg \max_{l=1...2^k} p(\mathbf{f}' | \mathbf{w}, \mathbf{b}_l).
\tag{6.159}
$$

Let us now indicate by $\mathbf{f}'_h$ and $\mathbf{w}_h$ the subsequences of $\mathbf{f}'$ and $\mathbf{w}$ corresponding to the h-th bit of $\mathbf{b}$ (i.e. $\mathbf{f}'_h$ is formed by the features in $\mathcal{S}_h$). By assuming that host features are independent of each other, and by remembering that the elements of $\mathbf{w}$ are independent as well, we can write:

$$
\hat{\mathbf{b}} = \arg \max_{l=1...2^k} \prod_{h=1}^{k} p(\mathbf{f}'_h | \mathbf{w}_h, b_{l,h}),
\tag{6.160}
$$

where $b_{l,h}$ is the h-th bit of the sequence $\mathbf{b}_l$. By assuming that channel coding is not used, bit-wise decoding of the transmitted sequence can be performed without losing optimality. Under this assumption, and by focusing on the h−th bit, the optimum decision criterion can be formulated as follows:

$$\hat{b}_h = \arg \max_{b_h \in \{-1,+1\}} p(\mathbf{f}'_h | \mathbf{w}_h, b_h) = \arg \max_{b_h \in \{-1,+1\}} \prod_{i | f_i \in \mathcal{S}_h} p(f'_i | w_i, b_h),$$

(6.161)

where we have exploited the fact that a given host asset is influenced only by the corresponding watermark component.

To go on, it is necessary that the embedding rule, the host feature statistics, and a proper attack model are defined. This will be the goal of next sections, where equation (6.161) will be specialized for various cases of theoretical and practical interest.

6.2.2 Binary signaling through AWGN channel

We start the analysis from the simplest case, namely the AWGN watermark channel. We remember that in the AWGN case, both host features and attack noise are assumed to be identical, independent, normally distributed random variables. Moreover, the watermark is simply added to the host features, and noise is added to the watermarked features. By using the same notation adopted in section 6.1.2 we express the host features under analysis as:

$$f'_i = f_i + n_i + \gamma w_i b_h,$$

(6.162)

where we focused on the h-bit of $\mathbf{b}$. Once again we introduce the new random variable x_i accounting for both f_i and n_i, i.e. [27]

$$x_i = f_i + n_i.$$

(6.163)

Decoder structure

Let us start by specializing equation (6.161) to the AWGN case. By neglecting the subscript h for simplicity, we have:

$$\hat{b} = \arg \max_{b \in \{-1,+1\}} \prod_{i=1}^{r} \frac{1}{\sqrt{2\pi\sigma_x^2}} \exp\left(-\frac{(f'_i - \mu_x - \gamma b w_i)^2}{2\sigma_x^2}\right),$$

(6.164)

where μ_x and σ_x^2 indicate the mean and variance of x respectively, and where we avoided to expressly indicate that only those i for which $f_i \in \mathcal{S}_h$ are included in the sum. By explicitly writing the above expression for

---

[27] We may also let $x_i = af_i + n_i$ where the coefficient $a \in [0, 1]$ accounts for host rejection at the embedder.

$b = 1$ and $b = -1$, and by passing to a logarithmic formulation, the ML criterion reduces to the following rule. If

$$\sum_{i=1}^{r} w_i(f_i' - \mu_x) > 0, \qquad (6.165)$$

then decide for $b = +1$, otherwise take $b = -1$. Before going on, some comments are in order. First we note that, as for the case of detection, the optimum decoding rule in the AWGN case relies on the correlation between the host features and the spreading sequence w_i. In this case, though, the mean value μ_x must be subtracted from f_i' before computing the correlation. Second, we note that decoding requires that the host feature mean μ_x is known. This may be a problem, since the exact statistic of $\mathbf{x}$ may not be known. A possible way to get around the problem is to choose the host feature set in such a way that $\mu_x = 0$. If such a solution is not viable, possibly because μ_x also depends on attacks, we may design w_i in such a way that

$$\sum_{i=1}^{r} w_i = 0, \qquad (6.166)$$

since in this way the term with μ_x is cancelled out. It is worth noting that if $\mu_w = 0$ the above condition is approximately true for large values of r. Nevertheless, if a strict equality must hold, the watermark has to be expressly designed to satisfy equation (6.166). As a last solution, if $\mu_w \neq 0$, then μ_x may be estimated directly on the asset at hand, since watermark insertion does not modify it.

Two cases of particular interest are obtained when $r = 1$ and when spreading is not applied, i.e. $w_i = 1$, $\forall i$. In the first case, the decoding rule reduces to comparing f_i' with μ_x, if $f_i' > \mu_x$ we conclude that $b = +1$, otherwise we let $b = -1$. In the second case, we simply have to sum all host coefficients together and look at the sign of the sum. If the sum is larger than $r\mu_x$ we have $b = +1$, otherwise we let $b = -1$. Finally, note that the use of the spreading sequence w_i adds security to the system, since, in this case, decoding is not possible without knowing w_i.

Error probability

We now calculate the error probability of the optimum decoder derived above. To do so, we note that the overall bit error probability is given by:

$$P_e = P(e|b = 1)P(b = 1) + P(e|b = -1)P(b = -1), \qquad (6.167)$$

where by $P(e|b = \pm 1)$ the probability of error conditioned to the sign of b is meant. By assuming that input bits are equiprobable, and by noting that for the symmetry of the problem we can argue that $P(e|b = 1) = P(e|b = -1)$, we obtain:

$$P_e = P(e|b = 1) = P(e|b = -1). \tag{6.168}$$

In order to calculate $P(e|b = 1)$ we introduce the observation variable ρ defined by[28]:

$$\rho = \frac{1}{r} \sum_{i=1}^{r} (f'_i - \mu_x) w_i. \tag{6.169}$$

The error probability, then, is given by:

$$P_e = P(\rho < 0|b = +1), \tag{6.170}$$

whose computation requires that the pdf of ρ is derived. We start by observing that, due to the normality of x_i's and hence of f''_i's, ρ follows a Gaussian pdf, whose mean and variance are given by:

$$\mu_{\rho|1} = E\left[\frac{1}{r} \sum_{i=1}^{r} (x_i + \gamma w_i - \mu_x) w_i\right] = \frac{\gamma}{r} \sum_{i=1}^{r} w_i^2 = \gamma \overline{w^2}, \tag{6.171}$$

$$\sigma_{\rho|1}^2 = \frac{1}{r^2} \sum_{i=1}^{r} w_i^2 \sigma_x^2 = \frac{\sigma_x^2 \overline{w^2}}{r} \tag{6.172}$$

where, according to the symbolism introduced in section 6.1.2, $\overline{w^2}$ is the sample mean square value of $\mathbf{w}$, calculated over the r features hosting the bit under analysis. Exploiting equation (6.171) and (6.172) to calculate P_e yields:

$$P_e = \frac{1}{2} \operatorname{erfc}\left(\sqrt{\frac{r\gamma^2 \overline{w^2}}{2\sigma_x^2}}\right). \tag{6.173}$$

Equation (6.173) gives the error probability for a particular watermark $\mathbf{w}$. If the overall performance of the decoder has to be calculated, such an equation must be averaged over the set $\mathbf{W}$ of admissible watermarks, nevertheless, if r is large enough, one may expect that $\overline{w^2}$ is approximately constant, thus making it possible to replace $\overline{w^2}$ with σ_w^2 [29]. By introducing the SNR parameter given by:

$$\mathrm{SNR} = \frac{\gamma^2 \sigma_w^2}{\sigma_x^2}, \tag{6.174}$$

---

[28]While not affecting decoder optimality, division by r will simplify the analysis later.

[29]It has to be observed that while for watermark detection $\overline{w^2}$ is averaged over the whole watermark length, in this case only $r = n/k$ samples are considered, thus increasing the dependence of $\overline{w^2}$ on $\mathbf{w}$.

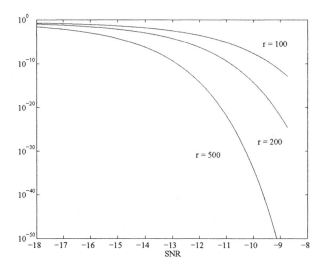

Figure 6.16: Error probability for an additive watermark immersed in white Gaussian noise.

the error probability can be given the more pleasant form:

$$\frac{1}{2} \operatorname{erfc}\left(\sqrt{\frac{r \, \mathrm{SNR}}{2}}\right). \tag{6.175}$$

Equation (6.173) reveals that in order to improve system performance, one may either increase the watermark strength γ or augment the number of host features assigned to each bit (thereby diminishing the payload). As to the dependence of P_e on σ_x^2, a common solution consists in estimating it directly on the asset at hand.

In figure 6.16, the error probability is plotted against SNR for several values of r.

Before leaving the AWGN channel, it is worth analyzing the decoder behavior when watermark spreading is not used, i.e. when $w_i = 1, \forall i$. In this case we have:

$$P_e = \frac{1}{2} \operatorname{erfc}\left(\sqrt{\frac{r\gamma^2}{2\sigma_x^2}}\right), \tag{6.176}$$

from which it is readily seen that no advantage in terms of error probability has to be expected from the adoption of a spread spectrum watermark. On the contrary, if $\overline{w^2} < 1$ then watermark spreading leads to poorer performance (this is not the case if $w_i \in \{-1, +1\}$). No need saying that watermark spreading may still be desirable to improve system secrecy, since in this way only authorized users are allowed to read the watermark.

6.2.3 Generalized Gaussian channel

As discussed in section 6.1.3, in some cases of practical interest the AWGN channel is not suitable to model the true pdf of host features. This is the case, for example, of block DCT coefficients used as host features in many image watermarking schemes. A more flexible model is provided by the Generalized Gaussian pdf, already introduced in section 6.1.3.

Watermark decoding in the presence of an additive Generalized Gaussian channel goes the same line followed for the AWGN case, the only difference being the pdf to be used in equation (6.161). More specifically, by letting β and c indicate the parameters of the Generalized Gaussian pdf used to model the host features, the analysis comes down to the following decision rule. If

$$\sum_{i=1}^{r} \beta^c (|f_i' + \gamma w_i|^c - |f_i' - \gamma w_i|^c) > 0, \qquad (6.177)$$

then take $b = 1$, otherwise let $b = -1$. Note that, as for 1-bit watermarking, the decision rule given above does not take attacks into account, i.e. we let $x_i = f_i$.

Computation of the bit error probability is analogous to the AWGN case, thus we leave it to the reader. It is only necessary to observe that fixing the watermark spreading sequence $\mathbf{w}$ and averaging over the host asset is too cumbersome, then P_e is usually calculated by fixing the host asset and averaging over $\mathbf{w}$. Moreover, in order to further simplify the analysis, w_i's are assumed to take value 1 or -1 with equal probability.

From a practical point of view, some implementation issues must be solved in order to apply the above analysis to real scenarios. These issues are similar to those encountered in the detectable case, hence similar solutions can be adopted, including estimation of pdf parameters on the asset at hand, and point elimination.

Of course, the effectiveness of the decoder derived under the Generalized Gaussian assumption depends on the accuracy with which this model fits the scenario at hand. Experimental results reported in the scientific literature show that a considerable improvement with respect to correlation based decoding are achieved whenever the host features exhibit a long-tailed behavior. This is the case, for example, of image watermarking schemes operating in the block-DCT domain, where the Generalized Gaussian model has been applied successfully.

6.2.4 Multiplicative watermarking with Gaussian noise

We now turn the attention to the decoding of a multiplicative watermark hosted by uncorrelated, normally distributed features, in the presence of additive white Gaussian noise. We also assume that both noise samples and host features have a zero mean, and that both the pdf of noise and host features do not depend on the index i, i.e. they are stationary processes.

As we did for the detection of a multiplicative watermark immersed in Gaussian noise, we do not take into account host rejection due to informed embedding, hence, for each hidden bit b, the received marked signal is written as:

$$f'_i = f_i + \gamma b w_i f_i + n_i = f_i(1 + \gamma b w_i) + n_i, \qquad (6.178)$$

with i ranging from 1 through r. Note that to simplify notation, we assumed that b is hidden in the first r samples of $\mathbf{f}$.

Decoder structure

The analysis still starts from equation (6.161). To specialize such a equation to the case at hand, we must calculate the pdf of f'_i under the assumption that $b = 1$ and $b = -1$. By remembering that both f_i's and n_i's are normally distributed, and that the watermark spreading sequence $\mathbf{w}$ is assumed to be known by the decoder, we have:

$$p(f'_i | w_i, b = 1) = \mathcal{N}(0, \sigma^2_{+,i}) \qquad (6.179)$$

$$p(f'_i | w_i, b = -1) = \mathcal{N}(0, \sigma^2_{-,i}), \qquad (6.180)$$

where we let

$$\sigma^2_{+,i} = \sigma^2_n + (1 + \gamma w_i)^2 \sigma^2_f, \qquad (6.181)$$

$$\sigma^2_{-,i} = \sigma^2_n + (1 - \gamma w_i)^2 \sigma^2_f. \qquad (6.182)$$

By substituting the above pdf into (6.161), and by passing to a logarithmic formulation, we obtain the following decision criterion:

$$\hat{b} = \text{sign} \left[\frac{1}{r} \sum_{i=1}^{r} \left(\frac{\sigma^2_{+,i} - \sigma^2_{-,i}}{\sigma^2_{+,i} \sigma^2_{-,i}} \right) (f'_i)^2 \ - \ \frac{1}{r} \sum_{i=1}^{r} \ln \frac{\sigma^2_{+,i}}{\sigma^2_{-,i}} \right], \qquad (6.183)$$

which can be rewritten in the following, more compact, form:

$$\hat{b} = \begin{cases} +1, & \text{if } \dfrac{1}{r} \sum_{i=1}^{r} k_i (f'_i)^2 > T_z \\[2mm] -1, & \text{otherwise} \end{cases} \qquad (6.184)$$

where we let:

$$k_i = \frac{\sigma_{+,i}^2 - \sigma_{-,i}^2}{\sigma_{+,i}^2 \sigma_{-,i}^2}, \tag{6.185}$$

and

$$T_z = \frac{1}{r} \sum_{i=1}^{r} \ln \frac{\sigma_{+,i}^2}{\sigma_{-,i}^2}. \tag{6.186}$$

An interesting simplification of the decoder is obtained when we assume that watermark spreading is not applied ($w_i = 1 \ \forall i$). In this case, $\sigma_{+,i}^2$ and $\sigma_{-,i}^2$ no longer depend on i, thus yielding:

$$\hat{b} = \begin{cases} +1, & \text{if } \dfrac{1}{r} \sum_{i=1}^{r} (f_i')^2 > \dfrac{\sigma_+^2 \sigma_-^2}{\sigma_+^2 - \sigma_-^2} \ln \dfrac{\sigma_+^2}{\sigma_-^2} \\ -1, & \text{otherwise} \end{cases} \tag{6.187}$$

where we let:

$$\sigma_+^2 = \sigma_n^2 + \sigma_f^2 (1 + \gamma)^2, \tag{6.188}$$

$$\sigma_-^2 = \sigma_n^2 + \sigma_f^2 (1 - \gamma)^2. \tag{6.189}$$

Error probability

We now calculate the error probability of the optimum decoder derived in the previous section. As usual P_e is calculated by averaging the error probabilities corresponding to the cases $b = 1$ and $b = -1$. By assuming that input bits are equiprobable, we have:

$$P_e = \frac{1}{2} \Big[P(e|b = 1) + P(e|b = -1) \Big]. \tag{6.190}$$

In order to calculate $P(e|b = 1)$ and $P(e|b = -1)$, the pdf of

$$z = \frac{1}{r} \sum_{i=1}^{r} k_i (f_i')^2 \tag{6.191}$$

conditioned to $b = 1$ and $b = -1$ must be computed. At this point we consider two cases. We first derive an exact expression for P_e by assuming that watermark spreading is not used, then we will consider the more complicated case in which the presence of w_i is taken into account. In this second case, the exact derivation of P_e is not possible, hence, we will resort to numerical simulations.

If $w_i = 1, \forall i$, we can can consider the simplified decision rule given in equation (6.187)). Then it is sufficient to derive the pdf of $z = 1/r \sum_{i=1}^{r} (f_i')^2$

with each f_i' normally distributed with zero mean and variance $\sigma_{\pm}^2$. It is at once evident that z follows a χ_r^2 distribution with r degrees of freedom. More specifically, if we let $\chi_r^2(z)$ be the central χ^2 distribution obtained by summing r independent standardized Gaussian random variables, we have:

$$
\begin{aligned}
p(z|b=1) &= \frac{r}{\sigma_+^2}\, \chi_r^2\left(\frac{rz}{\sigma_+^2}\right) = \\[2mm]
&= \frac{r}{2^{r/2}\Gamma(r/2)\sigma_+^2}\left(\frac{rz}{\sigma_+^2}\right)^{n/2-1} \exp\left(-\frac{rz}{2\sigma_+^2}\right) u(z),
\end{aligned}
\tag{6.192}
$$

$$
\begin{aligned}
p(z|b=-1) &= \frac{r}{\sigma_-^2}\, \chi_r^2\left(\frac{rz}{\sigma_-^2}\right) = \\[2mm]
&= \frac{r}{2^{r/2}\Gamma(r/2)\sigma_-^2}\left(\frac{rz}{\sigma_-^2}\right)^{n/2-1} \exp\left(-\frac{rz}{2\sigma_-^2}\right) u(z).
\end{aligned}
\tag{6.193}
$$

The error probability can now be easily calculated by considering the cumulative χ_r^2 distribution. In figure 6.17, the error probability of the optimum decoder in the absence of watermark spreading is given. More specifically, in figure 6.17a, the error probability in the absence of noise is plot for $\sigma_f^2 = 0.25$ and for $r = 100, 200, 300$. Figure 6.17b reports the error probability when noise is added $\sigma_n^2 = 0.1$. As it can be seen, in the presence of noise P_e gets considerably higher thus calling for higher values of r. Alternatively, channel coding may be used to improve system performance.

We now consider the more general case of a spread watermark. In this case, the general form of the sufficient statistic z must be considered (equation (6.191)). As it can be seen, we still have to consider the sum of the square values of r independent normally distributed random variables. However, the variances of these variables are different thus making it impossible to adopt the χ^2 analysis we used before. As usual, an approximate expression of the error probability can be obtained by resorting to the central limit theorem, so to conclude that z is approximately normally distributed. More specifically, by remembering that the pdf f_i' conditioned to $b = 1$ and $b = -1$, are given by equations (6.179) and (6.180), it is easy to demonstrate that

$$
p(z|b=-1) = \mathcal{N}\left(\overline{k\sigma_-^2},\, \frac{2}{r}\,\overline{k^2\sigma_-^4}\right),
\tag{6.194}
$$

$$
p(z|b=1) = \mathcal{N}\left(\overline{k\sigma_+^2},\, \frac{2}{r}\,\overline{k^2\sigma_+^4}\right),
\tag{6.195}
$$

with

$$
\overline{k\sigma_+^2} = \frac{1}{r}\sum_{i=1}^{r} k_i\sigma_{+,i}^2; \qquad \overline{k\sigma_-^2} = \frac{1}{r}\sum_{i=1}^{r} k_i\sigma_{-,i}^2,
\tag{6.196}
$$

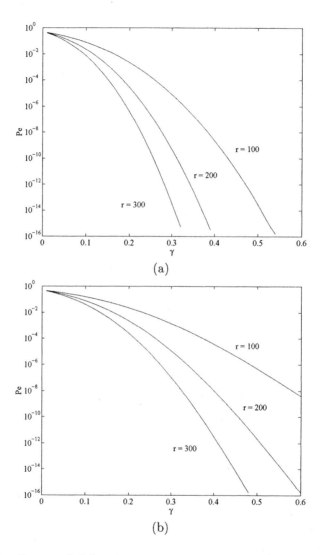

Figure 6.17: Error probability for a multiplicative watermark hosted by normal features when spreading is not used. Part (a) was derived by assuming that noise is not present, whereas in part (b) an additive white Gaussian noise with $\sigma_n^2 = 0.1$ was assumed. For both plots we let $\sigma_f^2 = 0.25$.

$$\overline{k^2\sigma_+^4} = \frac{1}{r}\sum_{i=1}^{r} k_i^2 \sigma_{+,i}^4; \quad \overline{k^2\sigma_-^4} = \frac{1}{r}\sum_{i=1}^{r} k_i^2 \sigma_{-,i}^4, \tag{6.197}$$

thus permitting us to conclude that

$$P_e = \frac{1}{4} \left[\text{erfc}\left(\sqrt{\frac{(T_z - \mu_{z|-1})^2}{2\sigma_{z|-1}^2}} \right) + \text{erfc}\left(\sqrt{\frac{(\mu_{z|1} - T_z)^2}{2\sigma_{z|1}^2}} \right) \right], \qquad (6.198)$$

where $\mu_{z|-1}$, $\mu_{z|1}$ and $\sigma_{z|-1}^2$, $\sigma_{z|1}^2$ are, respectively, the mean and variance of z conditioned to $b = -1$ and $b = 1$, as appearing in equations (6.194) and (6.195), and where T_z is given by equation (6.186).

As we already noted, the use of the central limit theorem introduces an approximation error that may be a significant one for low error probabilities. This is especially true in this case, and with multibit watermarking in general, since the sum leading to z does not contain so many terms as in the case of 1-bit watermarking. Typical values of r, in fact, are in the order of a hundred, as opposed to detectable watermarking where common values of n are in the order of several thousands. It is, then, of fundamental importance to develop alternative methods to evaluate the performance of a readable scheme. A practical possibility consists in the use of Montecarlo simulations[30]

In figure 6.18, the error probability obtained through Montecarlo simulations is given. The plot has been obtained by letting $\sigma_f^2 = 0.25$ and $\sigma_n^2 = 0.1$, for $r = 200$. In the same diagram the error probability in the absence of spreading is also shown. The plot consider only rather high bit error rates, to keep the computational burden of the simulations reasonably small. It is possible to observe how watermark spreading does not impact system performance significantly.

Implementation issues

With regard to the practical implementation of the decoder described in this section, it is worth pointing out that the separate knowledge of σ_f^2 and σ_n^2 is required. It is necessary, then, that suitable estimation procedures are adopted by the decoder. Alternatively, a suboptimum decoder may be adopted, by estimating σ_f^2 on the asset at hand, and by neglecting the presence of noise. It is also worth remembering that our analysis relies on the assumption that σ_f^2 and σ_n^2 do not depend on i. If this is not the case, a solution similar to that described in section 6.1.7 can be adopted, i.e. the feature vector hosting the watermark is split into a number of subregions within which σ_f^2 and σ_n^2 can be assumed to be constant. The derivation of the optimum decoder structure in this case follows the same guidelines described so far, and it is left to the reader.

---

[30]Note that as opposed to watermark detection in this case the estimation of the bit error rate is not needed to set the decoding threshold T_z, hence computational complexity is a less important concern.

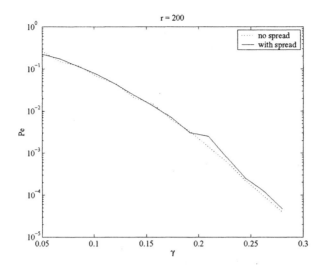

Figure 6.18: Error probability for an antipodal multiplicative watermark hosted by normal features in the presence of spreading (solid line), for $\sigma_n^2 = 0.1$, $\sigma_f^2 = 0.25$, and $r = 200$. The bit error rate in the absence of spreading is also given (dotted line). Both curves have been obtained through Montecarlo simulations.

6.2.5 Multiplicative watermarking of Weibull-distributed features

The last case we consider, before passing to the analysis of quantization-based schemes, regards multiplicative watermarking of Weibull-distributed features, a case typically corresponding to multiplicative image watermarking in the magnitude-of-DFT domain. The analysis is similar to that carried out in section 6.1.7 for the case of detectable watermarking.

Noticeably, in order to make the analysis mathematically feasible, we assume noise is not present, i.e. we let:

$$f_i' = f_i(1 + \gamma b w_i), \quad i = 1, 2 \ldots r \tag{6.199}$$

where, for simplicity, we focus on a generic bit b hidden within r consecutive host features.

Decoder structure

Let us observe that being f_i's distributed according to a Weibull pdf (see equation (6.114)), we have:

$$p(f_i'|\, b = 1) = \frac{\beta}{\alpha(1 + \gamma w_i)} \left(\frac{f_i'}{\alpha(1 + \gamma w_i)} \right)^{\beta - 1} \exp\left[-\left(\frac{f_i'}{\alpha(1 + \gamma w_i)} \right)^{\beta} \right],$$
(6.200)

and

$$p(f_i'|\, b = -1) = \frac{\beta}{\alpha(1 - \gamma w_i)} \left(\frac{f_i'}{\alpha(1 - \gamma w_i)} \right)^{\beta - 1} \exp\left[-\left(\frac{f_i'}{\alpha(1 - \gamma w_i)} \right)^{\beta} \right].$$
(6.201)

By inserting the above expressions into (6.161), and by adopting a logarithmic formulation, we obtain the following decision rule:

$$\hat{b} = \text{sign} \left[\sum_{i=1}^{r} \frac{(1 + \gamma w_i)^{\beta} - (1 - \gamma w_i)^{\beta}}{\alpha^{\beta}(1 + \gamma w_i)^{\beta}(1 - \gamma w_i)^{\beta}} (f_i')^{\beta} + \sum_{i=1}^{r} \beta \ln \frac{1 - \gamma w_i}{1 + \gamma w_i} \right].$$
(6.202)

which can be expressed in the following compact form:

$$\hat{b} = \begin{cases} +1, & \text{if } \sum_{i=1}^{r} v_i (f_i')^{\beta} > T_z \\ -1, & \text{otherwise} \end{cases}$$
(6.203)

where we let:

$$v_i = \frac{(1 + \gamma w_i)^{\beta} - (1 - \gamma w_i)^{\beta}}{\alpha^{\beta}(1 + \gamma w_i)^{\beta}(1 - \gamma w_i)^{\beta}},$$
(6.204)

and

$$T_z = \sum_{i=1}^{r} \beta \ln \frac{1 + \gamma w_i}{1 - \gamma w_i}.$$
(6.205)

Error probability

By relying on the optimum decoder structure derived so far, we now evaluate the error probability for the multiplicative Weibull channel. The analysis is similar to the one carried out for watermark detection on a multiplicative Weibull channel 6.1.7, thus we will only give the final results and some guidelines on how they can be obtained. More specifically, we start by observing that each term z_i of the sum forming z ($z = \sum_i v_i (f_i')^{\beta}$) follows an exponential distribution:

$$p(z_i|b) = |\lambda_i| \exp(-\lambda_i z_i) u(z),$$
(6.206)

with

$$\lambda_i = \frac{(1 - \gamma w_i)^\beta}{(1 + \gamma w_i)^\beta - (1 - \gamma w_i)^\beta}, \tag{6.207}$$

if $b = 1$, and

$$\lambda_i = \frac{(1 + \gamma w_i)^\beta}{(1 + \gamma w_i)^\beta - (1 - \gamma w_i)^\beta}, \tag{6.208}$$

if $b = -1$. Computation of $p(z|b)$ can now follow two distinct paths. The first one refers to the case in which watermark spreading is not applied, hence permitting to conclude that z_i's are identically distributed random variables. More specifically, for each i we have:

$$\lambda_i = \lambda = \frac{(1 \mp \gamma)^\beta}{(1 + \gamma)^\beta - (1 - \gamma)^\beta}, \tag{6.209}$$

where the sign at the numerator depends on the sign of b. By noting that the exponential pdf with $\lambda = 1/2$ corresponds to a χ_2^2 distribution with two degrees of freedom, we easily see that z follows a non-central χ_{2r}^2 pdf with $2r$ degrees of freedom, i.e.:

$$p(z|b) = 2\lambda \chi_{2r}^2(2\lambda z) = \frac{2\lambda}{2^r \Gamma(r)} (2\lambda z)^{r-1} \exp(-\lambda z) u(z). \tag{6.210}$$

Equation (6.210) can be used to calculate the bit error probability conditioned to $b = 1$ and $b = -1$ and hence derive the overall P_e of the decoder. In figure 6.19 the error probability resulting from equation (6.210) is plotted as a function of γ, for $\sigma_f^2 = 0.25$ (remember that throughout our analysis we assumed attack noise is not present).

When the watermark is multiplied by a spread spectrum sequence $\mathbf{w}$ prior to insertion, the analysis gets more complicated, however a closed form expression giving the decoder P_e can still be obtained. Once again we are led to an analysis which closely resembles the detection case. As a matter of fact, the pdf and cdf of z still have the form given in equations (6.145) through (6.148), with the only difference that now λ_i's are given by equations (6.207) and (6.208). Though the resulting expression may be rather complicated, the above equations permit to calculate the overall bit error probability of the decoder exactly. A problem with the exact analysis is that evaluation of the error probability requires that ad hoc numerical programs are used since the usual 64 bits double precision arithmetic is not enough accurate. Alternatively, Montecarlo simulations may be used. However, even in this case, the estimation of very low error probabilities is problematic due to computational complexity.

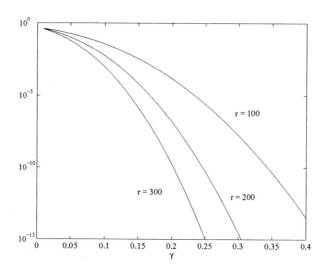

Figure 6.19: Error probability for a non-spread, multiplicative watermark hosted by Weibull-distributed features ($\sigma_f^2 = 0.25$). Results are obtained in the absence of noise.

Practical implementation of the decoder

Implementation problems of the optimum decoder for the Weibull channel are similar to those discussed in the Gaussian case. A major difference between the two cases is that for the Weibull-channel decoder the knowledge of σ_n^2 is not necessary. This is not surprising since we assumed the channel to be noise-free. A similar simplification could be applied to the Gaussian case by letting $\sigma_n^2 = 0$.

6.2.6 Quantization Index Modulation

In quantization-based schemes watermark decoding assumes a rather different meaning, because the host features are no more looked at as disturbing noise. This is a direct consequence of the informed embedding paradigm such techniques rely on. The most important consequence of the above property is that, in the absence of attacks, the error probability is equal to zero, which constitutes a dramatic improvement with respect to SS-based algorithms. Such an advantage is gradually lost when attack noise gets stronger, nevertheless, at least for scenarios where the necessity of a high payload is coupled with the presence of weak attacks, quantization-based techniques, significantly outperform SS methods. In this paragraph, we give a quantitative analysis of the achievable bit error rate for two of the simplest QIM schemes, namely DM with bit repetition and ST-DM (see section 4.3.2).

Dither Modulation watermarking

We start by analyzing DM watermarking without bit repetition. First of all, we have to derive the optimum decoder structure. This is a meaningful problem only if attack noise is taken into account, if attacks are not present, in fact, no source of uncertainty exists and decoding is a straightforward operation. In the sequel, we will assume that attacks take the form of additive white Gaussian noise.

Derivation of the optimum decoding rule still relies on equation (6.159). Given that we are analyzing DM without bit repetition, and under the usual assumption that host features are independent of each other, we can adopt a bitwise decoding scheme, i.e.:

$$\hat{b} = \arg \max_{b \in \{0,+1\}} p(f_i'|b), \qquad (6.211)$$

where, $f_i' = f_{w,i} + n_i$, and $f_{w,i}$ is the watermarked host feature obtained as described at the end of section 4.3.2. In order to evaluate $p(f_i'|b)$, we start by assuming that $b = 0$, then we introduce the probability that the host feature f_i is mapped into the codebook entry $u_{0,k}$. Let such a probability be $p(u_{0,k})$. The pdf of f_w prior to noise addition is[31]:

$$p(f_w|0) = \sum_k p(f_w|u_{0,k})p(u_{0,k}) = \sum_k \delta(f_w - u_{0,k})p(u_{0,k}), \qquad (6.212)$$

where the sum extends to all the codebook entries.

When noise is considered, the resulting pdf is obtained by convolving $p(f_w|0)$ and the normal pdf characterizing noise, leading to:

$$p(f'|0) = \sum_k \frac{1}{\sqrt{2\pi\sigma_n^2}} \exp\left(-\frac{(f' - u_{0,k})^2}{2\sigma_n^2}\right) p(u_{0,k}), \qquad (6.213)$$

where σ_n^2 is the variance of noise. By assuming $b = 1$ a similar result is obtained with the only difference that the entries of codebook $\mathcal{U}_1$ must be considered. The two pdf's conditioned to $b = 0$ and $b = 1$ are depicted in figure 6.20.

In order to decide whether $p(f'|0)$ is larger than $p(f'|1)$ or not, the following simplification is made: we assume that $p(u_{0,k}) = p(u_{1,k}) = 1/|\mathcal{U}_0| = 1/|\mathcal{U}_1|$ do not depend on k, a condition which is true only if the host features

---

[31]We neglect the subscript i for simplicity.

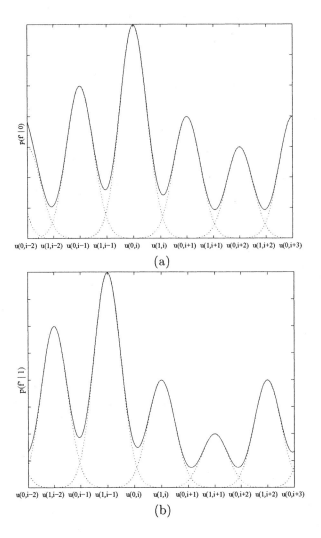

Figure 6.20: Probability density function of f' conditioned to $b = 0$ (a) and $b = 1$ (b).

follow a uniform pdf. Usually, this is not the case, however, by assuming that $p(f)$ varies smoothly with f, $p(u_{0,k})$ and $p(u_{1,k})$ are locally constant. Then, at least locally, the pdf of f' has the form reported in figure 6.21, and optimum decoding reduces to minimum distance decoding. More specifically, we have:

$$\hat{b} = \arg\min_{b=0,1} \left(\min_{u_{b,k} \in \mathcal{U}_b} |f' - u_{b,k}| \right), \qquad (6.214)$$

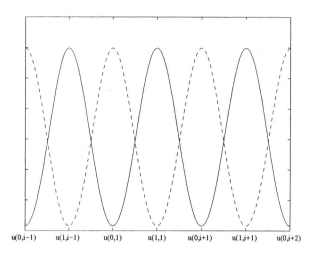

u(0,i−1) u(1,i−1) u(0,1) u(1,1) u(0,i+1) u(1,i+1) u(0,i+2)

Figure 6.21: Probability density function of f' conditioned to $b = 0$ (dashed line) and $b = 1$ (solid line) for equiprobable codebook entries.

and decoding regions can be easily achieved. For example, if codebook entries are defined as in section 4.3.2, i.e.[32]:

$$\mathcal{U}_0 = \{k\Delta, k \in \mathbb{Z}\}, \tag{6.215}$$

$$\mathcal{U}_1 = \{k\Delta + \Delta/2, k \in \mathbb{Z}\}, \tag{6.216}$$

we obtain (see figure 6.22):

$$R_0 = \bigcup_k \left[(4k - 1)\frac{\Delta}{4}, (4k + 1)\frac{\Delta}{4}\right], \tag{6.217}$$

$$R_1 = \bigcup_k \left[(4k + 1)\frac{\Delta}{4}, (4k + 3)\frac{\Delta}{4}\right]. \tag{6.218}$$

To calculate the bit error probability, we note that for the symmetry of the problem we have $P_e = P(e|1) = P(e|0)$. Eventually, $P(e|0)$ can be calculated as the probability that f' belongs to R_1 under the assumption that quantizer $\mathcal{Q}_0$ was used, leading to:

$$P_e = \sum_{k=0}^{+\infty} \left[\text{erfc}\left(\frac{(4k + 1)\Delta/4}{\sqrt{2\sigma_n^2}}\right) - \text{erfc}\left(\frac{(4k + 3)\Delta/4}{\sqrt{2\sigma_n^2}}\right)\right]. \tag{6.219}$$

---

[32]We assumed that $d = 0$ for simplicity, since translating the codebook by a fixed amount does not change the bit error probability.

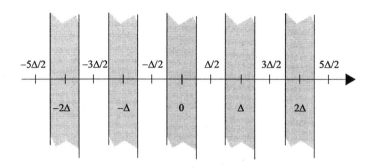

Figure 6.22: Decoding regions for scalar DM watermarking. Grey areas correspond to R_0 and white areas to R_1.

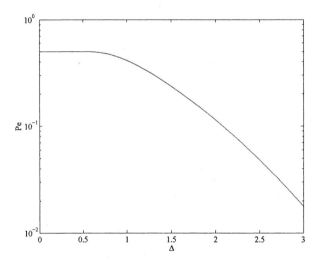

Figure 6.23: Error probability for scalar DM watermarking. Results have been
obtained by letting $\sigma_n^2 = 0.1$.

In figure 6.23, the bit error probability for DM watermarking is given for
Δ ranging from 0.1 to 3, and $\sigma_n^2 = 0.1$. We can observe how the bit error
probability does not depend on the host feature statistic, hence reaching
low values of P_e even with no repetition[33].

Interestingly, the bit error probability can be computed also for different
types of noise (even if the decoder structure may be no longer optimal).
For example, if the attack consists in the addition of uniformly distributed

---

[33]It should be observed that the results in the right part of the figure correspond to
very high Watermark to Noise Ratios, i.e. weak attack.

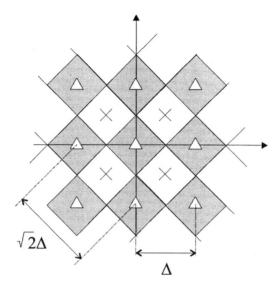

Figure 6.24: Decoding regions for DM with $r = 2$.

noise taking value in $[-\eta/2, \eta/2]$, the bit error probability is given by:

$$P_e = \begin{cases} 0 & \eta < \Delta/2 \\ 1 - \Delta/2\eta & \Delta/2 \leq \eta \leq 3\Delta/2 \end{cases} \tag{6.220}$$

By comparing the above equation and the error probability in the presence of Gaussian noise given in figure 6.23, it is readily seen that uniform noise addition turns out to be a more effective attack against DM watermarking than Gaussian noise addition.

Dither modulation with bit repetition

We now consider DM with bit repetition, in which the same bit is embedded in r host features. With regard to the optimum decoder structure, due to the independence of host features and noise samples, it can be easily demonstrated that optimum decoding is still achieved by a minimum distance strategy[34]. The geometric representation of decision region is now more complicated. For the simple case $r = 2$, such regions assume the form reported in figure 6.24.

The exact computation of bit error probability in the presence of Gaussian noise is now very complicated. Instead we will give an upper bound

---

[34]We assume soft sequence decoding is used.

of such a probability. Let us assume that $b = 0$ and that noise is added to the codebook entry $u_{0,0} = (0,0)$ in figure 6.24. If $\mathbf{f}'$ falls inside the gray sub-region centered in $(0,0)$, then b is correctly decoded. Conversely, if $\mathbf{f}$ falls outside such an area, then a decoding error is likely to occur. We will upper bound the error probability by assuming that in this second case an error is always made. We then have[35]:

$$P_e = 1 - P_c \leq 1 - P\left(f_1 \in \left[-\frac{\Delta}{2\sqrt{2}}, \frac{\Delta}{2\sqrt{2}}\right]\right) \cdot P\left(f_2 \in \left[-\frac{\Delta}{2\sqrt{2}}, \frac{\Delta}{2\sqrt{2}}\right]\right) =$$

$$= 1 - \left(1 - \mathrm{erfc}\left(\sqrt{\frac{\Delta^2}{16\sigma_n^2}}\right)\right)^2.$$

$$(6.221)$$

The generalization of the above analysis to the case $r > 2$ is not straightforward, since the shape of the detection regions becomes very complicated. For example, for $r = 3$ the set of points for which $\|\mathbf{f} - \mathbf{u}_{0,i}\| \leq \|\mathbf{f} - \mathbf{u}_{1,j}\| \; \forall j$ is a truncated octahedron. However, an exact analysis of the bit error probability for DM with bit repetition is still possible. The resulting error probability is plotted in figure 6.25 for $\sigma_n^2 = 0.1$ and different values of Δ. As it can be seen the improvement with respect to the scalar case is evident. For further details on the computation of the exact bit error probability for DM with bit repetition, readers are referred to the further reading section at the end of the chapter.

Spread-Transform Dither modulation

In section 4.3.2 we anticipated that the performance of DM with bit repetition can be improved through a different way of exploiting the availability of r features for each bit, namely through ST-DM. We recall that in ST-DM the watermarked feature vector $\mathbf{f}_w$ is calculated as:

$$\mathbf{f}_w = \mathbf{f} - \rho_f \mathbf{w} + \mathcal{Q}_{0/1}(\rho_f)\mathbf{w}, \qquad (6.222)$$

where $\mathbf{w}$ is a unitary norm, spreading sequence assuming values $\pm 1/\sqrt{r}$, ρ_f is the projection of $\mathbf{f}$ over $\mathbf{w}$ and $\mathcal{Q}_{0/1}(\rho_f)$ is the quantized correlation value conveying the hidden bit. More specifically, the admissible values for ρ_w are obtained by quantizing ρ by means of two quantizers $\mathcal{Q}_0$ and $\mathcal{Q}_1$ corresponding to $b = 0$ and $b = 1$ respectively. Such quantizers are built as for scalar DM (see figure 4.27). In order to keep the distortion along each component of the host feature vector below Δ, the quantization step for ρ is set to $\Delta\sqrt{r}$.

---

[35]Here features f_1 and f_2 are obtained by rotating the axis in figure 6.24 by $\pi/4$.

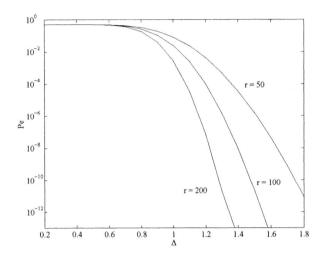

Figure 6.25: Bit error probability for DM with bit repetition, for various values of r ($\sigma_n^2 = 0.1$).

Watermark decoding is straightforward, it only needs to project $\mathbf{f}'$ over $\mathbf{w}$ and apply the minimum distance scalar DM decoder to such a projection. By assuming that $\mathbf{f}_w$ is corrupted by additive white Gaussian noise, we have:

$$\mathbf{f}' \cdot \mathbf{w} = \rho_w + \mathbf{n} \cdot \mathbf{w} = \rho_w + \nu, \tag{6.223}$$

where ν is still normally distributed with zero mean and variance:

$$\sigma_\nu^2 = \sum_{i=1}^{r} w_i n_i = \sum_{i=1}^{r} w_i^2 \sigma_n^2 = \sigma_n^2. \tag{6.224}$$

The error probability, then, can be calculated by applying equation (6.219), yielding

$$P_e = \sum_{k=0}^{+\infty} \left[\mathrm{erfc}\left(\frac{(4k+1)\sqrt{r}\Delta/4}{\sqrt{2\sigma_n^2}} \right) - \mathrm{erfc}\left(\frac{(4k+3)\sqrt{r}\Delta/4}{\sqrt{2\sigma_n^2}} \right) \right]. \tag{6.225}$$

A plot of the bit error rate of ST-DM is given in figure 6.26 for various values of r. The improvement with respect to the bit repetition strategy is evident. Basically, such an improvement may be explained by observing that whereas in DM with repetition every noise component impairs the watermark, with ST-DM those components orthogonal to $\mathbf{w}$ do not have any negative impact on the decoder.

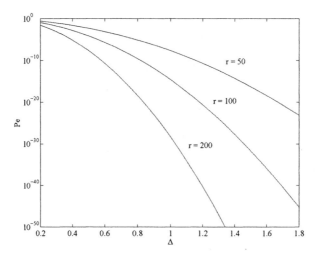

Figure 6.26: Bit error probability of ST DM watermarking for various values of r $(\sigma_n^2 = 0.1)$. The improvement with respect to DM with bit repetition is evident.

6.2.7   Decoding in the presence of channel coding

In general, the performance of a data hiding system can be improved through the use of channel coding. The benefits achievable in this way include lower bit error probability, and hence higher robustness, power saving, i.e. reduced watermark obtrusiveness, increased payload.

In section 3.4, we already described how channel coding may be used to code the watermark message prior to its insertion within the host asset. More specifically, we briefly outlined the main coding typologies that can be encountered in digital watermarking and discussed their advantages and drawbacks. Our review included conventional block and convolutional codes, as well as a number of data-hiding-specific techniques, among which a prominent role is played by informed coding, exemplified in section 3.4.5 by dirty-paper trellis coding.

We now take a look at the decoder side, by considering watermark reading in the presence of channel coding. From a very general perspective, two distinct approaches are possible: hard and soft decoding. With hard decoding each bit is decoded separately, then channel decoding is applied to the extracted bit sequence, possibly correcting, or revealing, decoding errors. In this case, no particular difference exists between traditional communication systems and data hiding, since both the communication and the watermark channel are looked at as binary channels characterized by their own bit error probability. It does not matter, then, if the underlying system operates in the asset rather than in the DFT domain, or whether an additive or a multiplicative embedding rule is adopted. The only parameter of interest is the bit error rate achieved by the system in the absence of coding. Noticeably, the decoder structure does not change due to the presence of channel coding, since the hidden information is retrieved on a bit by bit basis. As to the performance improvement brought by channel coding, it depends on the error correcting/detection capabilities of the code in a way that is in all similar to what happens in digital communication systems.

When soft decoding is adopted the situation is rather different. In this case, in fact, sequence decoding is used to take into account the relationship between coded bits. We start the analysis from block codes. To this aim, let $\mathbf{c}$ be the n-bit long codeword associated to a block of k bits of $\mathbf{b}$. For the sake of clarity, we will assume that $\mathbf{c}$ is an antipodal sequence, i.e. $c_i = \pm 1$. Then, let us focus on a single codeword and let $\mathcal{S}_j$ be the set of host features associated to the j-th bit in $\mathbf{c}$. Optimum ML decoding of $\mathbf{c}$ amounts to choosing $\hat{\mathbf{b}}$ according to the rule:

$$\hat{\mathbf{b}} = \arg\max_{\mathbf{c}_l \in \mathcal{C}} \prod_{j=1}^{n} \prod_{i \in \mathcal{S}_j} p(f_i'|c_{l,j}), \qquad (6.226)$$

where $c_{l,j}$ indicates the j-th bit of $\mathbf{c}_l$, and $\mathcal{C}$ is the set of admissible codewords. By passing to a logarithmic formulation, equation (6.226) comes down to:

$$\hat{\mathbf{b}} = \arg\max_{\mathbf{c}_l \in \mathcal{C}} \sum_{j=1}^{n} \sum_{i \in \mathcal{S}_j} \ln(p(f_i'|c_{l,j})). \qquad (6.227)$$

By following the same analysis carried out in the previous sections to derive the optimum decoder structure, it is easy to show that ML decoding corresponds to decide for the codeword $\hat{\mathbf{c}}$ such that:

$$\hat{\mathbf{c}} = \arg\max_{\mathbf{c}_l \in \mathcal{C}} \sum_{j=1}^{n} c_{l,j} r_j, \qquad (6.228)$$

where r_j is the sufficient statistic for the j-th bit of $\mathbf{c}$. More specifically, we have:

$$r_j = \sum_{i \in \mathcal{S}_j} w_i f_i', \qquad (6.229)$$

for the AWGN case;

$$r_j = \sum_{i \in \mathcal{S}_j} \beta^c (|f_i' + \gamma w_i|^c - |f_i' - \gamma w_i|^c), \qquad (6.230)$$

for the additive, Generalized Gaussian, case;

$$r_j = \sum_{i \in S_j} \left[\left(\frac{\sigma_{+,i}^2 - \sigma_{-,i}^2}{\sigma_{+,i}^2 \sigma_{-,i}^2} \right) (f_i')^2 - \ln \frac{\sigma_{+,i}^2}{\sigma_{-,i}^2} \right], \qquad (6.231)$$

for the multiplicative Gaussian case, and

$$r_j = \sum_{i \in S_j} \left[\frac{(1 + \gamma w_i)^\beta - (1 - \gamma w_i)^\beta}{\alpha^\beta (1 + \gamma w_i)^\beta (1 - \gamma w_i)^\beta} (f_i')^\beta + \beta \ln \frac{1 - \gamma w_i}{1 + \gamma w_i} \right], \qquad (6.232)$$

for multiplicative watermarking of Weibull-distributed features. Note that, while theoretically optimal, the above decoding strategies may not be feasible for the number of admissible codewords in $\mathcal{C}$ may be extremely large.

With reference to QIM schemes, the use of soft decoding is complicated by the fact that, prior to coding, each bit is associated to an entire codebook $\mathcal{U}$. For this reason, most of the systems proposed so far relies on hard decoding.

Convolutional codes are known to achieve better results than block codes, hence they are extensively used in data hiding systems. Even in this case, both hard or soft decoding may be used. When hard decoding is adopted, the peculiarities of data hiding applications need not be taken into account, since the single bit error probability is sufficient to characterize the watermark channel. One of the main advantages of convolutional codes over block codes is the availability of efficient soft decoding algorithms, based on the well-known Viterbi algorithm. Such an algorithm, which can be applied for any channel with i.i.d. noise, requires that a suitable metric is defined to *measure* the likelihood of all possible transmitted codewords. In digital watermarking applications, such a metric can be easily obtained by following the ML arguments used so far. More specifically, by focusing on the decoding of the h-th bit [36], the metric $m_{h,l}$ associated to the l-th branch is computed as[37]:

$$m_{h,l} = \ln p(\mathbf{f}_h' | \mathbf{c}_l), \qquad (6.233)$$

where $\mathbf{f}_h'$ is the vector with the features hosting the h-th bit, and $\mathbf{c}_l$ indicates the coded bits associated to the l-th branch of the Trellis.

Finally, it is worth remembering that due to the particular nature of the watermarking problem, informed coding techniques may be conveniently used to improve system performance. A systematic analysis of watermark

---

[36]We assume for simplicity that a convolutional code encoding one bit at a time is used.

[37]Note that since we are using a metric directly related to the likelihood of each branch, Viterbi's algorithm will try to maximize such a metric instead of minimizing it.

detection in the presence of informed coding is outside the scope of this book, all the more that research in this field is still on going, and new results are likely to appear in the years to come. Interested readers may refer to the simple examples we gave in section 3.4.5, to the general analysis carried out in chapter 9 and to the additional references given at the end of this chapter.

6.2.8 Assessment of watermark presence

Decoding of a readable watermark always results in a decoded bit stream, however, if the asset at hand is not marked, decoded bits are meaningless. Even with readable watermarking, then, it is desirable to have the possibility of assessing whether a given asset is watermarked or not. This possibility assumes an even greater importance when decoding is accomplished by looking for the watermark exhaustively by changing the geometrical configuration of the host asset. This is rather a common approach to cope with geometrical manipulations, e.g. rotations, where a rough estimate of the original geometrical configuration is first obtained, then the watermark is searched exhaustively in a neighborhood of the recovered geometry.

In the following, we briefly review the strategies that can be adopted to verify whether the host asset contains a watermark or not.

Optimum detection

As with detectable watermarking, the assessment of watermark presence can be formulated as a binary hypothesis test, with H_1 corresponding to the presence within the host asset of any message watermarked by using a given key K. The key may correspond to the spreading sequence $\mathbf{w}$ in spread spectrum watermarking, or to the particular codebooks used by a QIM scheme, or to any information distinguishing the watermarks belonging to a particular user, or class of users, from the others. Hypothesis H_0 amounts to saying that a watermark generated with K is not present in A[38]. With respect to detectable watermarking the situation is now complicated by the necessity of averaging $p(\mathbf{f}'|H_1)$ over the set of possible messages $\mathbf{B}$.

To go on, some assumptions must be made on the watermarking algorithm. For sake of brevity, we will not give a detailed analysis covering the most common additive watermarking algorithms, on the contrary, we will only consider the case of spread spectrum watermarking in AWGN noise, for its exemplificative value. In this case, the spreading sequence $\mathbf{w}$ plays

---

[38] H_0 also accounts for the possibility that A contains a watermark generated with a key K' other than K.

the role of K, and hence we can write:

$$\ell(\mathbf{f}') = \frac{p(\mathbf{f}'|H_1)}{p(\mathbf{f}'|H_0)} = \frac{p(\mathbf{f}'|\mathbf{w})}{p(\mathbf{f}'|H_0)}. \tag{6.234}$$

Let us focus on the numerator of $\ell(\mathbf{f}')$. We have:

$$p(\mathbf{f}'|\mathbf{w}) = \prod_{j=1}^{k} \sum_{b_j} p(\mathbf{f}'_j|\mathbf{w}_j, b_j)p(b_j), \tag{6.235}$$

where $\mathbf{f}'_j$ is the vector with the features hosting the j-th bit of $\mathbf{b}$ (i.e. S_j), $\mathbf{w}_j$ is the corresponding subset of coefficients in $\mathbf{w}$, and b_j is the j-th bit of the hidden message sequence. We now assume that hidden bits take value ± 1 with equal probability, thus permitting us to write:

$$p(\mathbf{f}'|\mathbf{w}) = \prod_{j=1}^{k} \frac{p(\mathbf{f}'_j|\mathbf{w}_j, 1) + p(\mathbf{f}'_j|\mathbf{w}_j, -1)}{2}. \tag{6.236}$$

In the AWGN case f'_i is given by:

$$f'_i = f_i + \gamma b_j w_i + n_i, \tag{6.237}$$

where the same bit b_j is used for several (say r) indexes i, and where both n_i and f_i are i.i.d. normally distributed random variables with $\sigma_x^2 = \sigma_n^2 + \sigma_f^2$. We have:

$$\ell(\mathbf{f}') = \frac{1}{2^k} \prod_{j=1}^{k} \left[\prod_{i \in S_j} e^{\frac{(f'_i)^2 - (f'_i - \gamma w_i)^2}{2\sigma_x^2}} + \prod_{i \in S_j} e^{\frac{(f'_i)^2 - (f'_i + \gamma w_i)^2}{2\sigma_x^2}} \right]. \tag{6.238}$$

By further assuming that $w_i = \pm 1$, we obtain:

$$\ell(\mathbf{f}') = \frac{1}{2^k} \prod_{j=1}^{k} e^{-\frac{r\gamma^2}{2\sigma_x^2}} \left[\prod_{i \in S_j} e^{\frac{2\gamma w_i f'_i}{2\sigma_x^2}} + \prod_{i \in S_j} e^{-\frac{2\gamma w_i f'_i}{2\sigma_x^2}} \right]. \tag{6.239}$$

A logarithmic formulation may be, equivalently, adopted, yielding:

$$\mathcal{L}(\mathbf{f}') = \sum_{j=1}^{k} \left(-\frac{r\gamma^2}{2\sigma_x^2} - \ln 2 + \ln \left[\prod_{i \in S_j} e^{\frac{2\gamma w_i f'_i}{2\sigma_x^2}} + \prod_{i \in S_j} e^{-\frac{2\gamma w_i f'_i}{2\sigma_x^2}} \right] \right). \tag{6.240}$$

The above expression determines the structure of the optimum detector for an additive readable watermark hosted by Gaussian features immersed in additive Gaussian noise, however, to completely specify the detector the detection threshold Λ must be specified. To this aim, we can resort to the Neyman-Pearson criterion thus defining Λ on the basis of the target false detection probability. This, in turn, requires that the statistics of $\mathcal{L}(\mathbf{f}')$ (or $\ell(\mathbf{f}')$) are derived. Unfortunately, this is a very complicated task, due to the complicated form of $\mathcal{L}(\mathbf{f}')$. A possibility consists in invoking the central limit theorem to assess that $\mathcal{L}(\mathbf{f}')$ is normally distributed. However, due to the form of the terms of the sum in (6.240), and to the rather small number of terms such a sum consists of[39], the central limit theorem should be applied with care, by keeping in mind that the actual false detection probability may deviate significantly from that predicted through the normality assumption. As an alternative Montecarlo simulations may be used to better estimate the actual error rate, especially if error rates are not too small.

A further difficulty with optimum watermark detection, is that it is very difficult to account for the use of channel coding. In this case, in fact, the numerator of $\ell(\mathbf{f}')$ should be averaged over valid codewords only, thus making the analytic analysis very cumbersome.

Verification after decoding

In many cases a suboptimal approach may be conveniently used. First, ML decoding is applied to get an estimate $\hat{\mathbf{b}}$ of the embedded message $\mathbf{b}$. Then, a ML hypothesis test is applied to verify whereas the host asset contains $\hat{\mathbf{b}}$ or not. In this way, the assessment of watermark presence is considerably simplified, however, we must keep in mind that the actual false detection probability may be higher than that set by the Neyman Pearson criterion, since each time we are looking for the most likely message. In some cases, analytical tools are available to get a more accurate bound on the actual P_f, however we will not detail them here. Additional bibliographical sources are given in the further reading section at the end of this chapter.

ML estimation of $\mathbf{b}$ in the presence of block channel coding may be a difficult task, thus the above approach may be further simplified by operating on a hard decision, bit by bit, decoding of the hidden message.

Checksum-based verification

In large payload applications watermark presence verification may be further simplified. In many applications, in fact, the set of meaningful messages does not cover the whole set $\mathbf{B}$. It is, thus, likely that decoding an unwatermarked asset, or using a wrong key, results in a meaningless

---

[39]The number of terms of the sum corresponds to the number of hidden bits, which only rarely is larger than a couple of thousands (usually some hundreds).

bit string, thus making it possible to recognize the host asset as a non watermarked one.

Alternatively, we can reserve some of the bits in **b** to contain a message checksum. If the decoded bits do not match the checksum we conclude that the asset is not marked. If the number of checksum bits is large enough, an arbitrarily small false detection probability may be achieved (though at the expense of payload reduction). In fact, it is known from coding theory that with n_r checksum bits, a false detection probability as low as 2^{-n_r} is obtained. For example, by letting $n_r = 30$, a false detection probability lower than $2^{-30} \approx 10^{-9}$ is obtained, which may be sufficient in many practical applications.

Watermark presence assessment for PPM and orthogonal watermarking

For some watermarking schemes, the verification of watermark presence is a conceptually easy task. This is the case of systems in which the hidden message is encoded in the position of one or more SS detectable watermarks (PPM watermark encoding), and systems based on orthogonal signalling, where the hidden information is encoded by choosing one out of 2^k watermark signals, e.g. an SS signal.

Let us consider first the PPM case, and let us assume that the watermark message is encoded in the position of M replicas of the same SS signal (see section 3.2.2). A straightforward way to verify watermark presence could be to check whether the detector response is above the detection threshold for at least one watermark position. Actually, many possibilities exist here, since one may wonder if the detection of just 1 watermark out of M is sufficient to decide that the asset at hand is watermarked. To devise the best detection strategy, let us consider the case in which the presence of the watermark is revealed if the detector response is above the detection threshold Λ for at least r different positions, with $r \leq M$. To determine the best choice of r, we must first fix Λ. For sake of clarity, let us indicate the false detection probability of the PPM algorithm by $P_{f,PPM}$. Note that $P_{f,PPM}$ must not be confused with P_f, which refers to the probability of falsely detecting the watermark at a single position. Similarly, let us $P_{m,PPM}$ denote the missed detection probability. Setting Λ results in a given P_f, then the false detection probability $P_{f,PPM}$ can be calculated through the formula:

$$P_{f,PPM} = \sum_{j=r}^{n} \binom{n}{j} P_f^j (1 - P_f)^{n-j}. \qquad (6.241)$$

For the values of r and n commonly adopted in practice, e.g. $r \in [1,5]$ and $n = 16384$, $P_{f,PPM}$ can be approximated with the first term of the sum in

Table 6.1: False alarm probabilities ensuring $P_{f,PPM} = 10^{-6}$. results have been obtained by letting $n = 16384$.

r	P_f
1	$6.12 \cdot 10^{-11}$
2	$1.2 \cdot 10^{-7}$
3	$1.85 \cdot 10^{-6}$
4	$7.98 \cdot 10^{-6}$
5	$2.06 \cdot 10^{-5}$

(6.241), hence:

$$P_{f,PPM} \simeq \binom{n}{r} P_f^r (1 - P_f)^{n-r}. \tag{6.242}$$

If we adopt the Neyman-Pearson criterion, we start by fixing $P_{f,PPM}$, thus permitting us to derive P_f and hence Λ. The P_f's obtained by letting $P_{f,PPM} = 10^{-6}$ and r vary between 1 and 5 are shown in table 6.1.

Once $P_{f,PPM}$ has been fixed, the choice of r can be made by selecting the value resulting in the lowest missed detection probability. To evaluate $P_{m,PPM}$, let us split the set of possible watermark positions in two sets: set A with true watermark positions, and set B with wrong positions (note that B is much larger than A, since A contains only M positions whereas B is composed by $n - M$ elements). For any r, the probability of missing the watermark can be written as ($P_{d,PPM}$ denotes the probability of detecting the watermark):

$$P_{m,PPM} = 1 - P_{d,PPM}, \tag{6.243}$$

$$P_{d,PPM} = P\{n_{d,A} = r\} + P\{n_{d,A} = r - 1, n_{d,B} \geq 1\} \ldots$$
$$+ P\{n_{d,A} = 0, n_{d,B} \geq r\}, \tag{6.244}$$

$$P_{d,PPM} = \sum_{j=0}^{r} P_m^j (1 - P_m)^{r-j} \sum_{i=j}^{n} \binom{n}{i} P_f^i (1 - P_f)^{n-i}, \tag{6.245}$$

where $n_{d,A}$ and $n_{d,B}$ indicate the number of watermarks detected when the positions contained respectively in sets A and B are considered. Though cumbersome in principle, the minimization of $P_{m,PPM}$ (or equivalently, the maximization of $P_{d,PPM}$) over all possible r, can be greatly simplified if we note that, usually, P_f and P_m are very small quantities, hence the following approximations hold:

$$P_{d,PPM} \simeq (1 - P_m)^r \simeq 1 - rP_m. \tag{6.246}$$

Exploitation of the above equation to set r requires that P_m is known, which in turn requires that a particular watermarking algorithm is specified, along with its operating characteristics. Once P_m is known equation (6.246) easily permits to choose the best value of r.

Equation (6.246) also permits to calculate the theoretical error probability. More specifically, the probability of missing the watermark is $P_{f,PPM}$ regardless of whether the image was attacked or not. As to $P_{m,PPM}$, in the absence of attacks or in the presence of moderate attacks, we can assume that P_m is much smaller than 1, leading to $P_{m,PPM} \simeq rP_m$. When, as a consequence of attacks, P_m increases significantly, the more exact expression given in (6.245) must be used. Note that, often, in the presence of attacks, P_m must be evaluated experimentally.

Watermark detection for systems based on orthogonal signaling goes the same path followed for the PPM case. Instead of looking for the same watermark at different positions in the feature space, we look for several different watermarks always at the same position. The results, then, are in all similar to those obtained for the PPM case and will not be detailed further.

6.3   Further reading

The bulk of theory of watermark detection as a statistical decision problem, has been developed by J. R. Hernandez, F. Perez-Gonzalez et al. for the case of additive watermarking of Gaussian [95] and Generalized Gaussian [94]. Such an analysis mainly applies to watermarks hosted by 2D-DCT coefficients. For an accurate analysis of the various models proposed to characterize those coefficients, readers are referred to the works by R. J. Clarke [44] and K. A. Birney and T. R. Fischer [31].

The extension of watermark detection theory to the multiplicative case is described in the works by M.Barni, F.Bartolini, V.Cappellini, A.Piva and A. De Rosa [9, 10, 12]. More recently such an analysis has been refined by Q. Cheng and T. S. Huang [40].

For a more detailed introduction to the theory of signal detection readers may refer to a number of excellent books on this topic, see for example the books by L. L. Scharf [195]), H. L. Van Trees [221], and S. M. Kay [118].

The analysis in the presence of signal-dependent noise we gave in this chapter is based on the fundamental work by I. J. Cox et al. [57], where more details about the mathematical derivation of the formulas contained in section 6.1.4 can be found.

The same authors also discuss the problem of accurately estimate the error probability of a detector based on normalized correlation [152]. A more in-depth analysis of the statistics of normalized correlation, may be found in the classical work by C. P. Cox [52] or M. G. Kendall [120]. The same book by M. G. Kendall contains a detailed discussion about the validity of the central limit theorem in various operating conditions, as well as a description of several methods to improve the accuracy of the analysis developed in this way.

Application of informed embedding principles to watermarking systems relying on normalized correlation detection has been outlined very briefly in this chapter. A much more comprehensive analysis is contained in [57, 153].

Multichannel watermark detection has received only a limited attention by researchers, with most of the works focusing on watermarking of color images. Among the papers published on this topic, two of the few addressing the problem from a statistical decision theory, and trying to take cross-band correlation into account are the papers by M. Barni, F. Bartolini et. al [19, 13].

Watermark decoding is usually casted into the framework of classical communication theory. In this framework, J. R. Hernandez, F. Perez-Gonzalez et al. have written a number of pioneering works, addressing the problem of optimum watermadk decoding into an additive Gaussian or Generalized Gaussian framework [95] and Generalized Gaussian [94]. The same authors also addressed the problem of watermark decoding in the presence of channel coding [96, 171], a problem which we only touch briefly in this book. In [96] the problem of watermark presence assessment in a readable, channel-coded, framework is also addressed

Watermark decoding of a multiplicative watermark is considered in [11] for the Gaussian case, and in [14] for Weibull distributed host features. In the latter work, the problem of optimum watermark presence assessment is also considered.

Detection of QIM watermarks is treated in great details in the works by B. Chen and G. Wornell [37, 39] and J. Eggers, B. Girod, R. Bauml and R. Tzschoppe [70, 71]. In addition to the above works, more details about decoding of QIM watermarks in the presence of channel coding may by found in [71, 42, 183, 184, 155].

The procedure for the exact computation of the bit error probability of DM with bit repetition was first introduced by F. Perez-Gonzalez et al. [169, 170]. This procedure has been used to plot the curves given in figure 6.25.

We concluded this chapter by considering the problem of watermark presence assessment for the case of readable watermarking. In addition to the papers already mentioned before [96, 171, 14], readers may refer to the paper by R. Baitello et al. [5], in which detection of a PPM modulated watermark is addressed. In the same paper an algorithm for fast exhaustive detection of a multiplicative watermark is described.

7

Watermark impairments and benchmarking

One of the advantages of watermarking technology is that the information embedded into the asset is resistant to format conversions that can occur to the watermarked data for storage, transmission or fruition purposes. This would not happen if the information was tied to a header, or to a separate file. The drawback is that some of these transformations (e.g. lossy compression) can cause a loss of information in the asset, and thus affect the embedded data. This issue, i.e. if and to which extent the watermarked data can survive asset transformations, is thus very important: one of the objectives of this chapter is just to analyze and model the possible types of impairments that can affect the hidden data because of the manipulations undergone by the host asset. Usually impairments are referred to as attacks.

As we have seen in the previous chapters, data hiding system designers may choose from a wide variety of different methodologies and algorithms, each of them optimizing one particular facet of the watermarking problem. While no technique exists which outperforms the others from all points of view, a comparative analysis of the algorithms proposed so far is of outmost importance. Such an analysis may be useful to highlight the merits and drawbacks of different algorithms/approaches, to guide system designers in the choice of the most suitable technique for a given application, to guide research towards new more efficient schemes, to clarify the ultimate limits of a given algorithm or class of algorithms. Unfortunately, due to the many perspectives data hiding systems may be looked at, and to the lack of a well established theoretical framework, a thorough comparison between different schemes is still a matter of research. Generally speaking two different approaches can be followed: theoretically compare the performance of various systems, or establish the merits and drawbacks of each algorithm through a set of well defined experiments aiming at judging performance from different points of view such as obtrusiveness, robustness, security and so on. The latter approach is referred to as benchmarking, since it usually comes out with a global score summarizing the overall system performance[1]. A description of the main approaches that have been proposed up to now for benchmarking watermarking systems is the other important objective of this chapter.

The two issues described above are evidently related one to the other, the main benchmarking goal is, in fact, to understand if and to which extent a given watermarking system is able to resist to a certain class of attacks, or to evaluate which of a given number of watermarking systems is able to better deal with some manipulations: thus before a reliable benchmarking can be defined it is important to understand which are the possible attacks, and which are their effects on different watermarking systems.

The chapter starts (section 7.1) with a classification of the various attacks. It continues (section 7.3) by examining a very simple, although not very common, attack, namely Gaussian noise addition, which allows (as we have seen in chapter 6) to theoretically evaluate the performance of watermarking systems, and thus to theoretically compare them. Then more commonly occurring manipulations are analyzed, such as sample values manipulation and filtering (section 7.4), compression (section 7.5), geometric transformations (section 7.6), editing (section 7.7), analog to digital and digital to analog conversions (section 7.8), and finally manipulations explicitly attempting at removing the watermark or making it unrecoverable (section 7.9). A few considerations are also made (section 7.10) about a very important topic of research, which is attack estimation. The last section (7.11) is devoted to the description of the most popular benchmarking approaches developed so far.

7.1 Classification of attacks

Before starting to describe and model the effects of the most common asset manipulations, it is useful to outline a possible classification of the various attacks. On this subject it is convenient, first of all, to distinguish between malicious and non malicious attacks:

Non malicious We classify as non malicious those attacks that can occur during the normal use of the asset. Their nature and strength is strongly dependent on the application for which the watermarking

---

[1] Alternatively, a pool of scores may be given summarizing different aspects of system performance, e.g. robustness, invisibility, complexity and so on.

system is devised; among these we have lossy compression, geometric and temporal manipulations, digital to analogue conversion, extraction of asset fragments (cropping), processing aimed at enhancing asset quality (e.g. noise reduction), etc. .

Malicious We say that an attack is malicious if its main goal is just to remove or make the watermark unrecoverable.

Malicious attacks themselves are better described by considering two distinct classes:

Blind A malicious attack is said to be blind if it tries to remove or make the watermark unrecoverable without exploiting any knowledge of the particular algorithm used for watermarking the asset. Among these we have, for example, the copy attack that estimates the watermark signal with the aim of adding it to another asset.

Informed Malicious attacks are said to be informed if they attempt to remove or make the watermark unrecoverable by exploiting the knowledge of the particular algorithm used for watermarking the asset. These attacks first try to extract some secret information about the algorithm from publicly available data, and then, based on this information, to nullify the effectiveness of the watermarking system.

The last class of attacks is very important for some applications because it is widely known that algorithm secrecy cannot be assumed. On the contrary, according to the Kerckhoff's principle, security must to be based solely on some parametric information (e.g. a key).

Actually, the above taxonomy has some limitations (as it always happens to taxonomys) because the distinction between malicious and non malicious, blind and informed attacks is sometimes a fuzzy one. Let us think, for example, to some noise reduction algorithms commonly used for enhancing image or audio quality, these could also be applied with the explicit aim of removing the watermark treated as noise. Similarly, some malicious attacks performing some random local geometric manipulation could not be considered really blind as they exploit some knowledge of the structure of the most common watermarking algorithms (in particular their inability to deal with complicated geometric transformations).

The importance of the different types of attacks is highly dependent on the application. For example, as an extreme case, data hiding for error recovery in multimedia transmissions does not have to be resistant to any kind of transformations if embedding is performed directly in the compressed domain. As another example, labelling application will not have to care about malicious attacks, but only about the most common asset manipulations.

Also the effects of the attacks on the watermark are very different: if, on one side, geometrical manipulations do not deteriorate very much the watermark signal, but solely make it very difficult to be found, on the other extreme, informed manipulations aims at removing the watermark, in such a way that no trace of it remains into the asset. Attention will thus be given in this chapter to the possibility of inverting the effects of the attack: if such an inversion is possible, in principle, a watermarking system can be designed that, with the aid of some additional pilot information, attempts to estimate the attack, and to invert it; this is, for example, the approach commonly used to deal with geometric manipulations. If inversion is impossible, different solutions have to be found.

In this chapter we deal only with non malicious and with blind malicious manipulations. The problem of informed malicious manipulations will be analyzed in chapter 8. After having introduced, in section 7.2, a set of figures of merit measuring some general parameters such as watermark obtrusiveness and attack strength, we consider (from section 7.3 to 7.8) all those manipulations that can normally occur to a multimedia asset, these should be normally considered as non malicious, but we will see that they, or some of their sophisticate developments, can also be used maliciously. Next (section 7.9) we analyze some important malicious blind manipulations.

7.2   Measuring obtrusiveness and attack strength

In order to compare systems based on different embedding and recovery rules, and operating in different host domains, a set of common, objective, parameters must be defined. In the following, we will use the false and missed detection probabilities, and the bit error rate, as the basis for our comparison. As to obtrusiveness, we will measure it through the Data to Watermark Ratio (DWR), expressing the ratio between the power of the host features and that of the watermark. The rationale for using the ratio between watermarked-induced distortion and host signal power, instead of an absolute measure such as the MSE, relies on the widely diffused opinion that a stronger signal can accommodate a stronger watermark without compromising invisibility[2]. To be specific, let $\mathbf{f}$ be the set of host features

---

[2]Results similar to those presented in this chapter can be obtained by using an absolute degradation measure such as MSE or PSNR.

and $\mathbf{w}'$ the embedded watermark signal[3], the DWR figure is defined as:

$$\text{DWR} = \frac{\frac{1}{n} \sum_{i=1}^{n} (f_i^2)}{\frac{1}{n} \sum_{i=1}^{n} (w_i')^2}. \tag{7.1}$$

As defined in the above equation, DWR depends on the asset at hand and the particular watermark we are embedding. In order to get a global measure of algorithm obtrusiveness, the average power is considered for both the asset features and for the watermark signal, by averaging the numerator and the denominator over all possible host assets and watermarks, yielding:

$$\text{DWR} = \frac{\sum_{i=1}^{n} E[f_i^2]}{\sum_{i=1}^{n} E[(w_i')^2]}. \tag{7.2}$$

If $\mathbf{f}$ and $\mathbf{w}'$ can be assumed to be stationary sequences, equation (7.2) can be simplified, leading to

$$\text{DWR} = \frac{E[f^2]}{E[w'^2]} = \frac{\sigma_f^2}{\sigma_{w'}^2}, \tag{7.3}$$

where, the second equality follows holds under the simplified assumption that both the host features and the watermark have a zero mean. It is worth noting that DWR does not coincide with the true asset to watermark power ratio, since the average power of the host features does not necessarily coincide with asset power: this, for example, is the case of systems embedding the watermark in a subset of the host features, e.g. a subset of DFT coefficients.

The strength of attacks is sometimes measured by the Watermark to Noise Ratio (WNR), giving the ratio between the power of $\mathbf{w}'$ and that of the noise introduced by attacks. Specifically, by letting $\mathbf{f}_w'$ indicate the attacked feature sequence [4], and $\mathbf{n} = \mathbf{f}_w' - \mathbf{f}_w$ be the attack noise, we have:

$$\text{DWR} = \frac{\sum_{i=1}^{n} E[(w_i')^2]}{\sum_{i=1}^{n} E[n_i^2]}, \tag{7.4}$$

that can be simplified to:

$$\text{WNR} = \frac{E[(w')^2]}{E[n^2]} = \frac{\sigma_{w'}^2}{\sigma_n^2}, \tag{7.5}$$

if both $\mathbf{w}'$ and $\mathbf{n}$ are stationary (zero mean) sequences.

---

[3]We refer here at the actual signal added to the host features, i.e. $\mathbf{w}' = \mathbf{f} - \mathbf{f_w}$.

[4]By using the symbol $\mathbf{f}_w'$ for the attacked feature sequence, we implicitly assumed that the feature sequence is watermarked, an assumption that can not be made in the case of detectable watermarking.

Another possibility is to consider the Data to Noise Ratio (DNR)[5]:

$$\text{DNR} = \frac{\sum_{i=1}^{n} E[f_i^2]}{\sum_{i=1}^{n} E[n_i^2]}, \tag{7.6}$$

or

$$\text{DNR} = \frac{\sigma_f^2}{\sigma_n^2}, \tag{7.7}$$

for stationary, zero mean, sequences. An advantage of DNR with respect to WNR is that it better reflects real scenarios where the maximum allowable distortion a pirate may introduce to remove the watermark depends on the energy of the host signal, rather than the energy of the watermark.

In the following sections the above definitions will be specialized to the cases of some particularly important watermarking systems and attacks.

7.3 Gaussian noise addition

Gaussian noise addition is not a very likely manipulation (although, as we will see in section 7.8, some noise can be assumed to be added to the asset during digital to analog and analog to digital conversion), however its effect is quite easy to model thus allowing a theoretical comparison between different watermarking systems. In particular, we consider here the direct addition of white Gaussian noise to the watermarked features: this is not a strong limitation given that any kind of linear transformation (e.g. full frame DFT or DCT, block based DCT, DWT) does not change the pdf of an additive white Gaussian noise, and thus if this kind of noise is added in the asset domain, the same type of disturb can be assumed to be present in most of the transformed domains used in practice.

7.3.1 Additive vs multiplicative watermarking

We start by analyzing the performance of additive and multiplicative spread spectrum watermarking in the presence of an additive Gaussian attack. First, we must specialize the definition of DWR, WNR and DNR to this case. By remembering the embedding rule defining additive and multiplicative watermarking, and by assuming that the host features form a zero mean, i.i.d. sequence, it can be easily shown that:

$$\text{DWR}_A = \frac{\sigma_f^2}{\gamma^2 \sigma_w^2}, \tag{7.8}$$

---

[5]It is immediate to verify that $\text{DNR} = \text{DWR} \cdot \text{WNR}$

$$\text{WNR}_A = \frac{\gamma^2 \sigma_w^2}{\sigma_n^2},$$ (7.9)

$$\text{DNR}_A = \frac{\sigma_f^2}{\sigma_n^2},$$ (7.10)

for the additive case, and

$$\text{DWR}_M = \frac{1}{\gamma^2 \sigma_w^2},$$ (7.11)

$$\text{WNR}_M = \frac{\gamma^2 \sigma_w^2 \sigma_f^2}{\sigma_n^2},$$ (7.12)

$$\text{DNR}_M = \frac{\sigma_f^2}{\sigma_n^2},$$ (7.13)

for the multiplicative case.

Watermark detection

Let us consider first additive and multiplicative detectable watermarking. A theoretical analysis of both these cases is possible only in the Gaussian case, e.g. when both the host features and attack noise are white, Gaussian processes. Under this assumption the relevant error probabilities are given by equations (6.36) and (6.41) for the additive case, and equations (6.103) and (6.110) for the multiplicative case. In order to compare the two classes of algorithms, we fix the DWR and plot the ROC curves for the additive and multiplicative optimum detectors. The results are given in figures 7.1 through 7.3. More specifically, in figure 7.1 we assumed noise is not present and DWR = 20dB. We considered the case of antipodal watermarking for which $\sigma_w^2 = 1$. As it can be seen the multiplicative scheme significantly outperforms the additive one. It has to be remembered, though, that equations (6.103) and (6.110) give only an approximation of the true false and missed detection probabilities, with a relative error which is higher for low values of P_f and P_m. More accurate results may be obtained by correcting the ROC curves as explained at the end of section 6.1.6, or by means of Montecarlo simulations.

In order to analyze the impact of noise on the above comparison, the ROC curves given in figure 7.1 have been redrawn for several values of DNR. The corresponding ROC curves are reported in figures 7.2 and 7.3. Upon inspection of the plots, it is at once evident that noise presence reduces that advantage of multiplicative watermarking, up to a point (DNR $\approx$ 0db) where the additive system starts achieving the best performance.

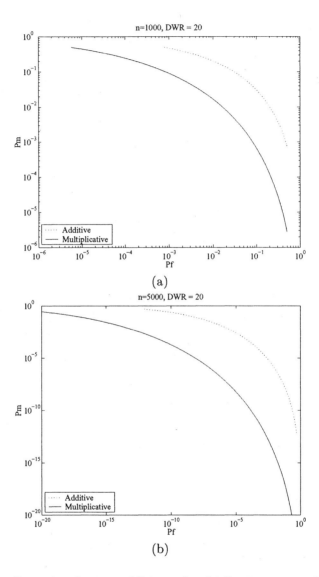

Figure 7.1: Comparison between additive and multiplicative watermarking in the absence of noise for $n = 1000$ and $n = 5000$. For both figures we let DWR = 20dB, and $\sigma_w^2 = 1$ (antipodal watermark).

It has to be noted, though, that DNR values for which additive watermarking outperforms the multiplicative scheme are very low, meaning that the attack results in a very strong degradation, e.g. when DNR = 0db (WNR

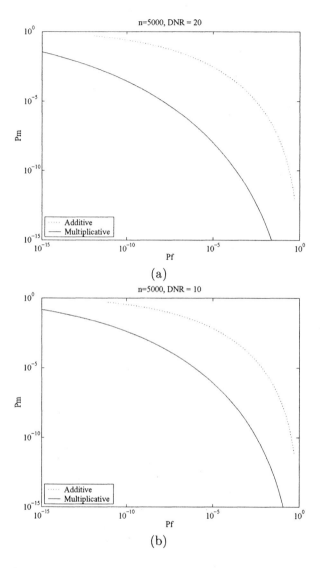

Figure 7.2: Comparison between additive and multiplicative watermarking in the presence of moderate noise (DNR = 20dB in part (a) and 10dB in part (b)). For both figures we let DWR = 20dB, and $\sigma_w^2 = 1$ (antipodal watermark).

= -20db) the attack has the same strength as the host features.

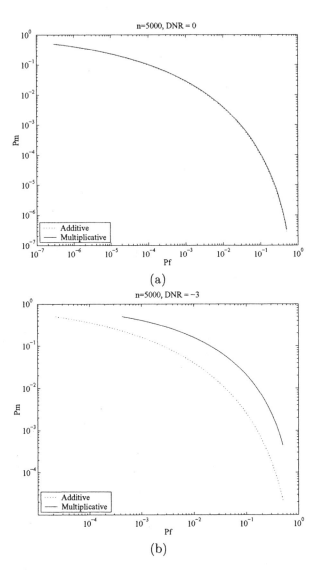

(a)

(b)

Figure 7.3: Comparison between additive and multiplicative watermarking in the presence of very strong noise (DNR = 0dB in part (a) and -3dB in part (b)). For both figures we let DWR = 20dB, and $\sigma_w^2 = 1$ (antipodal watermark).

Watermark decoding

We now turn the attention to multibit watermarking. As before we only consider the case of an antipodal watermark embedded into Gaussian fea-

tures in the presence of additive Gaussian noise. In this case, the relevant
equations are those derived in sections 6.2.2 and 6.2.4. As opposed to watermark detection two cases are possible, according to whether watermark
spreading is used or not. In the latter case, an exact expression of the
error probability has been derived for both the additive and the multiplicative case, whereas in the first case, only an approximation of the true bit
error probability is available for the multiplicative case. For this reason,
our comparison only considers the case of a non-spread watermark, in such
a way to avoid that the approximation error biases the analysis. All the
more that no difference of performance has been found between spread and
non-spread watermarks[6].

As for the detection case, we start the analysis by letting $\sigma_n^2 = 0$. In
figure 7.4a, the bit error probability is plotted as a function of DWR for a
fixed value of r ($r = 200$). Upon inspection of the plot it comes out that
multiplicative watermarking permits to save about 3dB with respect to the
additive case. In figure 7.4b we fixed the DWR (DWR = 20dB) and varied
r. The results shown in the figure permit us to conclude that the advantage
of the multiplicative scheme is retained for all r, with higher advantages
for higher values of r.

We now must consider the case in which attack noise is present. To
do so, we redraw the plot in figure 7.4a for DNR = 10dB and DNR -3dB,
thus obtaining the diagrams shown in figure 7.5. As we already noted in
the case of watermark detection, a more pronounced sensitivity to noise
of multiplicative watermarking is revealed, even if additive watermarking
starts performing better only for very strong noise. In order to get more
insight into this particular aspect, in figure 7.6 the error probability for a
fixed DWR (DWR = 17dB and 20dB) is plotted against DNR. The behavior
noted previously is confirmed, with a performance threshold around 0dB:
for DNR > 0dB multiplicative watermarking ensures better performance,
whereas for negative DNR's the additive approach is preferable.

7.3.2   Spread Spectrum vs QIM watermarking

One of the most fundamental distinctions among readable watermarking techniques regards whether a spread spectrum or a QIM approach is
adopted. The former approach relies on the basic assumption that spread
spectrum communication is a preferred communication technique in the
presence of very noisy channels where the possible presence of a jammer
has to be taken into account. Additionally, SS communications provide
a good degree of security against non-allowed access to the data. On the

---

[6]Such a difference does not exist at all for the additive case, whereas it can be argued
to be very small in the multiplicative case, see figure 6.18.

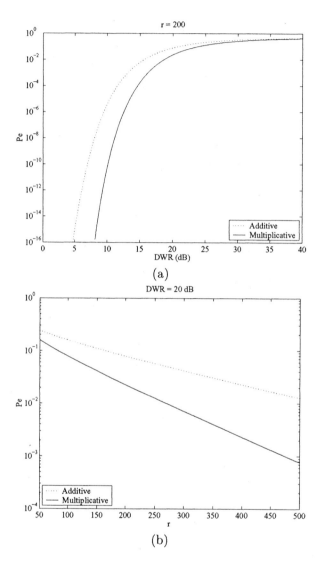

Figure 7.4: Bit error probability for additive and multiplicative watermarking in the absence of noise for various values of DWR (a) and r (b).

other hand, QIM-like schemes rely on the informed embedding principle, in that they exploit the fact that the embedder knows part of channel noise, and hence it can take proper countermeasures. It is the purpose of this section to compare SS and QIM techniques, from the point of view of com-

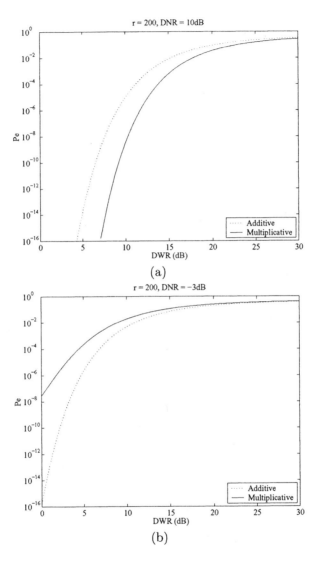

Figure 7.5: Bit error probability for additive and multiplicative watermarking for DNR = 10dB (a) and DNR = -3dB (b).

munication reliability, i.e. achievable bit error rates. The comparison will rely on the results we derived in chapter 6. The analysis in this section follows an empirical approach, in that, once the bit error rates are given, we use them to compare the various systems for several working environments.

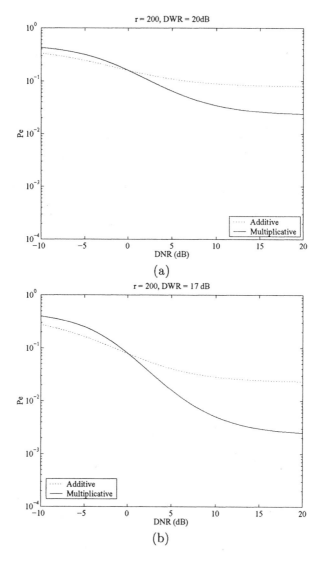

Figure 7.6: Plot of bit error probability vs DNR, for additive and multiplicative watermarking. In part (a) we let DWR = 20dB, whereas in part (b) we have DWR = 17dB.

For a more theoretical comparison between the two different approaches, readers are referred to chapter 9, where a framework for the evaluation of the watermark channel capacity is described.

In section 7.3.1, we observed that a definitive judgement on whether the best choice for an SS watermark consists of an additive or a multiplicative embedding rule can not be given, since the superiority of a system on the other depends on the operating parameters. For this reason, when comparing spread spectrum and QIM systems, we will consider both the additive and the multiplicative schemes. As to QIM, in section 6.2.6 we found that ST-DM always outperforms DM with bit repetition, hence in the following we will take into account only the performance of the ST-DM scheme.

We start our analysis by specializing the definition of DWR, DNR and WNR to the case of ST-DM watermarking.

The computation of DWR for DM with bit repetition is usually made under the assumption that the host feature variance is much larger than Δ thus permitting to conclude that quantization noise is uniformly distributed in the quantization interval. Whereas this is a reasonable assumption in the DM case, its application to ST-DM case is not justified. To be specific let us start by noting that, due to equation (4.89), we have

$$E[w^2] = \frac{1}{r}E[(\rho_w - \rho_f)^2], \tag{7.14}$$

that is, DWR is proportional to the quantization noise affecting ρ. In addition, due to the normality of $\mathbf{f}$, ρ_f is a normally distributed random variable with variance σ_f^2. To go on with the computation of $E[w^2]$, it is necessary that the value of the shift d appearing in equations (4.79) and (4.80) is fixed. To this aim, it can be argued that a lower quantization error is achieved by letting $d = \Delta/4$, since in this way two symmetric codebooks are obtained[7]. Under this assumption, and by assuming bits 0 and 1 are equiprobable, we have:

$$rE[w^2] = E[(\rho_w - \rho_f)^2|b = 0] = \sum_{i=-\infty}^{\infty} \int_{i\Delta-\Delta/4}^{i\Delta+3\Delta/4} (x - \Delta/4 - i\Delta)^2 f_{\rho_f}(x)dx,$$

$$\tag{7.15}$$

with f_{ρ_f} corresponding to a zero mean Gaussian pdf with variance σ_f^2. By applying some simple algebra to the above expression, $E[(\rho_w - \rho_f)^2]$ and DWR can be given the following general form:

$$E[(\rho_w - \rho_f)^2] = \alpha\Delta^2, \tag{7.16}$$

$$\mathrm{DWR} = \frac{r\sigma_f^2}{\alpha\Delta^2}, \tag{7.17}$$

---

[7]The truthfulness of this assertion can be easily verified by following the same analysis reported below.

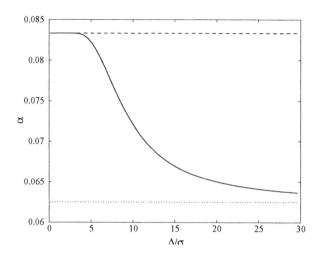

Figure 7.7: Plot of α vs Δ/σ_f. The dashed line indicates the value of α for a uniform quantization error ($\alpha = 1/12$), whereas the dotted line gives the limit of α for $\Delta/\sigma_f \to \infty$ (1/16).

where α is a factor depending on Δ/σ_f. The values of α deriving from equation (7.15) are plotted in figure 7.7 as a function of Δ/σ_f. Interestingly, the quantization error is always smaller than that obtained by assuming a uniform distribution ($\alpha = 1/12$, dashed line), and tends to $1/16$ when Δ/σ_f increases. This gives ST-DM an extra advantage with respect to DM with bit repetition, since a larger Δ can be used for a given DWR.

It is worth noting that a completely different result is obtained if we assume that d is a random variable uniformly distributed in $[0, \Delta)$ and average the quantization error over d as well, since in this case $\alpha = 1/12$ regardless of the value assumed by Δ/σ_f[8]. For the sake of simplicity, in the sequel we will always assume that $d = 1/4$.

The computation of WNR and DNR is by far simpler, since we immediately have

$$\text{WNR}_Q = \frac{\alpha\Delta^2}{r\sigma_n^2}, \qquad (7.18)$$

and,

$$\text{DNR}_Q = \frac{\sigma_f^2}{\sigma_n^2}. \qquad (7.19)$$

We are now ready to compare the performance of SS and ST-DM in the presence of an additive Gaussian attack. In fact, when attack noise is

---

[8]Interested readers can find more details about dither quantization in [196].

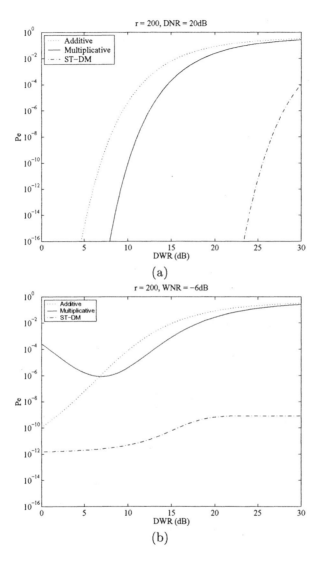

Figure 7.8: Comparison between SS and ST-DM watermarking. The curves have been obtained for a fixed DNR (DNR = 20dB - part (a)) or WNR (WNR = -6dB, part (b)), and by letting $r = 200$.

not present, the comparison does not make sense, since being QIM schemes characterized by a null bit error probability they largely outperform those based on blind embedding. We start by considering a case in which a

moderate amount of noise is present, let us say DNR = 20 dB. In figure
7.8a, the corresponding bit error rate is plotted against DWR ($r = 200$).
As expected, the ST-DM scheme largely outperforms both the additive
and multiplicative SS algorithms, where the noise component due to host
features is still predominant. One may wonder why the ST-DM plot is not
flat at all, due that ST-DM bit error probability should does not depend
on host signal power. Actually this is only partially true, since due to
the presence of α in equation (7.17), a higher quantization step can be
used for smaller values of DWR. In addition, having fixed DNR, it turns
out that WNR decreases with DWR hence the right part of figure 7.8a is
characterized by higher WNR values, hence justifying the behavior of the
ST-DM curve. For sake of clarity, in figure 7.8b, the bit error rate is plotted
by fixing WNR, thus leading to an approximately flat bit error rate for the
ST-DM scheme. Note that the non-monotic behavior of the bit error rate
for multiplicative watermarking can now be explained by noting that in
this case DNR is not constant. Note also that figure 7.8b refers to a much
more noisy channel than part (a) of the same figure.

In order to highlight the dependence of the various schemes upon noise
in figure 7.9, the bit error rate is given as a function of DNR for a fixed
value of DWR (DWR = 15dB). As it can be seen the distance between ST-
DM and SS reduces for increasing noise levels, up to a point that for DNR's
lower than approximately 0 dB SS schemes tend to perform better[9]. This is
not surprising, since for low values of DNR the importance of host features
as disturbing noise decreases, hence diminishing the positive impact of the
informed embedding approach. Note also that an interval may exist in
which the multiplicative scheme gives the best performance, whereas for
very large and very small values of DNR the best results are always achieved
by the ST-DM and Additive SS schemes respectively.

To summarize, we can say that the QIM approach (noticeably ST-DM)
has to be preferred in most practical applications, whereas the additive SS
scheme is more appropriate only when the attack strength is expected to
the very high. This, in turn, makes the QIM approach preferable when
a high payload, possibly at the expense of robustness, is a primary need.
Conversely, when only a few, very robust, bits have to be hidden, the SS
approach may still be a convenient choice.

The above analysis holds only for normally distributed host features
and additive Gaussian noise, thus an ultimate decision about the superiority of one scheme on the others can not be taken. For instance, ST-DM
watermarking is intrinsically less robust to radiometric scaling of the host
features, whereas multiplicative schemes are almost not affected by this

---

[9]Indeed DNR = 0db is a very low value (WNR = -15dB in figure 7.9) since it means
that the noise has approximately the same strength as the host features.

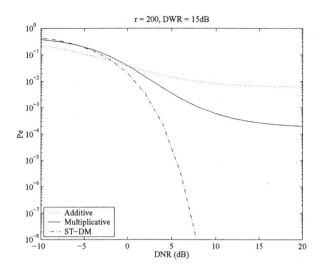

Figure 7.9: Comparison between SS and ST-DM watermarking for varying DNR's. The curves have been obtained by letting $r = 200$ and DWR $= 15$dB.

kind of attack (see the analysis presented later in section 7.4.1).

Finally, we would like to mention a bunch of alternative perspectives watermark effectiveness may be seen from. These include watermark presence assessment, multiple watermarking, security, conditional access to the hidden data. To date no detailed analysis of the above items is available, it is reasonable, though, to think that no good-for-all solution will ever be developed, thus leaving room for the coexistence of a number of different approaches each optimizing system performance from a different point of view.

7.4 Conventional signal processing

There are many manipulations that are commonly performed on multimedia assets. Let us start by considering the case of sample-by-sample (pointwise) transformations. A common example of this case is contrast enhancement for still images and videos. This is usually achieved on the basis of a previous analysis of the color histogram, by identifying the minimum and maximum color values, and by applying a linear function mapping them to the extrema of the available range. Thanks to the linearity of the most commonly used signal transformations (DCT, DFT, DWT), the same linear function is applied to the watermarked features regardless of the watermarking domain. A similar linear process can be applied to audio samples, e.g. to modify the loudness.

It is thus important to understand how watermark recovery is affected by a linear modification of the watermarked features. In particular, a question naturally arises about the behavior of the receivers developed in chapter 6 when this kind of processing is applied. For this reason, in the first part of this section, we analyze the performance of some of the systems described in chapter 6 in the presence of the so called gain attack. Where by gain attack we mean the scaling of host features by a constant factor plus noise addition. For simplicity we limit ourselves to the case of Gaussian distributed features.

7.4.1   The gain attack

Hereafter we analyze how the performance of the optimum decoders derived in the previous chapter degrade in the presence of a gain attack. Our analysis will regard additive and multiplicative SS watermarking and ST-DM watermarking.

Additive SS watermarking

Let us start by considering the expression of WNR and DNR for the gain attack. By focusing on multibit watermarking, we have:

$$f'_{w,i} = gf_i + g\gamma bw_i + n'_i, \qquad (7.20)$$

with $b = \pm 1$, and n'_i the additive Gaussian part of the attack. According to the definitions given at the beginning of the chapter, the computation of WNR and DNR should pass through the computation of the attack noise $\mathbf{n} = \mathbf{f}'_w - \mathbf{f}_w = (g-1)\mathbf{f} + (g-1)\gamma b\mathbf{w} + \mathbf{n}'$. However, we can observe that, in most cases, a simple scaling of the host features, not accompanied by noise addition, does not cause any perceptual deterioration of the host asset. Indeed, at very low, and very large gain values the effects of quantization of asset samples, and of their limited dynamic range, may become problematic, but in principle, a change of the scale of the asset samples does not cause any annoying effect. This means that it is not realistic to let g contribute to computation of DNR and WNR. Under this assumption, we have:

$$\text{WNR} = \frac{\gamma^2 \sigma_w^2}{\sigma_{n'}^2}, \qquad (7.21)$$

$$\text{DNR} = \frac{\sigma_f^2}{\sigma_{n'}^2}. \qquad (7.22)$$

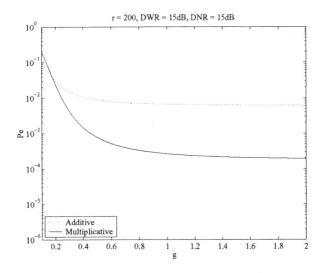

Figure 7.10: Bit error probability for additive and multiplicative spread spectrum watermarking in the presence of a gain attack.

In practice the bit error probability is equal to that obtained for the AWGN attack, with the only difference that both the host features and the watermark are scaled by a factor g, i.e.

$$P_e = \frac{1}{2}\mathrm{erfc}\left(\sqrt{\frac{r}{2}\frac{g^2\gamma^2\sigma_w^2}{g^2\sigma_f^2 + \sigma_{n'}^2}}\right). \qquad (7.23)$$

Multiplicative SS watermarking

In this case the watermarked and attacked features assume the form

$$f'_{w,i} = gf_i(1 + \gamma bw_i) + n'_i, \qquad (7.24)$$

with $b = \pm 1$, and n'_i the additive Gaussian part of the attack. As for the additive case, we will assume that only n'_i contributes to form the WNR and DNR figures. Under this assumption, the evaluation of the bit error probability goes the same line we followed in the AWGN case, with the only difference that now the host features are scaled by a factor g. Hence, the bit error probability is still given by equation (6.198), by replacing σ_f^2 with $g^2\sigma_f^2$.

The performance of additive and multiplicative spread spectrum watermarking in the presence of a gain attack are sketched in figure 7.10 for

$r = 200$, DWR = DNR = 15db. As it can be seen, spread spectrum techniques are almost insensitive to this kind of attack, since the host features and the watermark are affected in the same way by the presence of the scale factor g factor. The slight dependence on g is due to the different importance assumed by the additive noise part of the attack which is not influenced by g. As a matter of fact, spread spectrum systems enhance their performance for increasing values of g, since for high g's the strength of the watermark with respect to the fixed amount of noise is increased.

ST-DM watermarking

Let us now consider the behavior of ST-DM, as described at the end of section 6.2.6, in the presence of the gain attack. Even in this case we have:

$$\mathbf{f}'_w = g\mathbf{f}_w + \mathbf{n}'. \tag{7.25}$$

Then we must consider the WNR and DNR figures. As for the SS case, the gain g does not have any impact on the host asset quality, hence permitting us to write:

$$\text{WNR} = \frac{\alpha\Delta^2}{r\sigma_{n'}^2}, \quad \text{DNR} = \frac{\sigma_f^2}{\sigma_{n'}^2}. \tag{7.26}$$

We must now calculate the bit error probability as a function of g and $\sigma_{n'}^2$. To this aim, it is immediate to verify that the correlation between the attacked features and the reference direction $\mathbf{w}$ is given by:

$$\rho' = g\rho_w + n_\rho = g\mathcal{Q}_{0/1}(\rho_f) + n_\rho, \tag{7.27}$$

with n_ρ normally distributed with variance $\sigma_{n_\rho}^2 = \sigma_{n'}^2$. The bit error conditioned to $b = 0$ is:

$$P_{e|0} = \sum_{i=-\infty}^{\infty} p(u_{0,i})P_{e|u_{0,i}}, \tag{7.28}$$

with

$$P_{e|u_{0,i}} = P\left(\rho' \in \bigcup_j [(j+1/2)\Delta, (j+1)\Delta] \,|u_{0,i}\right)$$

$$= \sum_{j=-\infty}^{\infty} \int_{(j+1/2)\Delta}^{(j+1)\Delta} \frac{1}{\sqrt{2\pi\sigma_{n_\rho}^2}} e^{-\frac{(\rho'-g(i+1/4)\Delta)^2}{2\sigma_{n_\rho}^2}} \, d\rho' \tag{7.29}$$

$$= \sum_{j=-\infty}^{\infty} \int_{(j+1/2-g(i+1/4))\frac{\Delta}{\sigma_{n_\rho}}}^{(j+1-g(i+1/4))\frac{\Delta}{\sigma_{n_\rho}}} \frac{1}{\sqrt{2\pi}} e^{-x^2/2} \, dx,$$

and

$$p(u_{0,i}) = P\left(\rho_f \in [(i - 1/4)\Delta, (i + 3/4)\Delta]\right) = \int\limits_{(i-1/4)\frac{\Delta}{\sigma_f}}^{(i+3/4)\frac{\Delta}{\sigma_f}} \frac{1}{\sqrt{2\pi}} e^{-\frac{x^2}{2}} \, dx.$$

$$(7.30)$$

Finally, let us note that for the symmetry of the problem (remember that we let $d = \Delta/4$) we have $P_{e|1} = P_{e|0}$, yielding $P_e = P_{e|1} = P_{e|0}$.

The error probability resulting from the above equations is plotted in figure 7.11 for several values of DWR, DNR, and g. Specifically, part (a) of the figure shows the bit error probability for DWR = 25db, $r = 30$ and WNR = -3db, 0db and +3db (DNR = 22dB, 25db and 28db). It is readily seen that the performance of ST-DM decrease rapidly as soon as g departs from 1, up to a point that for $g < 0.9$ and $g > 1.1$, the error probability is unacceptably high. At the same time, is can be seen that the influence of WNR (DNR) on this behavior is negligible. This is not the case when DWR is varied, since, as it is shown in figure 7.11b, for lower values of DWR, the range of admissible g's is wider. This is an interesting result, since, as opposed to the AWGN case, where the performance are almost insensitive to DWR, robustness against the gain attack may be improved by increasing the watermark strength. In figure 7.12, the error probability for different values of r is shown. Even in this case the range of admissible g's increases for higher values of r, however the improvement is less evident than that obtained by varying DWR.

7.4.2 Histogram equalization

A more general example of pointwise manipulation which is sometimes used for images and video is histogram equalization. The goal of this manipulation is to transform the colour value of each image/video sample in such a way to obtain an image having a given histogram (usually uniform). The theoretical basis of this kind of processing is the relation between the pdf of a function of a random variable and that of the variable itself, in particular, if X is a random variable having cdf (cumulative density function) $F_X(x)$, and if $Y = g(X)$ is a function, that we assume here invertible and monotonicaly increasing for simplicity (the result is easily generalizable), the cdf of the random variable Y is related to that of X by the relation:

$$F_X(x) = F_Y(g(x)). \qquad (7.31)$$

Thus given the distribution of X (easily obtainable from the histogram of X) and the distribution of Y (easily obtainable from the desired histogram)

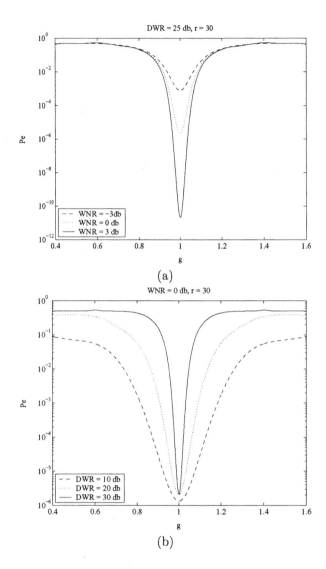

Figure 7.11: Bit error probability for ST-DM watermarking in the presence of a gain attack. Different values of WNR (part (a)) and DWR (part b) are considered.

it is simple to build the needed transformation $g(x)$: for each y look for the value x for which $F_X(x) = F_Y(y)$, then set $g(x) = y$. In practice, then, this attack is a generalization of that analyzed in section 7.4.1 where instead of a simple linear transform, a more general non linear function is considered.

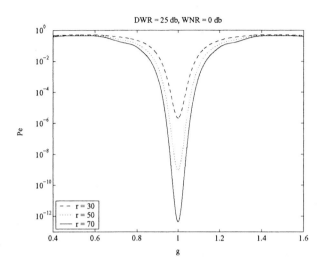

Figure 7.12: Bit error probability for ST-DM watermarking in the presence of a gain attack. Different values of r are considered.

A general theoretical analysis of this attack is very cumbersome, hence, to evaluate its impact on different watermarking algorithms it is necessary to resort to experimental or simulation results.

Even more complicated is the analysis of watermarking systems operating into some transformed domain (or hybrid ones) due to the fact that the effects of a non linear transformation in the asset domain can not be modelled easily in the transformed domain.

7.4.3 Filtering

Another type of manipulation commonly applied to multimedia assets is linear filtering. Linear filtering consists of the convolution of a point spread function (kernel) with asset samples. The effects of such a process on systems operating in different domains are quite different.

Let us consider first systems working in the asset domain: the effect of linear filtering is to correlate the watermarked asset samples, i.e.:

$$\mathbf{f}' = \mathbf{h} \otimes \mathbf{f}_w, \qquad (7.32)$$

where $\mathbf{h}$ is the kernel of the linear filter and $\otimes$ represents the convolution operation. Given that embedding is performed directly in the asset domain, a consequence of this attack is to correlate the watermark signal. This can be a problem for those systems, as those based on spread spectrum communication theory, that exploit the characteristics of uncorrelated watermark signal during the recovery phase. As an example let us examine the case of 1D signal (e.g. and audio file) watermarking[10] with an additive 1-bit watermark (see equation (4.33)). If a linear filter is applied, the attacked watermarked features result to be:

$$f_i' = \sum_k h_k f_{w,i-k} = \tilde{f}_i + \gamma \sum_k h_k w_{i-k}, \tag{7.33}$$

where we have defined:

$$\tilde{f}_i = \sum_k h_k f_{i-k}, \tag{7.34}$$

i.e. the filtered original image, and the detector response (see equation (6.28)) is:

$$\rho = \frac{1}{n} \sum_{i=1}^{n} f_i' w_i = \frac{1}{n} \sum_{i=1}^{n} \tilde{f}_i w_i + \gamma h_0 \frac{1}{n} \sum_{i=1}^{n} w_i^2 + \gamma \frac{1}{n} \sum_{k \neq 0} h_k \sum_{i=1}^{n} w_{i-k} w_i. \tag{7.35}$$

While the statistic of the detector response does not basically change if the asset is not watermarked[11], the mean of ρ can be quite different from the Gaussian case under H_1 hypothesis, due to the third term of the right hand side of equation (7.35), thus affecting the missing error probability. In particular, watermark designers may limit the effects of filtering by choosing a watermark signal which is as white as possible (in such a way that the third term becomes negligible).

In the case of a multibit watermarking scheme, the filtering process also introduces a correlation among portions of the watermark signal carrying different bits, i.e. Inter Symbol Interference (ISI) appears.

For the ST-DM method, it is quite simple to see that the decision variable is given by:

$$\rho' = h_0 Q_{0/1}(\rho_f) + \sum_{k \neq 0} h_k \left[\sum_{i=1}^{r} f_{i-k} w_i - (\rho_f - Q_{0/1}(\rho_f)) \sum_{i=1}^{r} w_{i \ominus k} w_i \right] \tag{7.36}$$

where $\ominus$ indicates the modulo r difference. In this case, further to the interference given by the second term of the right hand side of the equation, the strong sensitivity exhibited by the quantization based schemes to the introduction of a gain (here given by h_0) can be quite critical.

We now pass to consider watermarking schemes operating in the transformed domain. For the most common signal transformations, i.e. DFT, it

---

[10]The example is easily generalizable to multidimensional signals.
[11]The mean and variance of ρ will be in this case estimated a posteriori as it is in the case of additive Gaussian noise.

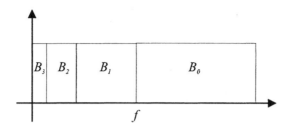

Figure 7.13: Example of subdivision of the spectrum in 4 bands of 1 octave width.

is valid the property that the convolution in the asset domain corresponds to a product in the transformed domain. We thus have that the attacked features will be:

$$f'_i = H_i f_{w,i} \qquad (7.37)$$

where H_i is the DFT of the filter kernel. By considering that transformed coefficients are complex variables, we can more precisely conclude that the process of filtering in the asset domain corresponds to the multiplication of the magnitude of the watermarked features by the magnitude of the DFT coefficients of the filtering kernel, and by the addition to the phase of the watermarked features of the phase of the DFT coefficients of the filtering kernel. Often, only the magnitude of DFT is watermarked, because of its invariance to translations in the asset domain, we have thus to deal with a varying multiplicative disturb. If the DFT of the filtering kernel varies smoothly enough to be considered piecewise constant, an analysis similar to that carried on for the gain attack can give some hints on the behavior of the watermarking method in the presence of filtering. More complicated is the case of DCT watermarking, where developing a tractable model is very difficult. It is anyway possible to demonstrate that if the filtering kernel is symmetric around some point[12], as it is often the case, the same equivalence between convolution in the asset domain and multiplication in the transformed domain is valid[13]. In any case, given that the transformed domain usually offers a frequency representation of the signal, and that it is possible to predict the effects of a linear filter on the frequency domain (thanks to Fourier analysis), useful indications on the deterioration caused by this kind of attack to the watermark signal can be always obtained.

Let us finally consider the case of hybrid techniques. What we have just told about full frame DCT transform is still valid for block based DCT methods, i.e. each DCT coefficient in a block is affected by a gain attack

---

[12]E.g. for the 2D case if $h(i, k) = h(-i, k) = h(i, -k) = h(-i, -k)$.

[13]In this case the filter affects in the same way identical absolute frequencies regardless of their sign, i.e. the filter has a symmetric behavior also in the frequency domain.

which is proportional to the filter response at that frequency. It is particularly interesting here to note that most of block based DCT watermarking methods treat separately each set of DCT coefficients having the same frequency, i.e. they model the asset as a set of parallel channels (one for each frequency) each one having the same statistical properties. For this approach the model of the gain attack that we have presented in section 7.4.1 is immediately applicable. The same can not be said about DWT: in this case it is useful to consider that DWT performs a sort of sub-band decomposition of the signal, i.e. it expresses the signal as the sum of different components each having a different frequency content. In particular, sub-bands are not uniform, i.e. they does not have the same bandwidth, which, on the contrary, decreases according to the frequency position (to be more precise bandwidths are constant if measured in a logarithmic scale, and have the value of 1 octave[14]). In figure 7.13 the subdivision of the spectrum into 4 bands 1 octave wide, as that performed by DWT, is exemplified[15]. The effect of a linear filter can thus be predicted by considering the band where the watermark signal is actually embedded, and the frequency response of the filtering kernel.

More than linear filters, non linear ones are commonly used for reducing the noise superimposed to images. A plethora of non linear filters exist, e.g. median and vector median, alpha trimmed and morphological filters, only to cite the most popular. Their effects on the watermark signal can not be precisely modelled, but some considerations can be useful. The main goal of this kind of filters is to eliminate from the asset those samples that appear to deviate from the statistic of the asset itself, i.e. the so called outliers. This suggests, then, that the watermark signal should not have this characteristic, i.e. should not modify too much the statistic of the original asset. Furthermore, although Fourier analysis can not be applied in this case, it is possible to affirm that non linear filters usually affects the highest part of the spectrum much more than the lowest one (i.e. they have a high pass behavior).

7.5 Lossy coding

Another very common manipulation multimedia assets may undergo is lossy coding. The goal of lossy coding is to compact as much as possible the representation of a the multimedia asset by discarding perceptually non relevant data. It is thus obvious how this process can negatively influence the performance of a data hiding system that, on its side, attempts to embed

---

[14]A band is 1 octave wide if $\log_2(f_u/f_l) = 1$ where f_u and f_l are the upper and lower frequencies of the band.

[15]Indeed the filters of contiguous DWT bands partially overlap.

useful information just into the perceptually non relevant parts of the host signal. The duality of the behavior of lossy coding (compression) and data hiding has already been discussed in chapter 5 where we concluded that, because of different complexity constraints, data hiding systems are much more effective than compression algorithms in exploiting the characteristics of the Human Perceptual Systems, thus allowing hidden data to survive also strong compression attacks. This is, in fact, what it is experimented with most of the watermarking tools available in the literature.

All lossy compression algorithms follow a three step scheme. The first step consists of the application of a mathematical transformation to the asset samples to project them into a space where the resulting coefficients can be considered almost independent, and where perceptual modeling is easier; the most common transformations are block DCT (e.g. in the JPEG standard) or DWT (e.g. in the JPEG2000 standard): this first step does not imply (at least in principle) any loss of information being the transformation reversible. The second step consists of the quantization of transformed coefficients: it is in this process that information (possibly the less perceptually relevant) is lost due to the non-reversible nature of the quantization process. Finally, in the third step, quantized coefficients are entropy encoded (e.g. by Huffman or arithmetic coding) without loss of information[16].

We now briefly describe the most common compression standards, paying particular attention to the phases where the information is lost: many other compression algorithms (e.g. SPIHT for images, those based on matching pursuit for video, etc.) are available, but being not standard (at least yet) they are only rarely encountered in consumer electronics applications, although they can be found in particular applications.

Let us start by considering the popular JPEG algorithm for still images. The images are first partitioned into 8×8 blocks that are then processed by DCT. Transformed DCT coefficients are quantized with a frequency dependent quantization step that can be chosen by the encoder (an example of quantization steps for each frequency are given by equation (5.59)). Quantized coefficients are then entropy encoded without loss of information. The quantization matrices are usually designed (similarly to matrix (5.59)) based on perceptual considerations. In particular the dependency of HVS sensitivity on frequency is usually considered, although more sophisticated masking effects can also be included for adapting the quantization matrix

---

[16]Indeed, as we will see, in the recent JPEG2000 standard the loss of information is not solely relegated to the second step but it is also occurring in the third one, anyway, at a logical level, the described three step process is completely suitable to explain the effects of compression.

on a block by block basis[17]. The most commonly observable effect is thus the attenuation of the high frequency components of the images.

The recently finalized JPEG2000 standard is, on the other side, based on DWT. After transformation, the coefficients in each band are quantized with a suitable quantization step (unique for each band) chosen on the base of perceptual considerations but usually quite fine. Quantized coefficients in each band are then partitioned into small blocks (e.g. 64×64 or 32×32) and entropy encoded with a bit-plane strategy: in practice the most significant bits are encoded first. The number of bit planes that are transmitted depends on the desired final bit rate and can be different from block to block (a rate distortion optimum strategy can be easily implemented that, given a desired final bit rate, chooses the number of bits to be assigned to each block in such a way to minimize the resulting distortion). This block-wise variable number of encoded bit planes actually makes the quantization step to be block-wise variable (in practice if the p LSBs are discarded in a block, the actual quantization step in that block is 2^p times higher than the one used during the quantization phase). As previously outlined, the loss of information is, in this case, occurring in both the two final steps, the final effect is anyway equivalent to having a block-wise variable quantization step in the second phase, followed by lossless entropy encoding in the third phase[18]. Another option made available by JPEG 2000 is the possibility of applying a power transformation to DWT coefficients before quantization, in such a way to produce a non-uniform quantization which well adapts to the properties of HVS (in particular to the fact that the HVS is sensitive to contrast more than to absolute luminance, see section 5.1.1).

Video coding standards also have a similar structure: the main difference is that, in order to reduce temporal redundancy, a motion compensation step is performed before the three phases described previously. In particular, each frame of a video sequence can be encoded in three different ways:

- I frames are encoded as still images, with a technique very similar to JPEG;

- P frames are first predicted based on the last I or P frame, in particular for each block of the frame the most similar block in the last I or P frame is looked for, the difference between these two blocks is then encoded as in the JPEG case (i.e. with a three step procedure: DCT, quantization, lossless entropy coding of quantized coefficients);

---

[17]We run the risk, in this latter case, to produce too much side information thus nullifying the effects of a more effective compression.

[18]Indeed the standard is a bit more complicated, but for our goals the present description is sufficient; interested readers can refer to [213]

- B frames are encoded in a similar way, the sole difference being that the prediction is now performed based on both the last I or P frame and the first next P frame.

Also in this case (as for JPEG), thus, the loss of information is restricted to the quantization phase. The DCT coefficients are quantized with a variable quantizer (see equation (5.59)) in the I frames and with an uniform quantizer in the P and B frames; usually quantization is coarser for B frames with respect to P, which, on their side, are quantized coarser than I frames.

Even for audio coding, the main responsible for loss of information is the quantization step. As an example, in the popular MP3 audio format, which corresponds to the Layer 3 coding mode of the MPEG-1 Audio standard, the coefficients in each subband are non-uniformly quantized, in practice a uniform quantization step is used after the coefficients are raised to a power of 3/4. Furthermore a different step size can be used from band to band, according to the allowed level of noise (estimated for each band as presented in section 5.4.5) and to the required bit rate.

7.5.1 Quantization of the watermarked features

From what we have said, it is evident that the loss of information caused by lossy coding algorithms is ultimately due to the quantization stage. The effects of quantization on data hiding systems is not always easy to understand, given that quantization is performed in a domain which, in general, does not coincide with the host feature domain. However, even in this case it is useful to refer to a frequency framework. As we have seen, in fact, quantization is performed by relying on perceptual criteria that cause higher frequency components to be quantized coarser than those at low frequencies. As a rule of thumb, then, we can say that lossy coding deteriorates the high frequency components of the watermark signal more heavily than the low ones.

In spite of the above observations, it is of great interest to investigate the case in which quantization just regards the watermarked features (e.g. JPEG compression of still images watermarked in the block-DCT domain, or JPEG 2000 compression of images watermarked in the DWT domain). Again, for simplicity, we will restrict our analysis to the case of zero mean Gaussian distributed features. Moreover we will only treat two cases of particular interest, namely 1-bit (detectable) additive SS, and ST-DM watermarking.

Additive spread spectrum watermarking

Let us start by noting that, after that the host features are quantized with a quantization step Δ_a, we have[19]:

$$f_i' = \begin{cases} f_i + q_{0,i} = x_{0,i} & H_0 \\ \gamma w_i + f_i + q_{1,i} = \gamma w_i + x_{1,i} & H_1 \end{cases} \tag{7.38}$$

where we can observe that, as opposed to equation (6.18), the additive noise (represented by $q_{0,i}$ and $q_{1,i}$) depends on the host features, and hence it follows a different statistic under the hypotheses H_0 and H_1. By assuming that a correlation based detector, equation (6.28), is used and that the threshold is chosen based on the desired value of the false alarm probability P_f, the following expression is obtained for the probability of missing the watermark[20]:

$$P_m = \frac{1}{2}\mathrm{erfc}\left(\sqrt{\frac{(\mu_{\rho|H_1} - \mu_{\rho|H_0} - \sqrt{2}\mathrm{erfc}^{-1}(2P_f)\sigma_{\rho|H_0})^2}{2\sigma_{\rho|H_1}^2}}\right), \tag{7.39}$$

with:

$$\begin{aligned}
\mu_{\rho|H_0} &= \frac{1}{n}\sum_{i=1}^{n} w_i \mu_{x_{0,i}}, \\
\mu_{\rho|H_1} &= \gamma\overline{w^2} + \frac{1}{n}\sum_{i=1}^{n} w_i \mu_{x_{1,i}}, \\
\sigma_{\rho|H_0}^2 &= \frac{1}{n^2}\sum_{i=1}^{n} w_i^2 \sigma_{x_{0,i}}^2, \\
\sigma_{\rho|H_1}^2 &= \frac{1}{n^2}\sum_{i=1}^{n} w_i^2 \sigma_{x_{1,i}}^2.
\end{aligned} \tag{7.40}$$

In order to estimate $\mu_{x_{0,i}}$, $\mu_{x_{1,i}}$, $\sigma_{x_{0,i}}^2$, and $\sigma_{x_{1,i}}^2$, it is convenient to refer to the pdf of the quantization noise, which can be easily verified to be given by:

$$p_q(q) = \mathrm{rect}\left(\frac{q}{\Delta_a}\right)\sum_k p_x(q + k\Delta_a), \tag{7.41}$$

---

[19]We use the symbol f_i' instead of $f_{w,i}'$ to expressly indicate that we do not know whether the asset at hand is watermarked or not.

[20]We have assumed that it is possible to apply the central limit theorem and, thus, that the detector response has a Gaussian distribution.

where $p_x(x)$ is the pdf of the to-be-quantized variable. We can then observe that if the to-be-quantized variable has a symmetric pdf (and thus zero mean), quantization noise follows a symmetric pdf as well (thus having zero mean too). From this it follows that given that f_i's are normally distributed with zero mean, the quantization noise $q_{0,i}$ also has zero mean, and thus $\mu_{x_{0,i}} = 0$ and $\mu_{\rho|H_0} = 0$. The same can not be said for $\mu_{\rho|H_1}$ given that in this case the noise $q_{1,i}$ results from the quantization of the variable $f_i + \gamma w_i$ that does not follows a symmetric pdf (it is a Gaussian centered at γw_i). For estimating the other parameters needed to compute P_m, we can observe that $x_{0,i}$ is nothing but the quantization of f_i, and thus:

$$\sigma^2_{x_{0,i}} = \Delta_a^2 \sum_k k^2 \int_{k\Delta_a - \Delta_a/2}^{k\Delta_a + \Delta_a/2} p_f(u) du, \qquad (7.42)$$

which (as it may be expected) does not depend on the index i. Similarly, if we define $y_i = x_{1,i} + \gamma w_i$, we have

$$\mu_{x_{1,i}} = \mu_{y_i} - \gamma w_i, \qquad (7.43)$$

and

$$\sigma^2_{x_{1,i}} = \sigma^2_{y_i} = E[y_i^2] - \mu^2_{y_i}, \qquad (7.44)$$

from which, given that y_i is nothing else that the quantization of $f_i + \gamma w_i$, we obtain:

$$\mu_{y_i} = \Delta \sum_k k \cdot \int_{k\Delta_a - \Delta_a/2}^{k\Delta_a + \Delta_a/2} p_f(u - \gamma w_i) du, \qquad (7.45)$$

and

$$E[y_i^2] = \Delta^2 \sum_k k^2 \int_{k\Delta_a - \Delta_a/2}^{k\Delta_a + \Delta_a/2} p_f(u - \gamma w_i) du. \qquad (7.46)$$

In figure 7.14 the ROC curve obtained for the quantization attack is reported. The curve has been obtained by using a binary watermark signal, and by imposing $n = 1000$ and DNR = DWR = 15 dB[21]. By comparing these results with those obtained for the case of Gaussian noise addition, it comes that the two types of attacks behave in the same way (in particular almost identical ROC curves are obtained by setting the same experimental parameters)[22]. This can be explained by the fact that in both cases the detection function (ρ) ultimately behaves as a Gaussian random variable with the same parameters.

---

[21]The DNR has been computed by referring to considerations similar to those that bring to equation (7.17), with the only difference that now the shift d is not present.

[22]Small differences start to appear only for low DWR values.

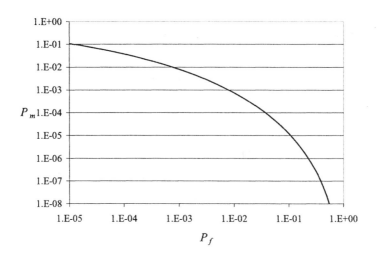

Figure 7.14: Plot of the ROC curve obtained with the correlation detector for the case of the quantization attack. The curve has been obtained by using a binary watermark signal $w_i = \pm 1$, and by imposing $n = 1000$, and DNR = DWR = 15 dB.

ST-DM watermarking

The computation of the bit error probability for ST-DM watermarking in the presence of host feature quantization with step Δ_a is rather cumbersome, and can be carried out analytically only by assuming that r is large enough. To be specific, let us consider the following model of the quantization attack:

$$f'_{w,i} = \mathcal{Q}_{\Delta_a}(f_{w,i}) = f_{w,i} + q_i, \qquad (7.47)$$

where $\mathcal{Q}_{\Delta_a}$ is a uniform quantizer with step Δ_a, and q_i is the corresponding (feature dependent) quantization noise. In this case the correlation between the attacked features and the reference direction $\mathbf{w}$ is given by:

$$\rho' = \sum_{i=1}^{r} f_{w,i} w_i + \sum_{i=1}^{r} q_i w_i = \rho_w + q_a. \qquad (7.48)$$

For the symmetry of the problem, we can restrict our analysis to the case $b = 0$. We have:

$$
\begin{aligned}
P_{e|0} &= \sum_{k=-\infty}^{\infty} p(u_{0,k}) P_{e|u_{0,k}} = \\
&= \sum_{k=-\infty}^{\infty} p(u_{0,k}) P\left(\rho' \in \bigcup_{j} [(j + 1/2)\Delta, (j + 1)\Delta] \,|u_{0,k} \right) = \\
&= \sum_{k=-\infty}^{\infty} p(u_{0,k}) \sum_{j=-\infty}^{\infty} \int_{(j+1/2)\Delta}^{(j+1)\Delta} f_{\rho'|u_{0,k}}(\rho') d\rho' = \\
&= \sum_{k=-\infty}^{\infty} p(u_{0,k}) \sum_{j=-\infty}^{\infty} \int_{(j+1/2)\Delta}^{(j+1)\Delta} f_{q_a|u_{0,k}}(\rho' - u_{0,k}) d\rho' = \\
&= \sum_{k=-\infty}^{\infty} p(u_{0,k}) \sum_{j=-\infty}^{\infty} \int_{(j-k+1/4)\Delta}^{(j-k+3/4)\Delta} f_{q_a|u_{0,k}}(\rho') d\rho',
\end{aligned}
\tag{7.49}
$$

where we have exploited the fact that $u_{0,k} = k\Delta + \Delta/4$. In order to evaluate the error probability, we need the pdf of the random variable q_a conditioned to *transmission* of the codebook entry $u_{0,k}$. Let us start by observing that q_a is the weighted sum of the quantization noise values q_i affecting each watermarked feature. As such, each q_i will depend on the corresponding $f_{w,i}$. The analysis is complicated by the fact that, at least in principle, the watermarked features $f_{w,i}$ are not independent. To see this, let us decompose the row vector $\mathbf{f}$ into a component $\mathbf{f}^{\perp}$ orthogonal to $\mathbf{w}$ and a component parallel to $\mathbf{w}$. Of course, ST-DM only affects the parallel part by replacing it with $\rho_w \mathbf{w}$, i.e.:

$$
\mathbf{f}_w = \mathbf{f}^{\perp} + \rho_w \mathbf{w}, \tag{7.50}
$$

where the components of $\mathbf{f}^{\perp}$ are a linear combination of the original features f_i that we assumed to be normally distributed i.i.d. random variables. The distribution of $\mathbf{f}^{\perp}$ will thus be a multivariate Gaussian with zero mean and covariance matrix C to be calculated. As to $\mathbf{f}_w$ it is immediate to see that, it still follows a multivariate Gaussian distribution, with:

$$
E[\mathbf{f}_w | u_{0,k}] = u_{0,k} \mathbf{w}, \tag{7.51}
$$

and covariance matrix:

$$
C = E[\mathbf{f}_w^t \mathbf{f}_w] = E[(\mathbf{f}^{\perp})^t \mathbf{f}^{\perp}]. \tag{7.52}
$$

In order to calculate $E[(\mathbf{f}^{\perp})^t\mathbf{f}^{\perp}]$, let us introduce the projection operators that project the vector $\mathbf{f}$ over the space orthogonal to $\mathbf{w}$. From linear algebra we know that such an operator has the form $P_w^{\perp} = I - \mathbf{w}^t\mathbf{w}$, hence we can write:

$$
\begin{aligned}
E[(\mathbf{f}^{\perp})^t\mathbf{f}^{\perp}] &= E[(I - \mathbf{w}^t\mathbf{w})\mathbf{f}^t\mathbf{f}(I - \mathbf{w}^t\mathbf{w})] \\
&= E[\mathbf{f}^t\mathbf{f}] - \mathbf{w}^t\mathbf{w}E[\mathbf{f}^t\mathbf{f}] - E[\mathbf{f}^t\mathbf{f}]\mathbf{w}^t\mathbf{w} + \mathbf{w}^t\mathbf{w}E[\mathbf{f}^t\mathbf{f}]\mathbf{w}^t\mathbf{w} \quad (7.53) \\
&= \sigma_f^2 I - 2\sigma_f^2\mathbf{w}^t\mathbf{w} + \sigma_f^2\mathbf{w}^t\mathbf{w}\mathbf{w}^t\mathbf{w} = \sigma_f^2(I - \mathbf{w}^t\mathbf{w}),
\end{aligned}
$$

where in the last equality we exploited the fact that $\mathbf{w}\mathbf{w}^t = \|\mathbf{w}\|^2 = 1$. By remembering that $w_i = \pm 1/\sqrt{r}$ we can conclude that:

$$
C_{ij} = E\left[f_i^{\perp} f_j^{\perp}\right] = \begin{cases} \left(1 - \dfrac{1}{r}\right)\sigma_f^2 & \text{if } i = j \\ \pm \dfrac{1}{r}\sigma_f^2 & \text{if } i \neq j \end{cases} \quad (7.54)
$$

which proves the dependency between $f_{w,i}$ coefficients. If r is large enough, though, we can assume that the $f_{w,i}$ are independent, thus permitting us to consider the q_i terms independent of each other. Furthermore, if r is large enough, we can exploit the central limit theorem, and approximate q_a, as given by equation (7.48), by a Gaussian random variable with mean (we avoid to explicitly indicate the conditioning to $u_{0,k}$ for notation simplicity):

$$
\mu_{q_a} = \sum_{i=1}^{r} w_i \mu_{q_i}, \quad (7.55)
$$

and variance:

$$
\sigma_{q_a}^2 = \sum_{i=1}^{r} w_i^2 \sigma_{q_i}^2 = \frac{1}{r}\sum_{i=1}^{r} \sigma_{q_i}^2. \quad (7.56)
$$

It now remains to estimate the mean μ_{q_i} and the variance $\sigma_{q_i}^2$ resulting from the quantization with a step size Δ_a of a Gaussian random variable having mean $u_{0,k}w_i = (k\Delta + \Delta/4)\,w_i$ and variance (approximately) σ_f^2. It is easy to demonstrate that

$$\mu_{q_i} = \sum_{l=-\infty}^{\infty} \int_{\Delta_a(l-1/2)}^{\Delta_a(l+1/2)} (q - l\Delta_a) \frac{1}{\sqrt{2\pi\sigma_f^2}} e^{-\frac{(q-u_{0,k}w_i)^2}{2\sigma_f^2}} dq$$

$$= \sum_{l=-\infty}^{\infty} \int_{\frac{\Delta_a(l-1/2)-u_{0,k}w_i}{\sigma_f}}^{\frac{\Delta_a(l+1/2)-u_{0,k}w_i}{\sigma_f}} (\sigma_f t + u_{0,k}w_i - l\Delta_a) \frac{1}{\sqrt{2\pi}} e^{-\frac{t^2}{2}} dt \qquad (7.57)$$

$$= u_{0,k}w_i - \Delta_a \sum_{l=-\infty}^{\infty} l \int_{\frac{\Delta_a(l-1/2)-u_{0,k}w_i}{\sigma_f}}^{\frac{\Delta_a(l+1/2)-u_{0,k}w_i}{\sigma_f}} \frac{1}{\sqrt{2\pi}} e^{-\frac{t^2}{2}} dt$$

Similarly the variance $\sigma_{q_i}^2$ can be computed based on the Mean Square Value of q_i that results to be:

$$\text{MSV}_{q_i} = \sum_{l=-\infty}^{\infty} \int_{\Delta_a(l-1/2)}^{\Delta_a(l+1/2)} (q - l\Delta_a)^2 \frac{1}{\sqrt{2\pi\sigma_f^2}} e^{-\frac{(q-u_{0,k}w_i)^2}{2\sigma_f^2}} dq$$

$$= \sigma_f^2 + \frac{1}{r} u_{0,k}^2 + \Delta_a^2 \sum_{l=-\infty}^{\infty} l^2 \int_{\frac{\Delta_a(l-1/2)-u_{0,k}w_i}{\sigma_f}}^{\frac{\Delta_a(l+1/2)-u_{0,k}w_i}{\sigma_f}} \frac{1}{\sqrt{2\pi}} e^{-\frac{t^2}{2}} dt$$

$$\qquad (7.58)$$

$$- 2\sigma_f \Delta_a \sum_{l=-\infty}^{\infty} l \int_{\frac{\Delta_a(l-1/2)-u_{0,k}w_i}{\sigma_f}}^{\frac{\Delta_a(l+1/2)-u_{0,k}w_i}{\sigma_f}} \frac{t}{\sqrt{2\pi}} e^{-\frac{t^2}{2}} dt$$

$$- 2u_{0,k}w_i \Delta_a \sum_{l=-\infty}^{\infty} l \int_{\frac{\Delta_a(l-1/2)-u_{0,k}w_i}{\sigma_f}}^{\frac{\Delta_a(l+1/2)-u_{0,k}w_i}{\sigma_f}} \frac{1}{\sqrt{2\pi}} e^{-\frac{t^2}{2}} dt.$$

It is worth observing that the MSW_{q_i} value also affects the strength of the attack, and hence the computation of WNR and DNR:

$$\text{WNR} = \frac{\alpha\Delta^2}{\sum_k p(u_{0,k}) \sum_{i=1}^{r} \text{MSV}_{q_i}}, \qquad (7.59)$$

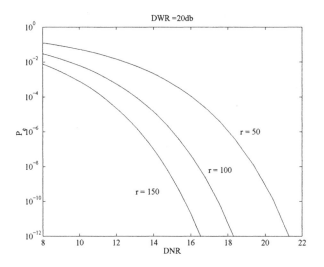

Figure 7.15: Bit error probability for STDM watermarking in the presence of a quantization attack.

$$\mathrm{DNR} = \frac{r\sigma_f^2}{\sum_k p(u_{0,k}) \sum_{i=1}^{r} MSV_{q_i}}, \qquad (7.60)$$

where we have used the more general definitions of DNR and WNR given in equations (7.6) and (7.4).

An exemplificative sketch of the bit error probability of ST-DM in the presence of a quantization attack is given in figure 7.15, where P_e is plotted against DNR for DWR = 20db and for various values of r.

The bit error probability characterizing the quantization attack can be compared to that of the additive Gaussian attack. As it happened in the additive spread spectrum case, it comes out that the two kinds of attack are equivalent, in that a very similar bit error probability is obtained for a wide range of r, DWR and WNR.

7.6   Geometric manipulations

The transformations that we have examined until now only affect the values of the samples of the asset: the goal of this section is to analyze those transformations affecting the position of samples, i.e. the so called geometric transformations. In general, the effect of this kind of transformations on the watermark depends on the domain in which the watermark has been embedded. In particular, systems operating in the asset domain will undergo exactly the same transformation that affects the watermarked asset, while the effects on transformed domain and hybrid systems will be very different.

7.6.1 Asset translation

Let us start by considering global transformations, i.e. geometric transformations that modify all asset samples positions according to a global rule. The simplest case is translation, i.e. all sample positions are modified by a fixed offset. If embedding is performed in the asset domain, the watermark signal will simply undergo the same translation as the asset, thus the problem for the watermark detector is to recover the synchronization of the embedded watermark signal which is not deteriorated, but simply translated to an unknown position (possibly at a fraction of the sampling step). The problem can be solved by an exhaustive search of some pilot watermark, i.e. a signal known by the detector and embedded solely for recovering the position of the actual watermark. The exhaustive search can be made more effective if the detection of the pilot signal is based on a correlation operation, in this case, in fact, we can exploit the correspondence between correlation in the asset domain and multiplication in the DFT domain, thus permitting us to resort to the FFT algorithm.

Let us then pass to see what happens with transformed domain techniques; for simplicity we will concentrate on a 1D signal, the phenomena being easily generalizable to higher dimensions. First of all let us notice that DFT magnitude is not affected by cyclic[23] translations in the asset domain, while a linear term is added to the phase. The validity of these properties is conditioned to the fact that the size of the DFT is not changed after the translation has occurred: this is granted if the translation is performed in a cyclic way. However, translation is usually associated (in particular for images) to cropping which reduces the size of the asset. In practice the translation is not performed in a cyclic way, but the samples that are disappearing at one extreme of the signal support are simply discarded. In this case it is important that a fixed size is used for computing the DFT during the embedding and recovery phases, regardless of the actual number of samples available. The size of the DFT is, in fact, inversely proportional to the frequency sampling step, given that a translation does not change the asset domain sampling frequency, maintaining constant the number of samples used for computing the DFT, is equivalent to keep the sampling step in the frequency domain constant. This is exemplified in figure 7.16,

---

[23]A cyclic translation consists of the replacement of the positions that are left free, because of the translation, at one extreme of the signal support, with the samples that, for the same reason, have disappeared at the other extreme, or in a more formal way: $A'(n) = A(n \ominus n_0)$ where $A'(n)$ is the translated signal, n_0 is the translation offset, $\ominus$ denotes modulo-N subtraction, and N is the size of the asset.

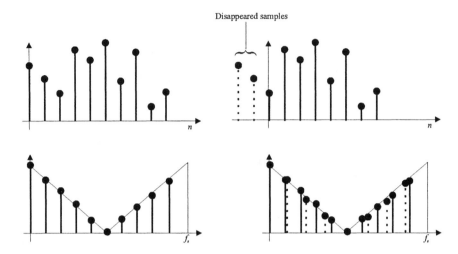

Figure 7.16: Effects of the translation of a 1D signal on the magnitude of its DFT. At the top left it is shown the original N-sample long signal; at the top right the same signal is reported after a 2 sample translation on the left; at the bottom left the DFT magnitude of the original signal is given, whereas at the bottom right the DFT magnitude of the translated signal is reported. In this later figure dotted samples correspond to the DFT computed on $N - 2$ samples, while solid ones correspond to the DFT computed on N samples and are the same as those of the original DFT.

where an N-sample signal is shown (top left), together with its version translated by 2 samples (top right), the magnitude of the DFT of the original signal (bottom left), and the magnitude of the DFT of the translated signal (bottom right). In the last figure dotted samples correspond to the DFT computed on $N - 2$ points while the continuous samples correspond to the DFT computed on N points. In the figure the loss of information due to cropping after translation has been neglected. We will see in following section 7.7 how cropping affects the spectrum of a signal; for now it is enough to say that if cropping is not too heavy, this loss of information is negligible. It is important, though, that missing (disappeared) points are replaced by zeroes so to sample the frequency spectrum exactly at the same points it was sampled before cropping.

Much more complex is the case of the DCT, the effects of the translation of the signal on its DCT coefficients are, in fact, not easy to model.

Considerations similar to those drawn for the transformed domain approach are still valid for hybrid techniques based on block-wise transforms. In this case we have also to consider that the asset translation makes it

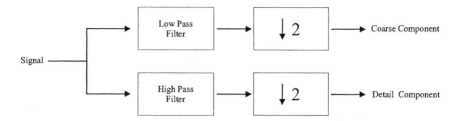

Figure 7.17: Scheme of a generic sub-band decomposition system.

impossible to identify the origin of the blocks. However, it has to be noted that for watermarking systems operating on the magnitude of DFT coefficients, this is a problem only if a different watermark is used in different blocks. With regard to systems working in the DWT domain (and in general exploiting some sort of sub-band decomposition) a particular analysis is worth to be performed. To this aim let us refer to figure 7.17 where a generic sub-band decomposition system is presented. The signal is first filtered with a low pass and a high pass filter, each filtered version is then down-sampled by a factor of two (this is possible because each filtered version has a bandwidth of half the original signal[24]). The critical step, from the point of view of signal translation, is just the down-sampling process. Down-sampling, in fact, is not a shift invariant operation, hence translation of the original signal may result in a completely different set of sample values[25]. To overcome this problem, two solutions can be envisaged:

- When the watermark is recovered, the sub-sampling process is performed twice, once by taking, the even samples, and the once the odd ones, the watermark is then looked for in both these sets.

- A correlated watermark is used.

The second solution is preferable because it also allows to partially deal with non integer pixel translation steps.

---

[24]Indeed, in general the filters used introduce some aliasing whose effects, for the case of a perfect reconstruction decomposition, can be compensated for during the reconstruction phase.

[25]This is the case, for example, when the signal is shifted one pixel left (or right) followed by down-sampling by a factor two. Subsampled pixels may either correspond to samples in even or odd position according to whether the input signal is translated or not. Thus, if an uncorrelated watermark signal was hosted by even (odd) samples, no trace of it would be found in the odd (even) translated samples.

7.6.2 Asset zooming

The analysis we carried out about the problems caused by simple signal translation can give us an idea on how bigger can be the effects of more complicated geometric manipulations. We pass, then, to consider zooming, which is basically a resampling process. First it is important to notice the different influence that resampling has in the temporal or spatial domain on our perception of the assets. This is reflected by the fact that, while the temporal sampling frequency (the frame rate for the case of video) has to be known for properly reproducing the digital content, the spatial sampling frequency is usually not known, and, in fact, the same image can be shown on displays having very different spatial resolution without any appreciable problem. Stating this in another way, if an audio segment originally acquired at 44.1 KHz sampling frequency, is down-sampled to, let us say, 16 KHz, the new sampling frequency has to be indicated instead of the old one in some field of the header of the file containing the segment, otherwise the reproduction of the audio would be slowed down. On the contrary, if a 1024 × 1024 image is down-sampled to, let us say, 600 × 600, the image can be displayed without having any knowledge of the new (or old) spatial sampling frequency. This is reflected by the fact that, while for audio files or videos the temporal sampling frequency (the frame rate for video) has always to be included in the audio or video file format, the same is not true for still images. The above considerations imply that the range of possible zoom values a watermarking system has to deal with in the temporal and the spatial domain are very different. In practice, if the zoom applied in the temporal domain is small enough, the nominal sampling frequency (i.e. the sampling frequency indicated into the audio or video file) does not need to be changed: this will result in a slightly slower or faster reproduction of the audio visual content, that is not perceptible. But if the zoom applied to the temporal domain is larger, then the change of sampling frequency has to be recorded into the file: based on this information, then, zooming can be inverted before watermark recovery, to obtain the same sampling frequency that was used during the embedding. Usually the amount of tolerable zoom for and audio segment is ±10%[26]. On the contrary in the spatial domain much larger factors can be found (it is common to deal with zooming factors ranging from -50% to +100%), effective interpolation techniques are, in fact, available, which allows not to deteriorate image quality in dramatic way. Furthermore, for still images a different zooming factor (anisotropic zooming) can be applied to the two dimensions: although very different factors would deteriorate the image aspect, factors differing by a

---

[26]This corresponds, if the nominal sampling frequency is not changed, to a decrease (or increase) of the reproduction speed by 10%.

small amount can make watermark detection quite difficult (consider, for example, the case of the Fourier-Mellin transform using a log-polar mapping that only accounts for isotropic zooming and rotations, see page 359). Indeed, something needs to be added with regard to video frames: the size of a frame can not, in fact, be whatever we want, international standard (ITU-T H.26x, ISO-MPEGx) have defined a set of possible dimensions to which every video has to adapt. This imply that if a frame is zoomed, let us suppose enlarged, it is likely that it is cropped as well[27] to fit again one of the standard sizes. If the area cut out by cropping is large (in practice this happens if the zoom is too large), then the loss of video information may not be tolerable.

Now that we have examined the different characteristics and influence of temporal and spatial zooming, we pass to analyze the effects that these processes have on different watermarking approaches. As usual we start by considering asset domain techniques. As we have seen the zoom implies a change of the sampling frequency, and, in practice, includes an interpolation operation. This causes a loss of synchronism between the watermark signal and the detector/decoder. Even in this case (as we have seen for the case of translations) an exhaustive search considering all admissible zooming factors is possible, but, if the range of allowed factors is large, such a search is likely to be computationally prohibitive. Furthermore, differently from the case of simple translation in which basically the watermark signal is not deteriorated, in the case of down-sampling a loss of information is likely to occur[28].

The effect of the zooming on watermark signals embedded in the frequency domain is easily modelizable; it is, in fact, well known that a zoom in the asset domain corresponds to a zoom by an inverse factor of the spectrum. Anyway, we must consider that discrete transforms (DFT and DCT) actually sample the spectrum with a step size that is inversely proportional to the number of points used for computing the transformation. Of course if a signal of N samples (for simplicity of notation we refer to the 1D case, the extension to higher dimensional signals being straightforward), whose DFT (DCT) is sampled in N equally spaced points, has been zoomed, passing to M samples, its DFT (DCT) is zoomed by a factor of M/N, and sampled in M equally spaced points. Then, if the original DFT (DCT) coefficients correspond to frequencies multiple of f_s/N ($f_s/2N$ for the DCT), where f_s is the sampling frequency, the DFT (DCT) coefficients of the

---

[27]Indeed some of these dimensions are scaled versions of others (e.g. the CIF 352×288 is doubled with respect to QCIF 176×144, and then some zooming factors can be applied without cropping.

[28]Down-sampling, in fact, usually implies a reduction of the spectral content of the signal.

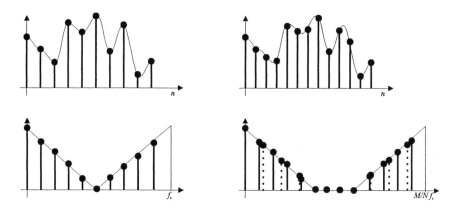

Figure 7.18: Example of the effects of the zooming of a 1D signal on its DFT. At the top left the original signal of N samples; at the top right the zoomed signal (of length M samples); at the bottom left the DFT of the original signal sampled at N equally spaced points, at the bottom right the DFT of the zoomed sampled at M equally spaced points. The dotted samples correspond to the DFT computed on the zoomed signal on N points.

zoomed signal correspond to frequencies multiple of $(f_s M/N)/M = f_s/N$ $((f_s M/N)/2M = f_s/2N$ for the DCT), i.e. exactly the same frequencies of the original signal. This is exemplified in figure 7.18, where at the top left the original signal consisting of N samples is sketched, its zoomed version of M samples is shown at the top right, the DFT of the original at the bottom left and the DFT of the zoomed signal at the bottom right: this last DFT is shown as computed on M points (continuous samples) and on N points (dotted samples). This conclusion should be contrasted to what we said with regard to cropping: in that case, in fact, to be sure to obtain the same DFT coefficients used during the embedding phase, when computing the DFT we had to use the same number of points used during embedding (regardless of the number of samples available at the detector), while in this case we must use just the number of points that are available at the detector. This would not be a problem if we knew the type of manipulations (cropping and translation or zooming) applied to the asset, but this is not possible in general. In addition, these attacks could have both been applied, thus further complicating the problem.

7.6.3   Image rotation

For images another common type of global geometric transformation is rotation. In particular a clockwise rotation by an angle of θ radiants can be expressed as:

$$\begin{bmatrix} x' \\ y' \end{bmatrix} = \begin{bmatrix} \cos\theta & \sin\theta \\ -\sin\theta & \cos\theta \end{bmatrix} \begin{bmatrix} x \\ y \end{bmatrix} \tag{7.61}$$

where (x', y') are the coordinates of the new sampling points, and (x, y) the original ones. The effect of this process on the watermark signal embedded in the spatial domain is not much different than that described with regard to translation: the watermark is moved to an unknown position, but no loss of information is occurring (at least if cropping is not applied). Similar considerations also apply to hybrid watermarking system.

On the contrary, a few words need to be said about transform domain techniques. It is well known that a rotation in the spatial domain corresponds to a rotation by the same angle of the Fourier transform. It has to be remembered, anyway, that the DFT is the Fourier transform of a periodic repetition of the signal: thus the DFT of a rotated image corresponds to the periodic repetition of the rotated image, and this is not exactly equal to the rotation of the Fourier transform of the periodic repetition of the original signal. This is exemplified in figure 7.19, where an image and its rotated version are displayed at the top, and the corresponding DFT's at the bottom. The main difference is constituted by the appearance of new frequency components, due to the different periodicitization of the signal. Similar considerations hold for the DCT.

Rotation and anisotropic zooming can be concatenated (i.e. applied one after the other) many times. In general it is possible to demonstrate that any sequence of any number of rotations and zooms can be obtained by simply concatenating one rotation, one anisotropic zoom, and another rotation. To demonstrate this, let us first show that any zoom can be expressed in matrix form as:

$$\begin{bmatrix} x' \\ y' \end{bmatrix} = \begin{bmatrix} a_x & 0 \\ 0 & a_y \end{bmatrix} \begin{bmatrix} x \\ y \end{bmatrix}, \tag{7.62}$$

i.e. by a diagonal matrix. Now let us suppose we have a sequence of zoom, rotation and zoom operations, that can be expressed by the product of the three matrices: $Z_1 R Z_2$, where (as shown in equation (7.61) and (7.62)) Z_1 and Z_2 are diagonal matrices, and R is an orthogonal matrix[29]. The product of the first two matrices can be expressed as: RA, where $A = R^{-1} Z_1 R$. It is easy to show that A is a symmetric matrix, in fact:

$$A^T = \left(R^{-1} Z_1 R\right)^T = R^T Z_1^T R^{-1^T} = R^{-1} Z_1 R = A. \tag{7.63}$$

---

[29]Let us recall that a characteristic of orthogonal matrices is that $R^{-1} = R^T$, i.e. the inverse matrix corresponds to the transposed.

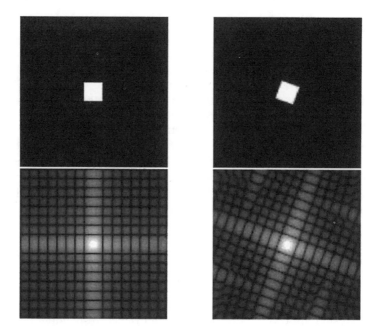

Figure 7.19: Exemplification of the difference between the DFT or a rotated image and the rotated DFT of the original image. It is evident that the DFT of the rotated image (bottom right) does not correspond to the rotation of the DFT of the original image (bottom left).

We have thus obtained that the combination of the three transformations can be expressed as:

$$Z_1 R Z_2 = R A Z_2. \tag{7.64}$$

Now let us observe that the product matrix $A Z_2$ is symmetric (as product of two symmetric matrices), and can thus be diagonalized and written as

$$A Z_2 = R'^{-1} Z' R', \tag{7.65}$$

where R' is a suitable orthogonal matrix (i.e. a rotation) and Z' is diagonal (i.e. an anisotropic zoom). By combining (7.64) and (7.65) we finally get:

$$Z_1 R Z_2 = R A Z_2 = R R'^{-1} Z' R' = R'' Z' R' \tag{7.66}$$

where $R'' R R'^{-1}$ is also orthogonal. If in addition to the three original transformations $Z_1 R Z_2$ we apply another zoom (which is combined with Z_1), or another rotation (which is combined with R'') the resulting combined transformation can always be expressed as a combination of one rotation, one zoom, and another rotation, which demonstrates our statement. In general, then, the combination of these three operations, can be described by a generic square matrix:

$$R''Z'R' = \begin{bmatrix} \cos\theta'' & \sin\theta'' \\ -\sin\theta'' & \cos\theta'' \end{bmatrix} \begin{bmatrix} z'_x & 0 \\ 0 & z'_y \end{bmatrix} \begin{bmatrix} \cos\theta' & \sin\theta' \\ -\sin\theta' & \cos\theta' \end{bmatrix}$$

$$= \begin{bmatrix} t_{11} & t_{12} \\ t_{21} & t_{22} \end{bmatrix} \tag{7.67}$$

7.6.4 More complex geometric transformations

With reference to images more complex geometric transformations can be tolerated by the Human Visual System. For example, the simple transformation described by equation (7.67), can be generalized by adding a constant translation term, and a term depending on the product of the coordinates x and y:

$$\begin{aligned} x' &= t_{10} + t_{11}x + t_{12}y + t_{13}xy \\ y' &= t_{20} + t_{21}x + t_{22}y + t_{23}xy. \end{aligned} \tag{7.68}$$

This transformation corresponds to move the four corners of the image into four new positions, and to modify coherently all the other sampling positions (see figure 7.20). If the new positions are not too far away from the original ones, the image deterioration is not perceptible, but the effects on watermark recovery can be dramatic. By the way, this is the first step of the popular Stirmak random bending attack, where the positions of the new corners are chosen randomly. In this attack, other two coordinates transformations are following this. The second step is given by:

$$\begin{aligned} x'' &= x' + d_{max}\sin\left(y'\tfrac{\pi}{M}\right) \\ y'' &= y' + d_{max}\sin\left(x'\tfrac{\pi}{N}\right) \end{aligned} \tag{7.69}$$

where M, and N are the vertical and horizontal dimensions of the image. This transformation applies a displacement which is zero at the border of the image, and maximum (of value d_{max}) in the center. Also in this case, if the maximum displacement d_{max} is not too large, the image deterioration is not perceptible, but the effects on watermark recovery can be dramatic, because a loss of syncronization which is spatially varying is introduced. The third step of the Stirmark geometric attack is expressed as:

$$\begin{aligned} x''' &= x'' + \delta_{max}\sin\left(2\pi f_x x''\right)\sin\left(2\pi f_y y''\right)\mathrm{rand}_x(x'',y'') \\ y''' &= y'' + \delta_{max}\sin\left(2\pi f_x x''\right)\sin\left(2\pi f_y y''\right)\mathrm{rand}_y(x'',y'') \end{aligned} \tag{7.70}$$

where f_x and f_y are two frequencies depending on the dimension of the image (always smaller than 1/20), and $\mathrm{rand}_x(x'',y'')$ and $\mathrm{rand}_y(x'',y'')$ are

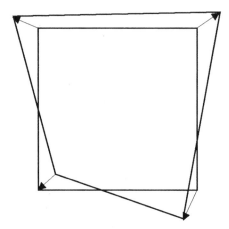

Figure 7.20: Example of the transformation that constitutes the first step of the popular Stirmark random bending attack.

random numbers in the interval $[1, 2)$. This last transformation introduces a random displacements at every pixel position. Again if δ_{max} is small enough (the default value for the Stirmark benchmark is 0.1), image deterioration is not perceptible, but the effects on watermark recovery are dramatic, because a random loss of synchronization is introduced. The Stirmark random bending attack, that is part of the Stirmark benchmark (see section 7.11.2), has demonstrated to be very effective in removing most of the watermarks developed so far. In addition, it can be seen as a first manipulation explicitly designed for removing the watermark (i.e. of malicious manipulation).

7.6.5 Countermeasures against geometric manipulations

While describing the effects of geometric manipulations on the hidden information, we have already mentioned some of the possible solutions that can be adopted to reduce the problems. Anyway, given the importance of this issue it is worth to dedicate to it more attention: we will thus analyze in this section in more detail some of the most popular countermeasures that can be adopted to get around this problem.

Our analysis will be as general as possible, nevertheless, some of the proposed solutions are explicitly designed to deal with image watermarking, since this is the case which is more heavily affected by this kind of attacks. Of course, we will not consider all kinds of geometric manipulations, on the contrary, we will focus on a class of global geometric transformations including cropping, translations, scaling and rotation[30]. For example, we will neglect more subtle attacks such as the random bending attack, or local geometric distortions which are rather common in the audio case.

We will assume that a blind detector/decoder is used, since with non-blind systems global geometric attacks are not a problem, given that the geometric transformation can be easily estimated and inverted by comparing the attacked and the original assets.

Before going on, two words are in order about cropping[31]. As it will be more deeply detailed later in section 7.7, the effect of cropping on the hidden information is twofold: on one hand the watermark content is damaged due to the loss of part of the information, on the other hand, watermark synchronization may be lost, in that the removal of part of the hosting signal results in a translation of feature space origin. In this section we only deal with the latter effect, since loss of information due to cropping can not be modeled as a geometrical transformation of the host asset.

Exhaustive search

By recognizing that the problem with geometric attacks is one of loss of synchronization, a possible solution consists of looking for the watermark at all possible translations, scaling factors and, in the case of images, rotation angles. Of course, the main problem with this approach is complexity, however several other problems exist making the exhaustive search of the watermark unfeasible in most applications.

To be specific, let us start the analysis by considering watermark detection. In this case, exhaustive search is conceptually straightforward. Let $\mathcal{D}(A, \mathbf{b}, K)$ indicate the detector operator, and let $A(\mathbf{d}, \mathbf{s}, \theta)$ be a copy of the host asset which has been translated by a displacement vector $\mathbf{d}$, scaled by a factor $\mathbf{s}$ and rotated by an angle θ, where the scale vector $\mathbf{s}$ possibly accounts for anisotropic scaling along the horizontal and vertical image axis. Exhaustive watermark detection amounts to the following decision rule. If a triplet $\{\mathbf{d}, \mathbf{s}, \theta\}$ is found such that

$$\mathcal{D}(A(\mathbf{d}, \mathbf{s}, \theta), \mathbf{b}, K) = \text{yes}, \qquad (7.71)$$

then decide for the presence of $\mathbf{b}$ within A. Computational complexity of the exhaustive search approach depends on several factors, including the number of geometric parameters to be taken into account and their

---

[30]Of course, rotation is only meaningful in the image watermarking case

[31]In audio and video watermarking, the hidden information is repeatedly embedded in proper subparts of the host signal. We speak about cropping when watermark recovery has to be performed on an asset segment which is smaller than the watermark repetition period.

quantization step. As to this point, the sensibility of the detector to small geometrical transforms must be evaluated. Suppose, for example, that an image watermarking system is capable of revealing the watermark even in the presence of 1 pixel displacement, an isotropic scale factor in the [0.9,1.1] range, and a rotation angle of 1 degree. By assuming that the input image size is 512×512, and that the minimum and maximum admissible scale factors are 0.5 and 2 respectively, it easily turns out that the detector has to consider $15 \times 15 \times 512 \times 512 \times 360 = 11,418,992,640$ different geometrical configurations, thus leading to an extremely high computational burden. Obviously, complexity is reduced if some of the parameters defining the geometrical transform can be neglected. This is the case, as we have seen, of an image watermark embedded in the DFT magnitude domain. Due to translation invariance, in fact, complexity decreases by a factor 512^2, thus reducing the search space to $43,560$ configurations. Similarly we have seen at page 345 that FFT can be used for effectively computing the correlation between the (possibly translated) watermarked asset and the watermarking signal.

A drawback of exhaustive watermark searching is that in this way the false detection probability considerably increases. More specifically, if the watermark is looked for at M different geometrical configurations, then the false detection probability is increased approximately by a factor M. In the common case that a given target $\hat{P}_f$ has to be granted, the detector must be designed by letting the false detection probability at each step be equal to $\hat{P}_f/M$, thus increasing considerably the probability of missing the watermark.

Exhaustive watermark search can not be directly applied to readable schemes, unless the decoder is also able to assess whether the host asset is marked or not. Watermark decoders, in fact, always result in an estimated bit sequence, thus it is impossible to use the decoder output to guess the right geometric transformation to be applied to the host asset. To circumvent this problem, a number of synchronization bits are usually added to the watermark, then the watermark is read at all possible geometrical configuration. Only the bit sequence read in correspondence of the, hopefully unique, configuration yielding the right synchronization bits is retained. No need saying that a sufficiently large number of sync bits must be provided for, to make the false detection probability negligible.

Synchronization template

One of the most common way to cope with geometric manipulations is to estimate the parameters of the transformation applied to the asset, and to invert it. To do so, a synchronization pattern is usually inserted at some fixed position in the frequency domain. The pattern may simply be a set of peaks, or more complicated patterns may be used. To improve security, the pattern may also depend on a secret key known to authorized users only. Once the synchronization template has been recovered, its position is used to estimate the parameters of the geometric transformation applied after watermark embedding.

To be more specific, let us focus on image watermarking (the audio case is a simpler one, since rotations has not to be considered). Translations are usually dealt with either by choosing a feature space which is invariant to spatial shifts, e.g. the frequency magnitude domain, or by applying a fast exhaustive search through FFT.

We have seen that any combination of spatial scaling (zoom) and rotations can be described by a simple matrix product as:

$$
\begin{bmatrix} x' \\ y' \end{bmatrix} = \begin{bmatrix} t_{11} & t_{12} \\ t_{21} & t_{22} \end{bmatrix} \begin{bmatrix} x \\ y \end{bmatrix},
\tag{7.72}
$$

to which corresponds, in the frequency domain, the following relation:

$$
\begin{bmatrix} f'_x \\ f'_y \end{bmatrix} = \begin{bmatrix} t'_{11} & t'_{12} \\ t'_{21} & t'_{22} \end{bmatrix} \begin{bmatrix} f_x \\ f_y \end{bmatrix}.
\tag{7.73}
$$

where we have that[32]:

$$
\begin{bmatrix} t'_{11} & t'_{12} \\ t'_{21} & t'_{22} \end{bmatrix} = \begin{bmatrix} t_{11} & t_{21} \\ t_{12} & t_{22} \end{bmatrix}^{-1}.
\tag{7.74}
$$

Assume, now, that the synchronization template consists of S peaks in the frequency domain. For each recovered peak, two equations like those in (7.73) can be written, where $(t'_{11}, t'_{12}, t'_{21}, t'_{22})$ are the unknowns, and (f_x, f_y), (f'_x, f'_y) known terms corresponding to the original and recovered peak coordinates. Then an MSE procedure can be activated to estimate $(t'_{11}, t'_{12}, t'_{21}, t'_{22})$. To correctly write the MSE system, correspondences between recovered peaks and those of the known sync template must be determined, e.g. by exploiting additional, geometrically invariant, information such as segment or area ratios. Alternatively, all possible matches may be considered and the one yielding the minimum residual square error retained.

Practical implementation of template-based synchronization must take into account that, due to unavoidable inaccuracies in peak localization, transformation parameters are always affected by error. Then it is necessary that, a reduced-extent, exhaustive search in the neighborhood of

---

[32]Note that the inverse is taken of the transposed matrix.

the recovered geometric configuration is performed, the exact extent of the search depending on the sensibility of the detector/decoder upon small geometric transformations. Additionally, in readable watermarking systems, provision must be made for synchronization bits, as described in the previous section.

Self-synchronizing watermarks

The use of a self-synchronizing watermark is an alternative solution (see section 3.1.4). The same watermark is inserted periodically either in the asset or the frequency domain, the repetition period being known. Such a period is estimated again at the detector side by looking at the peaks of the autocorrelation of the watermarked asset. By comparing the original period and the estimated one, the detector can trace back to the scaling factor and to the rotation angle applied to the asset after embedding. Such transformations are then inverted and the original asset configuration recovered. No need saying that, due to unavoidable inaccuracies of the estimate, a local exhaustive search is required. As to translations, an exhaustive search is usually performed, possibly exploiting fast correlation computation through FFT.

In order to analyze the impact of watermark periodicity on detector performance, let us consider the simple case of an additive watermark embedded in normally distributed features, in the presence of additive Gaussian noise.

Optimum watermark detection follows the analysis carried out in section 6.1.2. More specifically, optimum detection is still based on the comparison between the correlation:

$$\rho = \frac{1}{n} \sum_{i=1}^{n} f_i' w_i, \tag{7.75}$$

and a threshold T_ρ set by applying the Neyman-Pearson criterion. The probabilities of missing the presence of the watermark or falsely detecting it also remain the same. The analysis changes when calculating the false detection probability in the presence of a watermark $\mathbf{v} \neq \mathbf{w}$. To be specific, let us assume that the watermark sequence $\mathbf{w}$ consists of exactly M periods[33]. Let also assume that each period is formed by $n_p = n/M$ samples, drawn from an i.i.d. distribution. Since $w_i = w_{i+kn_p}$ for any k, correlation can be conveniently rewritten as:

$$\rho = \frac{1}{n} \sum_{k=1}^{M} \sum_{i=1}^{n_p} f_{(k-1)n_p+i}' w_i = \frac{1}{n} \sum_{k=1}^{M} \sum_{i=1}^{n_p} f_{k,i}' w_i, \tag{7.76}$$

---

[33]We neglect border effects for simplicity.

where in the last expression we introduced the symbol $f'_{k,i} = f'_{(k-1)n_p+i}$ to simplify notation. If the host asset contains a watermark $\mathbf{v} \neq \mathbf{w}$ we have:

$$f'_{k,i} = f_{k,i} + n_{k,i} + v_i, \qquad (7.77)$$

easily yielding $\mu_{\rho|\mathbf{v}} = 0$ and

$$
\begin{aligned}
\sigma^2_{\rho|\mathbf{v}} &= \mathrm{var}\left[\frac{1}{n_p M}\sum_{k=1}^{M}\sum_{i=1}^{n_p}(f_{k,i} + n_{k,i} + \gamma v_i)w_i\right] = \\
&= \mathrm{var}\left[\frac{1}{n_p M}\sum_{k=1}^{M}\sum_{i=1}^{n_p}(f_{k,i} + n_{k,i})w_i\right] + \mathrm{var}\left[\frac{1}{n_p M}\sum_{k=1}^{M}\sum_{i=1}^{n_p}\gamma v_i w_i\right] = \\
&= \frac{1}{n_p M}\sigma_x^2\overline{w^2} + \mathrm{var}\left[\frac{1}{n_p M}\sum_{i=1}^{n_p}M\gamma v_i w_i\right] = \\
&= \frac{\overline{w^2}(\sigma_x^2 + \gamma^2\sigma_v^2 M)}{n}.
\end{aligned}
$$

$$(7.78)$$

By comparing the above expression with the analogous expression we obtained for the classical SS AWGN case (see equation(6.49)), we note that in this case the term depending on σ_v^2 is increased by a factor M, thus quantifying the impact of the use of a periodic watermark on system reliability. As a matter of fact, system performance decreases only with reference to the possibility of distinguishing between different watermarks. From this point of view, though, the negative impact of watermark periodicity maybe rather heavy, especially if a small period (hence a large M) is used. In addition, watermark periodicity may be exploited by attackers to remove the watermark, since they can estimate the watermark signal period just as the detector/decoder does, and use this information to remove the watermark.

The use of a self-synchronizing watermark for readable watermarking systems is a more complicated piece of work, at least with QIM and binary SS signalling. The reason for such a difficulty is that in this case the watermark can not be made perfectly periodic, since it depends on the to-be-hidden bit sequence, and, in the QIM case, on the host signal. As a matter of fact, to date, a detailed analysis of self-synchronizing watermarking for this case is not available.

Trying to achieve geometric invariance

The most elegant solution to cope with geometric manipulations consists in the choice of a set of host features which are invariant to geometrical transformations. Unfortunately, it is not easy to find a set of features which are both invariant to geometric manipulations and capable of conveying a high payload, while resisting to conventional signal processing attacks. With reference to shifts in the asset space, we have already mentioned that the most common solution consists in inserting the watermark in the magnitude of DFT coefficients, for the well known invariance of DFT magnitude against spatio/temporal translations. As to scaling and rotation, a possibility consists in embedding the watermark in the Fourier-Mellin domain, as described in section 4.1.2. Note that such an approach does not ensure invariance to anisotropic scaling, where the horizontal and vertical axis are scaled by a different factor. To get around the problem, the watermark could be embedded in a log-log mapping of the DFT magnitude, however in this case robustness against rotation is lost[34]. Histogram-based watermarking is an alternative solution.

Though attractive, invariant based methods have not given satisfactory results yet, sometimes because what is gained from the point of view of geometric robustness is lost from the point of view of robustness against coding and filtering, sometimes because the capacity of invariant features is very limited.

Feature Based Geometric Normalization

The last approach to watermark detection/decoding in presence of geometric manipulations relies on Feature Based Geometric Normalization (FBGN) of the host asset. The idea FBGN relies on is very simple: always insert and recover (detect or decode) the watermark when the host asset assumes a reference geometric configuration, where by geometric configuration the scale factor and, in the image/video case, orientation angle are meant. To be meaningful, reference geometry must be given with respect to a coordinate system which is known both to the encoder and the detector/decoder. To achieve this, FBGN techniques define the reference geometric configuration with respect to a set of asset features, e.g. edges or corners in the image case. Hopefully, such reference features are chosen so that they are stable with respect to all the manipulations the watermark must survive.

To be more specific, let us consider an example in which geometric normalization is performed by relying on image edges (figure 7.21). As a first step, image edges are extracted, then geometric normalization is performed by calculating the central inertial axis of edge pixels and the cor-

---

[34]It is worth to highlight here that both behaviors can not be obtained simultaneously: this constitutes, in our opinion, the principal limitation of the use of Fourier-Mellin transform for watermarking applications, given that a very simple and almost imperceptible attack consists of applying a small isotropic zoom followed by a small rotation (or vice versa).

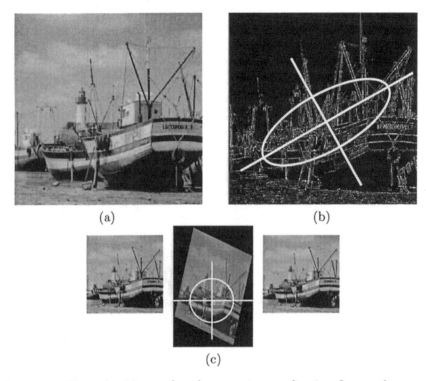

(a)   (b)

(c)

Figure 7.21: Example of feature-based geometric normalization. Image edges are first extracted (b). Then the image is rotated and scaled in such a way that the inertial axis of image edges assume a reference orientation and scale (c). After watermark insertion the image is brought back to its original geometry.

responding inertial moments. Before inserting the watermark, the image is rotated and scaled so that the central inertial axis and the corresponding moments assume given reference values. After watermark insertion the image is transformed back to its original format. The same operations are performed before attempting to recover the watermark, so that geometric transformations possibly applied after watermark insertion are automatically corrected. Indeed during embedding, to avoid the degradation due to the double interpolation required to pass from the original geometrical configuration to the reference one and backwards, the geometrically normalized host asset is used only to compute the watermark. The difference between the non-marked normalized asset and the watermarked one is then transformed back to the original geometrical configuration and added to the original asset.

In FBGN techniques, robustness ultimately relies on the stability of features used to normalize the image. As a matter of fact, it is very difficult to find a set of features which is robust against the wide variety of manipulations images may undergo. Possible solutions include the use of edges, corners, or image regions. FBGN algorithms tend to be very sensitive to image cropping as well, since when cropping occurs some of the reference features are likely to be lost, thus compromising the effectiveness of geometric normalization. Possible solutions consist of repeatedly applying geometric normalization to a set of subparts of the host asset, or in choosing the reference features in such a way to minimize the impact of cropping on the establishment of a geometrical reference.

7.7   Editing

Another class of common manipulations is related to editing operations, i.e. to those operations aimed at extracting part of the information from an asset, and recombining it with other parts.

The first case is cropping, i.e. the extraction of a part of the data constituting the asset. This operation has two effects: the first is that part of the watermark signal is lost, the second is that (in general) the watermark signal is also translated We already considered translations in section 7.6.1, thus here we concentrate on the effects of information loss. For the techniques working in the asset domain, cropping implies that part of the watermark signal is removed. The loss of watermark information depends on the way the information is associated to the watermark signal. In particular, during embedding, it is convenient to use some interleaving technique to spread the information bits all over the watermark signal, in such a way that, although a part of the signal disappears, the rest of it still contains information about all message bits. In other words, the information bits have to be repeatedly embedded at different locations inside the asset. Differently, transformed domain techniques are intrinsically more robust against cropping, given that each coefficient depends on all asset samples, thus, even if some of the samples are lost, the remaining ones can be sufficient to adequately recover all the transformed coefficients[35]. In this case, cropping can be modelled as the product of the signal by a rectangular window: this corresponds in the DFT domain to the circular convolution of the DFT of the signal with a sinc-like sequence of the form:

$$W(k) = \frac{\sin \pi k \frac{M}{N}}{\sin \pi k \frac{1}{N}}, \qquad (7.79)$$

---

[35]Of course all the aspects regarding the number of samples to be used, and invariance to translations as described in section 7.6.1 must be carefully considered.

where M is the size of the cropped segment, N is the number of points of the transformation, and for simplicity we have neglected the phase term which depends on the position where cropping occurred. In practice, cropping introduces a correlation of the DFT coefficients that, thus, can no longer be considered independent. This correlation also influences the watermarking message, by causing an Inter Symbol Interference (ISI) to appear among the different bits. It can be shown that a similar effect also occurs for the DCT case. These effects are more accentuated if the cropped area is small (i.e. if M is small with respect to N). The effects of cropping on hybrid techniques is not different from those described for asset domain embedding: part of the watermark signal is removed, and synchronization is lost (for block-based transform techniques we suppose that the cropped region is larger than block dimension, and thus the cropping does not affect the signal inside the single block (except for blocks at the border of the cropped region)).

Cropping is a very common non malicious manipulation, it is, in fact likely that a sub-image of interest is extracted from a large picture, or an audio segment from a longer track, or a scene from the whole film. Regardless of the application, it is then crucial that a watermarking system is able to deal with this kind of attack.

Similarly to what we have seen when talking about geometric manipulations, also with reference to editing we can find some processes whose aim is basically to confuse watermarking systems, thus making them unable to recover the embedded message. An example is the removal and/or duplication of single samples in an audio file: the removal introduces a loss of synchronization to which asset domain and hybrid techniques are quite sensitive, while duplication is applied solely (and not necessarily) to compensate for the removal (in such a way that the total audio length remains unchanged). A similar procedure can be applied to still images, where single columns or rows can be exchanged, deleted, duplicated. In general frequency domain techniques embedding the watermark in the magnitude of DFT coefficients are less sensitive to this kind of manipulations, thanks to the invariance of this embedding to signal translations. In video, further to the application of single frame attack, the frames themselves can be swapped, removed, or duplicated. This poses some problems to frame based systems: in particular, if the watermark depends on the frame index, its recovery can be very critical if the frame sequence has been manipulated. On the other side, if the watermark is independent on frame index, the watermarking system can be weak against the averaging attack described later in section 7.9.

A particular form of editing that has emerged recently with the release of the MPEG4 standard, is object manipulation. The video coding part of the MPEG4 standard allows, in fact, to code single scene objects instead of whole frames: in practice, each object present in the video can be encoded in a separate MPEG4 stream, thus allowing to manipulate it easily. Without decoding the stream, in fact, an object can be removed from a video and put into another one, or translated to a different position[36].

7.8   Digital to analog and analog to digital conversion

The possibility of recovering the embedded message even after conversion to an analog format is one of the most attractive peculiarities of data hiding systems. Data hiding is, in fact, a unique way to tightly associate some information to a multimedia asset, and make this information persistent against changes of the asset format. The analog version is just an extreme case of the possible format changes an asset can go through. Of course, to recover the embedded data, a final conversion from analog to digital is required, thus making these two processes (D/A and A/D) to be always considered together.

The model of the transformations applied to an asset as a consequence of D/A - A/D conversions can be considered as a collection of the attacks that we have seen up to now. In fact, first of all the geometry of the original asset is likely not to be recoverable (in general the asset will be slightly translated, resized, rotated, cropped), furthermore the value of asset samples will undergo some distortions, like linear and non linear modifications, and noise addition. As we will see, most of these manipulations are very light, at least if, as it often happens, the devices used for D/A and A/D conversion are of high quality.

For audio, recording after A/D conversion is almost surely not to start and stop exactly at the beginning and the end of the track, translation and cropping can thus be very heavy. On the contrary we can assume that sampling frequency has been fixed by the watermarking system, and thus re-sampling problems can be neglected (although the audio track could have been sub-sampled before reproduction, and thus the high frequency portion of the watermark signal missed, this does not regard correct sampling step recovery that can be quite accurate). On the other side loudness manipulations (being linear or slightly non-linear) will be present with high probability, and with unpredictable strength. An example of an application for which it is crucial that the watermarking system is able to survive this attack is given by broadcast radio monitoring, where an automatic system listens to radio programs to check if (and possibly how many times) some audio tracks (music or commercials) are actually transmitted.

---

[36]More complex object manipulations, e.g. rotation, zooming, requires that the object stream is first decoded.

The situation is almost complementary for still images: in this case, in fact, cropping and translation (at least if a small part of the image is not explicitly selected) have small values, and similar is the case for rotations. On the contrary, sampling frequency (or in other words the image dimension) can be very different from the original. It could, indeed, be possible to resize all images to a fixed size before watermark embedding (then restoring them to the original size[37]), and doing the same for A/D conversion: this would allow to always have approximately the same sampling frequency. Anyway, this approach fails as soon as the image has also been cropped. With regard to color level manipulations we can have, as for the audio case, unpredictable effects (both linear and non linear).

The case of D/A-A/D attack for video is not very likely to occur, anyway there are applications for which this it is the main manipulation that the watermarked data can suffer. An example is given by the digital cinema scenario: in this case it is foreseen that moving pictures are distributed, through a fast telecommunication network, to digitally equipped moving picture theaters. During projection, videos are watermarked with the identifier of the theater, in such a way that, if someone in the public records the movie with a video camera, the theater where this copyright violation has been perpetrated, can be identified by analyzing the illegal copy. In this case it is easy to imagine that geometric manipulations can be very heavy (a good model could be given by the transformation described by equation (7.68)), and the same can be said about color distortions.

7.9 Malicious attacks

Most of the attacks that we have seen up to now are common manipulations that can be applied to multimedia asset for enhancing their enjoyability (histogram manipulations, noise reduction filtering), for storing them in an effective way (compression), for adapting them to some presentation format (geometric manipulations). Anyway, all of these attacks can be used maliciously with the explicit objective of removing or making the watermark unrecoverable. In this section we will analyze those attacks explicitly devised to prevent watermark detection/decoding.

In general the idea is that the watermark can be considered as noise added to the host asset; many sophisticated noise reduction tools can thus be used for removing or making it so weak to be unrecoverable. For example mean (and in general linear) and median (and in general non linear)

---

[37]With this regard it is worth observing that it is not needed that the image itself undergoes this double resizing process, since such a process can be applied only to the watermark signal, as we suggested when discussing the FBGN approach against geometrical manipulations (see page 361).

filters, already analyzed in section 7.4.3, can effectively reduce watermark strength. The main limitation of this mean and median filters is that they do not make any assumption on the statistic of the cover data, and, furthermore, they assume that the watermark is additive and independent on the cover data itself: in fact, it can be demonstrated, that the mean and median filters are the ML estimators of the cover asset if we assume that the watermark is additive, Gaussian (Laplacian for the median filter), and independent from the cover data, i.e. they solve the problem:

$$\mathbf{f}' = \arg \max_{\mathbf{f}} p(\mathbf{f}_w | \mathbf{f}), \qquad (7.80)$$

where $\mathbf{f}_w$ and $\mathbf{f}$ are the watermarked and original feature vectors respectively. Furthermore, these attack are often applied without any knowledge of the watermarking algorithms, not even of the embedding domain: they are, in fact, applied in the asset domain, regardless of which are the features hosting the watermark[38]. Because of the strong and not realistic assumptions they make, these simple attacks are not very effective, in that, usually the asset is unacceptably degraded before the watermark is removed.

A more effective and sophisticated attack can be obtained by suitably modelling the cover data. In particular a MAP criterion can be used:

$$\mathbf{f}' = \arg \max_{\mathbf{f}} p(\mathbf{f}_w | \mathbf{f}) p(\mathbf{f}), \qquad (7.81)$$

for estimating the original non-watermarked features. Many solutions exist also in this case according to the different models adopted for the cover data and for the watermark. In particular, if the watermark is supposed to be additive, independent, identically distributed and Gaussian, and the cover features are modeled as non-stationary locally Gaussian, according to the following equation:

$$f_i = \overline{f}_i + \epsilon_i, \qquad (7.82)$$

where $\overline{f}_i$ indicate the local mean of the feature values computed around position i, and ϵ_i are Gaussian independent random variables with zero mean. Then the solution of the MAP problem coincides with the popular Lee filter:

$$f_i' = \overline{f_{wi}} + \frac{\text{Var}[f_w]_i}{\text{Var}[f_w]_i + \sigma_w^2} \left(f_{w,i} - \overline{f_{wi}} \right) \qquad (7.83)$$

where $\overline{f_{wi}}$ and $\text{Var}[f_w]_i$ are the local mean and variance of the watermarked features. When the local variance of the watermarked features is much higher than the variance of the watermark (σ_w^2), the filter has almost no

---

[38]In spite of this, in the following we will always refer to attacks directly performed on the watermarked features

effect, while when the local variance of the watermarked features is much lower than the variance of the watermark, the filter behaves like a mean filter. In practice, then, the watermarked features are left almost unchanged when $\frac{\text{Var}[f_w]_i}{\text{Var}[f_w]_i + \sigma_w^2} \approx 1$, i.e. when the feature variance is high. This suggests a possible countermeasure to this attack: it seems, in fact, natural to give to the watermark signal a strength just proportional to $\frac{\text{Var}[f_w]_i}{\text{Var}[f_w]_i + \sigma_w^2}$. By the way, this countermeasure applies to the watermark signal a weight which is very similar to that derived in equation (5.57) based on perceptive considerations: in practice, we get an agreement between concealment and robustness requirements.

Although developed on the basis of a quite precise model of the watermark embedding algorithm, this attack can also be (and it is) used in a blind way, i.e. without making any hypotheses about the watermarking system at hand (not even about the embedding domain). In this case the watermarked asset is modelled as:

$$A_{\mathbf{w}} = A + W(A, \mathbf{w}, K) \qquad (7.84)$$

where in general the added watermark $W(A, \mathbf{w}, K)$ depends on the original asset A, on the watermark signal $\mathbf{w}$, and on the key K[39]. The MAP noise removal filter (7.83) can thus be applied, although its effectiveness is reduced due to the fact that all the hypotheses made to derive it are not valid. Of course, if some knowledge of the watermark embedding system is available, a more effective MAP removal attack can be developed (informed attacks will be better discussed in chapter 8).

A more heuristic removal attack that has been devised particularly for still images is the so called Frequency Mode Laplacian Removal (FMLR) filter. It is based on the experimental observation that the Laplacian operator, whose 3×3 mask is:

$$\begin{bmatrix} 1 & 1 & 1 \\ 1 & -8 & 1 \\ 1 & 1 & 1 \end{bmatrix} \qquad (7.85)$$

is effective in highlighting the subtle modifications introduced by the watermark. It is then proposed to estimate an non-watermarked image as:

$$\hat{I} = I - \alpha \left(\nabla^4[I] - \nabla^2[I] \right) \qquad (7.86)$$

where by ∇^2 we have indicated the Laplacian operator, by ∇^4 the same operator applied twice, and α is a weighting parameter with values in the

---

[39] By referring to the notation of equation (1.2), we have that $W(A, \mathbf{w}, K) = \mathcal{E}(A, \mathbf{w}, K) - A$.

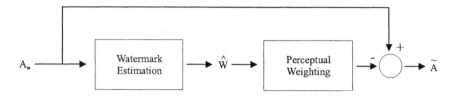

Figure 7.22: Functional scheme of the remodulation attack.

range $[0.05, 0.15]$. A more effective implementation of this attack is obtained by working in the block DCT domain (thus the name Frequency Mode Laplacian Removal); in practice the image and its Laplacian filtered version are partitioned in 8×8 blocks which are DCT transformed. The DCT coefficients of the attacked image blocks are obtained as:

$$
\begin{aligned}
\mathrm{DCT}\{\hat{I}\}(u,v) = {} & \mathrm{DCT}\{I\}(u,v) \\
& - \alpha \left(\mathrm{DCT}\{\nabla^4[I]\}^\gamma(u,v) - \mathrm{DCT}\{\nabla^2[I]\}^\gamma(u,v) \right) w(u,v),
\end{aligned}
\tag{7.87}
$$

where the parameter γ (having default value 0.3) is used to give more importance to low intensity values, while the weighting term $w(u,v)$, whose expression is given by:

$$
w(u,v) =
\begin{bmatrix}
0.0 & 0.2 & 0.3 & 0.4 & 0.5 & 0.6 & 0.7 & 0.8 \\
0.2 & 0.3 & 0.5 & 1.0 & 1.0 & 1.0 & 1.0 & 0.7 \\
0.3 & 0.5 & 1.0 & 1.0 & 1.0 & 1.0 & 1.0 & 0.6 \\
0.4 & 1.0 & 1.0 & 1.0 & 1.0 & 1.0 & 1.0 & 0.5 \\
0.5 & 1.0 & 1.0 & 1.0 & 1.0 & 1.0 & 0.5 & 0.4 \\
0.6 & 1.0 & 1.0 & 1.0 & 1.0 & 0.5 & 0.4 & 0.3 \\
0.7 & 1.0 & 1.0 & 1.0 & 0.5 & 0.4 & 0.3 & 0.2 \\
0.8 & 0.7 & 0.6 & 0.5 & 0.4 & 0.3 & 0.2 & 0.2
\end{bmatrix}
\tag{7.88}
$$

has the goal of concentrating the effects of the operator in the mid frequency range, i.e. where watermarking techniques usually embed most of the energy.

The watermark removal filters described so far can be used also for estimating the watermark from the asset. Let us suppose that we have obtained a non-watermarked copy of the asset A', we can then also get an estimate of the watermark $\hat{W}(A, \mathbf{w}, K)$ as:

$$
\hat{W}(A, \mathbf{w}, K) = A_{\mathbf{w}} - A'.
\tag{7.89}
$$

The availability of an estimated watermark (that can be obtained as we described before or in some other way), allows us to implement a more effective watermark removal strategy, named remodulation attack, which consists of the subtraction of the estimated watermark, possibly weighted by a perceptual mask, from the watermarked asset according to the scheme in figure 7.22. Obviously, if the watermark is estimated as in equation (7.89), and perceptual masking is not applied, the effect of this attack is exactly the same as for the denoising attack used to estimate A'. As in the other cases, this attack can increase its effectiveness if more details about the watermark embedding strategy are known, for example it can be applied directly in the embedding domain if this is publicized.

Another attack that relies on an estimate of the watermark, consists of adding such an estimate to another asset for producing a forged copy. Is this the so called copy attack, whose aim is different from that of the previously described attacks. In this case, in fact, the attacker's goal is a non-watermarked asset to seem watermarked.

When more than one watermarked asset is available collusion attacks become feasible. As a first example, let us suppose that we have multiple copies of an asset with different watermarks (this situation is common for fingerprinting applications, where, each distributed asset is marked with the identifier of the consignee), it is then possible averaging the different assets to obtain a copy where the different watermarks have a lower energy. In general, if we average N differently watermarked copies, the energy of each different watermark in the average asset is reduced by $1/N^2$, thus resulting more difficult to recover. A basic idea to build a countermeasure to this attack is to use different watermark signals which are not perfectly orthogonal, but that, on the contrary have some parts in common: in particular, each codeword associated to a consignee is composed by a certain number of sub-codewords, and each sub-codeword can be found in other codewords. When N assets are averaged it is very unlikely that all of them will have embedded codes having no equal sub-codewords, in general thus the common sub-codewords of the codes embedded into the N colluding assets, will have their energy reduced by less than $1/N^2$, it is then more probable to identify at least a subset of the consignee, i.e. those whose codewords contain the detected sub-codeword(s).

A more sophisticated collusion attack that can be applied is the so called collage attack. It consists of building an attacked (new) asset, by assembling different sub-parts taken from the differently watermarked copies of the asset. If the size of each sub-part is chosen to be smaller than the minimum dimension from which the watermarking system is able to recover the watermark, any watermark will not be recovered in the composed (attacked) asset. This attack can even be generalized to the case in which the attacker has available different assets watermarked with different watermarks (but also non-watermarked assets work well in this case): if he wants to forge one of these assets he can partition it into small sub-parts, then, for each sub-part, look for the sub-part of one of the other assets that is most similar to it, finally replace the found sub-part to the original one. The success of this attack is related to the availability of a large archive of (possibly non-watermarked) assets where to look for the sub-parts needed to build the collage. The resistance of the watermarking technique to this kind of attacks is ultimately related to its robustness: it depends in fact either on the minimum tolerable cropping size, or on the maximum amount of sustainable modifications.

The dual situation of the collusion attacks described above must be considered as well. We now have many different assets, containing the same watermark: in this case, by averaging the watermarked assets, it is possible to get an estimate of the watermark itself (at least it is easier to estimate it by means of one of the previously described watermark removal techniques) and then attempt to remove it from all assets (e.g. by a remodulation attack). The sole countermeasure to this kind of attack is to make the watermark signal as much dependent as possible on the cover asset, in such a way that it is impossible to have exactly the same watermark in different assets (this is achieved, for example, by perceptually adapting the watermark, or by using informed embedding/coding schemes).

We have seen in section 7.6.5 that some counter-measures against geometric manipulations, are based on the introduction of some peaks in the magnitude of DFT: either the peaks are directly embedded, as in template based techniques, or they appear because of the periodic structure of the watermark, as with self-synchronizing watermarking systems. In any case, the presence of the peaks in the DFT domain is crucial for making possible the estimation and subsequent compensation of geometric manipulations. This critical characteristic suggests the so called template removal attack, whose aim is just to make less detectable those peaks, and then apply a light geometric manipulation (e.g. a rotation by 1 degree, and/or a zoom by 1.05) that can not be recovered because the detector misses the synchronization pattern. For example, to remove DFT peaks, a simple local maxima detection algorithm can be applied, followed by the interpolation of the detected peaks, based on the neighboring DFT values. This attack has demonstrated to be very effective against template based techniques, where usually only a few high peaks are added, and less effective, although often still successful, against self-synchronizing systems, where the number of resulting peaks is much higher (they have usually a lower height).

7.10 Attack estimation

Up to now we have examined the effects of various attacks on different watermarking systems. Usually watermarking systems adopt a blind behavior with respect to attacks, i.e. the recovery strategy does not make any attempt to estimate the attack, in order to try to compensate for it. The notable exception is constituted by geometric attacks for which many estimation/compensation strategies have been developed as detailed in section 7.6.5. Certainly this blind approach is not effective. We have seen, in fact, that the watermark recovery algorithms are developed based on given models of the host features (namely Gaussian, Generalized Gaussian, or Weibull distributed) and of the occurred manipulation (namely Gaussian noise addition); almost always, anyway, these models do not match the actual attack. It is thus reasonable that performance could be improved if the attack is estimated and compensated for, in such a way to make the hypotheses regarding the models on which the recovery algorithm is based, more realistic. As a matter of fact it is very common in communication systems to include some technique for estimating the transmission channel: it is thus reasonable to assume that such an approach could bring some benefits even to data hiding systems.

Actually, a few attempts have been made for trying to estimate and compensate the attacks. The basic idea is to embed into the asset, further to the message bearing watermark, which is unknown to the decoder, a pilot watermark which is, on the contrary, known to the decoder. Given that this watermark is known, it is possible to estimate the manipulation(s) applied to it. Once the manipulation has been estimated it can be compensated for thus enhancing the quality of the unknown watermark. In order to implement such a strategy, first of all we need to define a model of the possible manipulations: given that such a model can not be too general, it can be adapted either to the most likely or to the most dangerous attacks. As an example, given the sensitivity to the gain attack of quantization based schemes, efforts have been devoted in the literature to try to estimate the possible scaling that the watermarked signal has undergone, thus making possible to compensate for it.

Research in this area is still in its infancy, but it will be very important in the future.

7.11 Benchmarking

As we have seen, a theoretical analysis and comparison of the performance of different watermarking algorithms can be performed only on a limited number of simple cases, that can not account for the complexity and variety of practical situations watermarking systems have to face with. As a matter of fact, a model fitting the actual distribution of host features does not exist. In the same way, attacks modelling is a very arduous, virtually impossible, task, due to the huge variety of manipulations the host asset may undergo.

A second deviation of real scenarios from theory is the way obtrusiveness is measured. Usually the degradation of the host signal is measured through classical metrics such as Signal to Noise Ratio (SNR), Peak Signal to Noise Ratio (PSNR), Mean Square Error (MSE), peak error, and so on. However, it is well known that such measures do not correlate well with the way degradations are perceived by a human observer.

The absence of watermark masking is a third important deviation from reality of the theoretical analysis. Watermark masking, in fact, in addition to improving unperceivability of the hidden information, also increases robustness, however such an improvement can not be predicted from a theoretical point of view.

In addition to these problems, several other practical issues, including synchronization effectiveness, algorithm complexity, finite precision of computer arithmetic, approximation of random numbers through pseudorandom sequences, make the validation of theoretical results through experimental analysis necessary. This is the scope of benchmarking systems: compare different watermarking algorithms by the light of objective experiments aiming at measuring a pool of properties which characterize the actual performance of the system. Commonly measured properties include robustness against a *standard* set of attacks, unobtrusiveness, and capacity.

As it can be easily argued, the design of an effective benchmarking system is a difficult task, due to the necessity of standardizing the attacks against which the algorithms must be tested, the desired level of unobtrusiveness, a set of reference host assets to be used in the experiments and so on. Up to date, several efforts have been made to design a good benchmarking system, however research in this field is still on going. In the sequel, we will briefly describe such efforts, and the solutions proposed so far.

7.11.1 Early benchmarking systems

The first problem to be addressed when designing a watermarking benchmark is the exact definition of the watermark properties to be measured. As a matter of fact, the validity of a certain algorithm depends on several factors including the application it has to be used for, the protocol exploiting the information conveyed by the watermark, the cover signal within which the message has to be hidden, and so on. It is not rare the case in which the effectiveness of a system relying on data hiding technology is flawed due to reasons other than the watermarking algorithm used to hide the data. Though all system aspects, ranging from the protocol level through purely implementation aspects deserve a careful analysis, in this chapter we are only dealing with signal processing aspects of information hiding. If a parallelism is made between data hiding and digital communication, we can say that here we deal only with the transport layer.

Even by restricting the analysis to the, so to say, transport layer, the target of evaluation may still be questionable, since it heavily depends on the intended application. It is undoubtable that evaluating the performance of an algorithm designed for authentication purposes is a completely different piece of work than analyzing the performance of a scheme designed for copyright protection applications. In any case, the effectiveness of a watermarking algorithm depends on the trade off it reaches among three major requirements, namely unobtrusiveness, robustness and payload[40]. The target application only determines whether the benchmarking procedure must focus on robustness, payload or unobtrusiveness.

As watermarking research was initially triggered by copyright protection applications, early benchmarking systems focused on robustness parameters. As a matter of fact, the structure of early benchmarking systems was rather simple. The to-be-tested algorithm was repeatedly used to mark a set of reference assets, then the host assets were attacked by means of several processing tools. Finally, the system tried to recover the embedded watermark resulting in an error if the watermark could not be revealed or if the decoded message was not equal to the original one. As to the attacks, when possible, the system applied them with increasing strength. The benchmark score, was simply a re-elaboration of the number of recovery errors occurred during the tests. Usually a final score summarizing the system performance as a whole was produced. Such a global score was computed by dividing attacks into a number of different categories, and by calculating the average of the scores obtained for each category.

In order to achieve a fair comparison between different schemes, a fixed level of obtrusiveness was used, e.g. by fixing the PSNR of the watermarked asset. No attempt was made to take into account the watermark payload into the benchmarking process. Thus 1-bit watermarking scheme were treated in the same way as multibit algorithms conveying some hundreds of bits. In the same way, no attempt was made to weight different attacks according to their importance or according to the distortion they introduced.

Despite their simplicity, the first generation of benchmarking systems

---

[40]The effectiveness of a watermarking algorithm also depends on parameters such as computational complexity, non-detectabily, security. However here we will not deal with such aspects.

has gained a wide popularity and have permitted to highlight most of the main drawbacks early watermarking systems suffered from. This is the case of the Stirmark benchmarking software whose usage strongly influenced the still image watermarking community, pushing researchers to continuously improve their systems so to obtain a higher score. Indeed, whereas first systems were not able to survive most of the attacks included in the Stir-Mark package, after a few years, many algorithms appeared which could claim a 0.9, or higher, score[41].

In the next section we give a brief overview of the StirMark benchmarking system.

7.11.2 StirMark

The first version of the StirMark watermark benchmarking package was released in the late nineties, and since then it has represented an important reference (arguably the only reference) for the whole watermarking research community. StirMark software was initially designed to deal with still image watermarking, however the StirMark project keeps on evolving and new releases have addressed audio watermarking too.

The structure of StirMark closely resembles the scheme described in the previous section. Images are marked with the strongest strength subject to invisibility. Then a set of transformations (attacks) with increasing strength is applied to the host images, to look whether the watermark survives them or not. Watermark recovery is treated as an on/off process in that, even for readable schemes, recovery is considered to be successful only if all the embedded bits are decoded without error. The benchmark score is produced by assigning 1 to correct watermark recovery and 0 to errors. The average score is then computed for each class of attacks and the overall score by averaging partial scores. The overall score is averaged on a set of standard images. In order to speed up the evaluation process, unobtrusiveness is not measured by human inspection. Instead, the PSNR is used as quality metric. For example, a commonly adopted approach consists in considering the quality satisfactory if the PSNR of the marked image is higher than 38dB.

Still images

Let us consider in more detail the still image StirMark benchmark. In the basic implementation (released in 1999), attacks are divided into seven categories, namely: signal enhancement, compression, scaling, cropping, rotation, random bending, geometric transformations other than those listed

---

[41]StirMark score ranges from 0 through 1.

above. Due to the importance of JPEG coding in everyday life applications, all the above attacks, but compression, are carried out with and without JPEG coding with 90 percent quality factor. The attacks each of the above categories consists of are listed below.

- *Signal enhancement*: this section includes Gaussian low-pass filtering with kernel

$$\begin{bmatrix} 1 & 2 & 1 \\ 2 & 4 & 2 \\ 1 & 2 & 1 \end{bmatrix} ;$$

  3 × 3 median filtering;
  linear sharpening with kernel:

$$\begin{bmatrix} 0 & -1 & 0 \\ -1 & 5 & -1 \\ 0 & -1 & 0 \end{bmatrix} ;$$

  and the Frequency Mode Laplacian Removal (FMLR) attack described in section 7.9.

- *Compression*: including color gamut compression to 256 values (GIF format), and JPEG coding with quality factors 10, 15, 25, 50, 60, 75, 80, 85, 90;

- *Zooming*: this section comprises uniform zooming (same factor along x and y directions) with factor 2.0, 1.5, 1.1, 0.9, 0.75 and 0.5;

- *Cropping*: in StirMark software cropping is achieved by simply removing a given percentage of pixels from image borders. Adopted cropping extents include: 1, 2, 5, 10, 15, 20, 25, 50 and 75% of the original image size;

- *Rotation*: several rotation angles are considered, including -2, -1, -0.5, 0.5, 1, 2, 5, 10, 15, 30, 45 and 90 degrees. The user may choose whether to crop or scale the rotated image so to retain the original image size;

- *Random bending*: the random geometric transformation described in section 7.6.4 was first developed for inclusion in the StirMark software, and for some years (and to some extent still today) it represented one of the major challenges for watermarking system designers.

- *Other geometric transforms*: this section includes simple geometric attacks such as line and column removal, horizontal flipping, and shearing (1% and 10%).

Upon inspection of the attacks included in the above list, it is at once evident that StirMark benchmark is strongly biased towards geometric transformations. Additionally, other kinds of attacks, such as histogram modification, synchronization removal, and estimation-based attacks, which may be particularly effective against certain classes of algorithms, are not considered at all. For these reasons, and to overcome several other weakness of the simple approach early benchmarking systems relied on, researchers has kept (and will continue to) developing more and more sophisticated algorithms. We will describe some of the newly proposed solutions in sections 7.11.3 and 7.11.4.

Audio

As watermarking research in general, even the development of benchmarking tools was initially focused on the watermarking of still images. StirMark software followed the same path: whereas several versions of the package for the benchmarking of image watermarking algorithms have already been released and widely diffused among the watermarking community, the development of an analogous software for audio watermarking is still at an early stage. The current implementation of the audio version of StirMark, attacks against audio watermarks are split into 9 categories: dynamic range modification, frequency filtering, ambience effects, format conversion, lossy compression, noise addition, modulation effects, time stretch and pitch shift, sample permutations. For each class provision is made for one or more attacks according to the following list:

- *Dynamics*: it includes reduction of signal peaks so to allow the reproduction of an overall louder signal, and denoising through thresholding (noise is removed by setting to zero, or any other fixed value, the value of samples below the threshold);

- *Frequency filtering*: high pass filtering removing all frequencies lower than a threshold, e.g. 50Hz; low pass filtering removing all the frequencies higher than a given threshold, e.g. 15KHz; equalization to change the power ratio between different frequency bands; L/R splitting to produce a stereo signal from a mono signal. In particular this last attack works by reducing certain frequencies in a channel and increasing them in the other and its effect on the watermark is similar to equalization;

- *Ambience*: this class comprises a pool of effects usually applied to simulate environment effects on the audio signal. It includes the introduction of a delayed copy of the signal to simulate a wide listening environment, and reverberation simulating shorter delays and reflections.

- *Format conversion*: it includes resampling, e.g. downsampling from 48kHz to 44.1KhZ for CD production, and sample inversion;

- *Lossy compression*: this class is of the outmost importance in internet-based applications where MP3 compression is always present;

- *Noise addition*: noise level may be defined absolutely, by specifying the maximum value of the noise signal, or as a percentage of sample values;

- *Modulation*: this class considers a number of effects which are available in most commercial packages for audio editing. Possible attacks include: chorus effect (addition of a modulated echo with varying delay), flanging (mixture of the signal with a delayed copy of itself, where the delay varies constantly), enhancing (increase of high frequencies to improve the audio signal);

- *Time stretch and pitch shift*: time stretching allows to adjust the duration of the audio signal so to fit a target time slot, pitch is not changed; pitch shift changes the base frequency of the signal without changing the speed;

- *Sample permutations*: this class includes attacks such as: insertion of zero crosses, whereby the duration of zero crosses is artificially modified, thus inserting small pauses in the signal; sample copy, whereby some randomly chosen samples are repeated several times; sample flipping, whereby the position of randomly chosen samples are exchanged; sample cutting, whereby a set of randomly chosen sample is removed from the signal

Attention is also paid to define a test set of audio signals presenting various characteristics: spoken text, classic music, pop music, jazz music, very loud rock music, urban ambience sounds.

Beyond StirMark

While StirMark represented, and still represents, a valuable tool for measuring the robustness of image watermarking algorithms, the need for more sophisticated benchmarking tools which take into account the various facets of the watermarking problem, not just robustness, is rapidly raising thus justifying the efforts of many research groups to develop new, more powerful, benchmarks. Though research in this field is still on going we summarize below the main streamline followed to create a, so-called, second generation watermarking benchmark[42]

7.11.3 Improving conventional systems

The simplest way to improve StirMark benchmark, consists in removing its main weakness without changing the overall benchmarking structure. We will refer to this class of benchmarks as *improved* conventional benchmarks, in that they still focus on robustness, and follow the basic StirMark approach according to which a watermarked asset is subject to a number of attacks and the number of successes and failures measured.

More powerful attacks

A straightforward improvement of StirMark consists in the introduction of new classes of attacks. According to a widely accepted opinion, attacks are divided into 4 main categories, namely: removal and interference attacks, geometrical attacks, cryptographic attacks, protocol attacks. Here we are only interested in attacks affecting the transport layer, hence we only considers the first two classes. Moreover we focus on the still image case. Among the proposed extensions, the most important differences with respect to the set of attacks considered by the first StirMark versions, consists in the introduction of the following types of attack for the interference removal category:

- *New types of noise*, including speckle, impulsive and multiplicative noise;

- *Histogram modification*, including equalization, stretching and quantization;

- *Denoising*, including Wiener filtering, soft and hard thresholding;

- *Remodulation*, whereby the watermarking signal is first estimated and used to remodulate the host signal in the attempt to remove the hidden watermark. A variant of this attack, called perceptual remodulation, takes perceptual masking into account to reduce the degradation of remodulation (for more details see page 369);

---

[42]We prefer not to make reference to any project in particular, since several important research institutions share this effort and it is not yet clear which approach, if any, will get the upper hand over the others.

- *Template removal*: this attack, which consists in the selective removal of peaks in the frequency domain, aims at fooling the synchronization mechanism adopted by many algorithms to cope with geometric manipulations. Template removal is usually followed by a geometric attack (for more details see page 370);

- *Wavelet-based compression*: this attack is inserted to account for the future diffusion of the JPEG2000 standard (for more details see page 336);

- *Multiple watermarking*: in some cases, noticeably some QIM-based schemes, the simple insertion of a second watermark is enough to remove the embedded information;

With regard to geometrical attacks, there is less room for improvement since StirMark already included a wide variety of attacks of this type. Possible extensions include projective transforms and non-uniform line removal, where removed columns or lines are not equally spaced.

Measuring perceptual distortion

A considerable improvement with respect to early benchmarking schemes can be obtained by adopting a perceptual metric to measure the degradation introduced by the watermark, or the attack, as perceived by a human observer. A first possibility consists of weighting the PSNR according to the sensitivity of the human eye to disturbs affecting different frequency bands. More specifically, the error signal is first transformed in the frequency domain, then its power is computed by weighting the magnitude of frequency coefficients by the CSF curve presented in section 5.1.2. At a more sophisticated level, Watson's model, described in section 5.4.2, may be used to account for visual masking, providing a more exact measure of the perceptual distortion introduced by the watermark (and the attack).

As an alternative to the adoption of theoretical metrics based on the HVS models described in chapter 5, a heuristic image-structure derived measure based on the image content in terms of edges, textures and flat regions may be defined. e.g. weighting disturbs occurring in different areas according to the content.

Watermark decoding vs detection

So far watermark detection and decoding have been treated in the same way by benchmarking algorithms, however they need to be treated differently. With regard to watermark detection, current systems focus on missed detection probability only. Ignoring the false detection rate, though, impairs seriously the value of the benchmark score, since it is rather obvious that a smaller missed detection probability can always be obtained at the expense of increasing P_f. For a fair comparison, then, robustness should be measured by fixing the probability of falsely deciding for watermark presence. This, in turn, poses serious implementation problems since P_f values adopted in practice are usually very small (10^{-6} through 10^{-12}) thus making virtually impossible their exact measurement. A possibility could be to derive experimentally the pdf of the detection statistic and use such a statistic to evaluate the false detection probability, however such an approach is strongly algorithm-dependent and may be difficult to automatize.

Watermark decoding is theoretically easier, nevertheless very small error probabilities may be hard to measure, thus making the comparison between different systems significative only for the heaviest attacks.

Application-dependent benchmarking

Another important issues to be addressed by second generation benchmarking systems, is the impact of the intended application on the benchmark score. As a matter of fact, the sole robustness is not enough to completely characterize a watermarking algorithm. It may well be the case that, for a given application, watermark capacity has a higher priority, as well as in some cases, the perceptual constraint may somewhat be relaxed. Though a satisfactory solution to this problem may require that the overall benchmarking architecture be revised, some possible approaches to account for the application scenario even in the context of conventional systems have been proposed. A possibility consists in weighting the results obtained for each class of attacks (or for each single attack) by a set of coefficients which depend on the application. Before starting benchmark routines the user is asked to select a scenario, then such an indication will be used to produce the final score.

Alternatively, one may define several operative conditions each one fitting a particular application, then the host asset is marked and the overall benchmark score produced. Operative conditions may include maximum allowable distortion, watermark payload, choice between detectable and readable watermarking, and all the parameters directly stemming from the application requirements.

Video watermarking

No benchmarking tool is currently (at least publicly) available which is expressly designed to deal with video watermarking. In spite of this, a list of attacks which should be taken into account to consider the peculiarities of the video case can be easily drawn. Of course, all the attacks considered in the image case still hold, since they can be applied separately to each frame of the video sequence. Video-specific attacks include:

- *MPEG compression*: of course when dealing with image sequences the superior compression capabilities of video coding schemes must be taken into account. Robustness should be tested against the major standards of the MPEG family, including MPEG2 and MPEG4. A possible difficulty may arise from the large number of coding parameters to be set. Just to mention some, they include: target bit rate, frame rate, structure of the coded sequence (GOP definition), choice between constant and variable bit rate, frame size. Other video coding standards such as those of the H.26x family may be taken into account for specific applications such as videoconferencing or video surveillance;

- *Format conversion*: it may include change of frame or bit rate, frame resizing, color format conversion;

- *Editing*: this considers all most common postproduction processing, including scene composition (especially suited to MPEG4 streams), addition of logos or subtitles, overlays and so on;

- *Temporal scaling*: including temporal resampling, frame insertion, temporal interpolation;

- *Geometric transformations*: in addition to classical image transformation applied frame by frame, video specific processing such as frame exchange and time jitter must be considered.

7.11.4 A new benchmarking structure

We already noted that a thorough evaluation of the performance of a watermarking algorithm can not rely on robustness only, on the contrary, such performance depend on how several characteristics including, among the others robustness, capacity, and imperceptibility, fit the application the watermark is intended for. By following this point of view, future benchmarking systems will be more and more configured as systems capable of measuring a set of merit figures and presenting them in a compact way. Some of the figures to be measured may correspond to those adopted for theoretical analysis, nevertheless, when measured by a benchmarking system they will have the advantage of reflecting the true properties of the watermarking algorithm rather then its supposed ones. Along with merit figures directly relating to the characteristics of the watermarking algorithm, other, so to say, environmental parameters must be set or measured.

This is the case, for instance, of attack-induced distortion. Instead of evaluating a generic robustness measure against a given attack, the system will provide a sequence of degradation/bit-error-rate pairs which completely characterize the algorithm performance from the point of view of that attack. In the end, it is up to the user to decide whether the performance of a given algorithm fits the application of his interest or not.

Though the list of to-be-measured characteristics may be rather long, parameters such as bit error rate, watermark induced degradation, watermark payload, false and missed detection probabilities will surely play a predominant role. As to the operating characteristics, attack-induced degradation will surely have to be taken into account.

Evaluation report

Given that the benchmark output will not consist of a global score, the question on how results should be reported arises. A first possibility is represented by the, so called, Raw Performance Measurement Plots (RPMP). Given k measured parameters and n measurements, e.g. each measurement deriving from a different test asset, the corresponding RPMP is obtained by showing the position of each measure in a k-dimensional space. In order to avoid the difficulties of visualizing points in k-dimensional space, only two parameters at a time may be taken into account. The other parameters are fixed, with each fixed value resulting in a cluster of measurements in the 2-dimensional RPMP.

Though extremely insightful, RPMP's may result too difficult to read, since no compaction effort is made. In order to simplify the analysis clusters may be approximated by ellipses, with the horizontal and vertical dimensions related to cluster spreading.

Finally, the possibility of summarizing all the measurements in a single overall score may be taken into account, due to the immediacy of such an overall score and to the similarity with other popular benchmarking systems, e.g. those measuring the performance of PC's hardware. In spite of this advantages, the feasibility of such an over-synthetic performance measure is questionable, since it inevitably oversimplifies the judgment losing the richness of the measurements taken by the system.

7.12 Further reading

All the manipulations that we have described in this chapter assume that any knowledge about the watermarking technique is not exploited. Nevertheless this should not support the idea that trying to keep the algorithm details secret can be a viable solution for assuring a sufficient degree of resistance to attacks. On the contrary it is widely acknowledged, since the publication in 1883 of the popular A. Kerckhoffs work [121], that algorithm secrecy can not be assumed, and that security should only rely on the secrecy of some parameter (usually referred to as the key).

Regarding image manipulations, many monographies are available, we only cite here the classical textbook of A. K. Jain [112] where more details about histogram processing, linear filtering and image transforms can be found. For non-linear filters, on the contrary, interested readers can refer to [179].

While it is very simple the relation existing between the correlation operation in the sample domain, and the multiplication in the DFT domain, the same is not true for the DCT: nevertheless, such a topic has been deeply investigated in [146].

The implications related to the contrasting objectives of watermarking and of lossy data compression have been analyzed in [232] and [125, 124, 77].

Readers interested in better understanding image coding standards can refer to the official documents, as for example [107] for JPEG and [108] for JPEG2000. A good tutorial introduction to JPEG2000 can be found in [187]. In particular it is worth here mentioning that the quantization and entropy encoding steps of JPEG2000 are more complex than those briefly outlined in section 7.5, although the basic principles are the same: readers interested to deepen this issue can see [213]). The official documents for MPEG video coding are [103] (MPEG1), [105] (MPEG2) and [106] (MPEG4). An excellent survey on MPEG4 object based coding features is given in [199]. A good overview on the MPEG audio encoding standard [102] can be found in [161]. A plethora of other compression techniques have been developed during the years, we are only citing, as examples, the SPIHT algorithm [192] for still images and the matching pursuit approach [3] for video.

To analyze the effects of signal translation on DCT coefficients it is again possible to refer to [146] given that a translation can be formally described as the convolution of the signal with a suitable delta function.

Details about the Stirmark random bending attack can be found in [176, 174].

An interesting approach for the optimization of the design of synchronization templates has been suggested by P. Moulin in [158]. The first to propose the use of the Fourier-Melling transform for dealing with zooming and rotations were J. J. K. Ó Ruanaidh and T. Pun in 1997 [162], and the technique has been refined by incorporating concepts from informed embedding theory by C. Y. Lin et al. in [135]. The first to propose self-synchronizing watermarks was M. Kutter [128], while the techniques exploiting the Feature Based Geometric Normalization approach have been presented in [4, 25, 20] (in particular in [25] the approach is generalized to work at a local level, thus achieving robustness also with respect to spatially varying geometric manipulations, as the random bending attack).

Techniques dealing with the particular issues related to the object manipulations capabilities of the MPEG4 standard have been presented in [8, 180, 26].

With reference to malicious attacks, a good review has been presented in [226]. Attacks based on stochastic signal models are described in more detail in [225]. The FMLR attack, later included in the Stirmark benchmark, was firstly proposed by B. Barnett and D. E. Pearson in 1998 [6, 7].

The principles on the use of anti-collusion codes for fingerprinting applications are described in [33, 67, 217].

The topic of attacks estimation for improving the performances of watermarking techniques is quite new, interested readers can find some attempts in this direction in [28, 69, 224, 127].

With regard to benchmarking the first systematic attempt to approach the problem is certainly due to F. A. P. Petitcolas that proposed the popular Stirmark benchmarking package in [175, 173, 130]: the website offering the Stirmark benchmarking service and a good database of reference images is still today active and can be found at the URL

- http://www.cl.cam.ac.uk/ fapp2/watermarking/stirmark/.

Details about the generalization of Stirmark to the audio case can be found in [206], and its implementation is available at the URL

- http://ms-smb.darmstadt.gmd.de/stirmark/stirmarkbench.html.

A newer benchmarking package, including more sophisticated attacks is available at the URL

- http://watermarking.unige.ch/Checkmark

and has been described in [226, 168]. Details about an effort to set up a public benchmarking service for watermarking technologies can be found at the URL

- http://www.certimark.org/.

8

Security issues

The greatest interest in data hiding literature has usually been turned to the robustness aspects of watermarking systems. On the contrary, security issues have often been neglected, and only recently the attention to this problem has grown, also because of the failure of some attempts to exploit watermarking systems in strongly non-secure environments, as it was the case of the SDMI affair.

Similarly to the case of robustness, security requirements are strongly application dependent. In particular the level of security that different applications need to satisfy can be highly variable; it is not difficult, for example, to find applications for which security is not even an issue (e.g. data hiding for error concealment, or content labelling). In this chapter we try to discuss the security aspects in all their implications, being as general and independent of the applications as possible. We will, anyway, from time to time, specify the application for which a particular issue is most important.

In section 7.1, we classified attacks as non malicious and malicious, furthermore we divided the malicious attacks into blind and informed. In this chapter we mainly concentrate on the latter class of attacks, i.e. on those attacks that exploit some knowledge that is available about the watermarking system. We assume, in fact, that security issues are related to all those attacks that can benefit from the knowledge of some details of the watermarking system functioning. As we will see, anyway, some considerations are due also with reference to the situation in which no knowledge is available about the watermarking system (see the *security by obscurity scenario* described in section 8.1), given that an attacker can always try to disclose any information that was supposed to be secret. In general, thus, this chapter also deals with the issues related to the problem of keeping secret the information not intended to be publicly known.

For the purpose of better specifying what 'knowledge' means in the watermarking field, it is useful to introduce a general security framework which can accommodate all types of watermarking systems. The framework is based on identifying the three types of information that can be kept secret or, on the contrary, made public regarding a watermarking system. These are:

- The embedding and decoding algorithms.

- The embedding parameters (we will refer to them as the embedding key).

- The recovery parameters (we will refer to them as the recovery key).

Given these types of information we can have four types of approaches on designing a watermarking system:

Type 1. Everything is kept secret. This approach is usually referred to as security by obscurity.

Type 2. The algorithms are public, but the keys are secret. The algorithms of this type are usually called symmetric.

Type 3. The algorithm and the recovery key are public, while the embedding key is secret. The algorithms of this type are usually called asymmetric.

Type 4. Everything is public. We will refer to this approach as the open cards approach.

At least in principle, from the point of view of the watermarking algorithm designer, the first case is the most favorable, as he has only to care about making the watermarking technique as robust as possible, while the other types of approaches present an increasing degree of difficulty, since the designer has to care also about all those attacks that exploit the information available about the algorithm. The attacker is in a complementary situation, since he can implement more and more effective attacks, as the information he has about the watermarking system increases. In practice, anyway, keeping information secret has always demonstrated to be very difficult, and thus it is preferable that not too much data are to be kept secret. It is likely that a tradeoff between these conflicting requirements offers the preferable solution.

Indeed the attacker has also the option to try to discover some of the information that was meant to be kept secret, in order to subsequently implement more effective attacks. It is then worth to distinguish between two kinds of attacks that can be brought to a watermarking system. In particular we will call *fair* those attacks which respect the rules of the game between the watermarking system designer and the attacker, i.e. which do not try to discover the information that is intended to remain secret, while we call *unfair* the attacks that first attempt to discover the secret information and then, based on this, actually attack the watermarked data.

Finally, similarly to what is done for cryptography, it is convenient to distinguish the attacks on the basis of which data is available to the attacker. More specifically, by following the Diffie-Helman classification usually adopted in cryptography, we can assume to have the following situations:

Only watermarked content. The attacker can only access one or more watermarked assets.

Chosen watermarked content. The attacker can choose one or more watermarked assets.

Original and watermarked pair. The attacker can access one or more pairs of original and corresponding watermarked assets.

Chosen original and watermarked pair. The attacker can choose one or more pairs of original and corresponding watermarked assets.

The first kind of attack is the most common one, every watermarking system has to deal with it, given that the availability of one or more watermarked assets to the attacker is certain. The second attack is related to the possibility for the attacker of having unlimited access to a watermark recovery device: by analyzing the response of this device to different inputs (watermarked and non-watermarked assets), the attacker can succeed in reaching its objectives. The third attack is not very common, given that original assets are not likely to be at anybody's disposal, they are usually stored in archives whose access is restricted, or they could even be discharged just after watermark embedding, if the watermarking system is blind as those we are mainly dealing with in this book. For some applications, anyway, it has been thought that the watermarking device is available to anybody, in such a situation then this attack makes sense. This is, for example, the case of the mechanism that has been devised for DVD copy control , where it is considered the possibility to change in the player the watermark embedded within a video[1]. Finally the fourth type

---

[1]This is needed for changing the status of the video from copy-once to copy-never (see section 2.1.3).

of attack is also related to the possibility for the attacker to access a water-mark embedding device, in such a way that he can generate as many pairs of original and corresponding watermarked assets.

The next four sections of this chapter are dedicated to examine in details the four cases of watermarking systems that we have identified, i.e. those based on the security by obscurity approach (section 8.1), the symmetric (section 8.2) and the asymmetric (section 8.3) ones, and finally those relying on the recently proposed playing with open cards approach (section 8.4). For each approach we will consider some possible systems presented in the literature, and, above all, the corresponding attacks (fair and unfair) and possible countermeasures. Finally in section 8.5 we give a short account of the possibility to make watermarking systems secure by integrating them with cryptographic tools.

8.1   Security by obscurity

This is the most common approach used by commercial products: it is, in fact, believed that letting the watermarking system details to be unknown can make the task of the attacker more difficult. Indeed this approach explicitly violates the Kerckhoff's principle which states that the security of a system can not be based on the secrecy of the algorithm, but solely on some secret parameters (keys), it is, in fact, very unlikely that the details of the algorithm can be kept secret for a long period. Nevertheless, for sake of completeness here we briefly analyze this scenario. Most of what we will say has already been analyzed in chapter 7 that was explicitly dealing with blind manipulations.

A fair attacker can only rely on blind manipulations as those described in chapter 7 and in particular in section 7.9. Thus if he has at his disposal many copies of the same asset, watermarked with different watermarks, he can simply average them to obtain a copy in which every watermark will be very weak. On the contrary, if he has different assets containing the same watermark, he can average them for obtaining an estimate of the watermark itself . Actually, in this way the attacker is not able to recover the actual watermark signal (as defined in section 1.1.1), since he does not know in which domain and how it was embedded, however he obtains an estimate of the modifications of the original asset introduced by the water-marking process to the original asset in order to embed the message: such modifications can thus be subtracted from other watermarked copies, prob-ably reducing the recovery capabilities (possible countermeasures against this attack have been presented in section 7.9). If the attacker has unlim-ited access to a watermark detector, he can blindly (and randomly) slightly modify the asset until the watermark is no longer recoverable: this attack can be very time consuming, and can produce an attacked asset of unacceptable quality. If pairs of watermarked and original assets are available, the modifications introduced by the watermarking process can be obtained in a straightforward way by simple subtraction (the same considerations just drawn with reference to the averaging attacks are valid also in this case).

An unfair attacker , on his side, can try to get some information about the watermarking process. First of all he can try to discover some details about the algorithm: the SDMI affair has, in fact, demonstrated that, though nothing is publicized about the system, an analysis of the literature and of the patents filed by the company producing the tool, can help the attacker to make sufficient hypotheses about the functioning of the algorithm, thus allowing him to optimize the attacks. This means that in practice, a complete obscurity about the watermarking system can not be reasonably assumed. Furthermore it is widely known which are the most important weaknesses of state of the art watermarking technology: as an example, the vast majority of image watermarking algorithms is not able to deal with local geometric manipulations as those caused by random bending; similarly, most algorithms rely, for resisting to global geometric transformations (scaling and rotation) to the embedding of some reference pattern (see page 356 and 358), the template removal attack (page 370) is then likely to be successfully applied. Some information about the functioning of the algorithm can also be extracted by analyzing the watermarked assets (e.g. by looking for anomalies in their spectral content), or by comparing them with their respective originals (i.e. by analyzing the modification induced by the watermarking process).

In conclusion we can say that it does not make too much sense to try maintaining the algorithms secret[2], both because this does not give significant advantages to the designer, and, mainly, because it is highly unlikely that this secret will remain completely inviolated.

8.2 The symmetric case

All the watermarking algorithms that we have analyzed in this book require that the embedding key is known for recovering the watermark. This constitutes a major weakness because it is very easy, for those having knowledge of the key, to remove (or make unrecoverable) the watermark without causing any relevant degradation of the watermarked asset. In particular, for the additive and the multiplicative cases the watermarking process can be perfectly reverted: we have in fact that the original features can be

---

[2]Besides, this conclusion is certainly not a novelty since it was first enounced in 1883 by Kerckhoff.

obtained by simple subtraction as:

$$f_i = f_{w,i} - \gamma w_i, \tag{8.1}$$

for the additive case (see equation (4.33)), and by simple division as:

$$f_i = \frac{f_{w,i}}{1 + \gamma w_i}, \tag{8.2}$$

for the multiplicative case (see equation (4.45)). Perfect reversibility can not be obtained for quantization based schemes, but the watermark can be effectively made unreadable if the key is available: for example for ST-DM we have that (see equation (6.222))

$$f_{w,i} = f_i - \rho_w w_i + \mathcal{Q}_{0/1}(\rho_f) w_i = f_i + q_{0/1}(\rho_f) w_i, \tag{8.3}$$

where $q_{0/1}(\rho_f) w_i$ indicates the actual difference between the i-th original feature and the marked one. Given that ρ_f can not be derived from the watermarked features (the quantization function $\mathcal{Q}_{0/1}()$ is not reversible), $q_{0/1}(\rho_f)$ can not nbe known with precision: thus f_i can not be obtained from $f_{w,i}$ and from the knowledge of w_i and of the embedded bit. It is anyway possible to erase the embedded code, by referring to the following equation:

$$f_{w,i} = f_i^{\perp} + \mathcal{Q}_{0/1}(\rho_f) w_i, \tag{8.4}$$

where $\mathcal{Q}_{0/1}(\rho_f)$ is known if the watermark is correctly decoded, and $f_i^{\perp}$, and $f_i^{\perp} = f_i - \rho_w w_i$ is estimated by the simple difference $f_{w,i} - \mathcal{Q}_{0/1}(\rho_f) w_i$. It is then sufficient to add to $\mathcal{Q}_{0/1}(\rho_f)$ a value $\pm \Delta_r$ (the sign can be chosen randomly with equal probability): in this way the projection of the feature vector in the direction of the vector $\mathbf{w}$ assumes a value which is exactly in the middle between the two quantizers. Of course the distortion with respect to the original is increased in comparison to that of the watermarked asset, however, by assuming that the quantization noise due to the watermark and the modification introduced to prevent its recovery are independent and both uniformly distributed, it is possible to demonstrate that such an increase is limited to about 2.4 dB.

From the above discussion it is clear that for this kind of schemes, usually referred to as symmetric, the recovery key is made available only to trusted players, i.e. only to players that almost surely are not willing to remove the watermark.

A fair attacker can then rely only on the available information about the functioning of the algorithm to implement more sophisticated watermark removal attacks as those described in section 7.9. For example he can identify the major weakness points of the algorithm (e.g. limited resistance

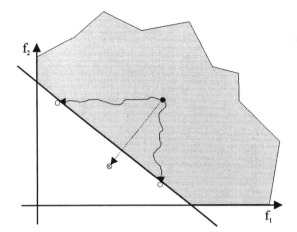

Figure 8.1: Sketch of the sensitivity attack when the boundary is an hyperplane (the case for $n = 2$ is depicted). The black dot is the watermarked features vector; the two white dots are the two estimated boundary points; the crossed dot represents the result of the closest point attack consisting in finding a feature vector which is sufficiently inside the non-detection region along the direction (dashed arrow) orthogonal to the detection boundary.

to geometric manipulations), and concentrate the attack on that direction. As another example he can perform directly the MAP or ML estimation of the original features given that the embedding domain is known[3]. Similarly, if he has at his disposal many watermarked assets, he can average directly the watermarked features, instead of the assets, thus having more effective results. In general the possible fair attacks correspond with those described in section 7.9.

The fact that the recovery key has to be maintained secret, does not imply that the detector can not be public. On the contrary, for some applications (e.g. copy control) it is needed that the watermark decoder is available in every user device. In this case the decoding engine should be enclosed into an antiforgery package, in such a way that access to the recovery key is almost impossible[4]. The possibility for the attacker of having unlimited access to a decoding device makes the chosen watermarked content attack feasible: a fair attacker can, in fact, iteratively (and slightly)

---

[3]This is not in general completely true, as some watermarking techniques have been proposed where the embedding domain itself is depending on a secret key [78, 148].

[4]If this is possible when the recovery device is implemented in hardware, it is hopeless when it is part of a software tool. Unfortunately it is likely that the latter will be the most common future scenario.

modify the watermarked features until he obtains an asset that the water-mark decoder fails to decode. This fair attack can be very time consuming, because the attacker is moving almost randomly, furthermore the final asset quality can be highly deteriorated. A more effective unfair version of the last attack can be performed. Let us initially suppose that we have a detectable watermarking scheme for which the detection region has the form depicted in figure 8.1 (i.e. the region boundary is an hyperplane in an n dimensional space). The attacker can iteratively modify the watermarked features until he finds a feature vector for which small modifications cause the watermark detector to switch between the non-watermarked and watermarked response, i.e. the attacker has found a feature vector that is very near to the detection boundary (more precisely, an approximation of a point of the boundary). By repeating this process he can obtain at least n boundary points, and then estimate the whole boundary (in this case the hyperplane). A possibility for obtaining the n boundary points is, for example, to iteratively modify, one at the time, all the components of the feature vector. Once the boundary is available, the point lying near to it on the non-detection region, and which is the nearest to the watermarked features can be selected to build the attacked asset. The attacker is then sure that the introduced distortion is the minimum possible. A sketch of this attack, that is usually referred to as 'sensitivity attack' is given in figure 8.1. The complexity of this attacks, that is very effective, is of the order of n. More complex detectors can have a more complicated decision boundary (this is for example the case of the asymmetric schemes that will be examined in section 8.3), and thus more points can be needed for estimating it. In figure 8.3 the time in years needed for estimating $n = 40000$ boundary points is plotted against the time in seconds needed for finding 1 boundary point for linear and quadratic detectors: it can be seen that if 1 point can be found in 1 s, only a few hours are needed to implement the attack on a linear detector (the time needed for estimating the boundary parameters is neglected). Instead of estimating the boundary description parameters, it is also possible to approximate it with a set of hyperplanes.

An effective countermeasure against this kind of attacks, is to use, as the decision boundary, a fractal curve, that can not be parametrized and can not be approximated by a finite set of hyperplanes. In practice a correlation based detection rule is used, thus resulting in a decision boundary that is an hyperplane: the correlation response is then split in two parts:

$$\rho = \frac{1}{n} \sum_{i=1}^{n} f_i' w_i = \frac{1}{n} \sum_{i=1}^{n/2} f_i' w_i + \frac{1}{n} \sum_{i=n/2+1}^{n} f_i' w_i = \rho_1 + \rho_2, \qquad (8.5)$$

and thus the detection boundary can be seen as the line in the 2D space

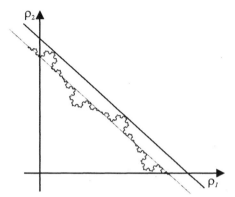

Figure 8.2: Example of a fractal detection boundary. The original linear detection boundary ($\rho_1 + \rho_2 = T_\rho$) is substituted with a fractal curve. The embedding phase needs to be modified in such a way that the watermarked asset continues staying into the detection region.

for which $\rho_1 + \rho_2 = T_\rho$ (dotted line in figure 8.2). Around this line a fractal curve (for example a Peano curve) can be built, and used as the actual detection boundary (see figure 8.2). Given that the detection boundary has been changed, the embedding phase should also be modified in order to be sure that the watermarked feature vector really lies inside the detection region: for example it would be possible to use an informed embedding approach for obtaining a given robustness to manipulations (this implies that the minimum distance between the watermarked feature vector and the detection boundary is fixed). Alternatively, the line enveloping the Peano curve and lying entirely into the detection region (bold line in the figure) can be used for making embedding easier. The boundary description must be kept secret inside the detection device. Given the non-parametric nature of the detection boundary, a sensitivity attack becomes impossible (all points of the curve are needed to describe it).

Having at disposal only one or more watermarked contents, the unfair attacker can try to estimate the watermark signal, e.g. by estimating through a MAP or ML attack the original features, and then inverting the embedding process, or through averaging. The watermark signal can then be used for implementing more effective attacks (e.g. the remodulation attack described at page 7.9), or to obtain the embedding key (for example, if a linear congruential random number generator is used to produce the watermark signal, a few samples of the pseudo random sequence are sufficient to estimate the generator parameters, and thus to get the used seed). This attack is even simpler if the unfair attacker can access pairs of original and

corresponding watermarked assets. The estimation of the watermark signal
is not useful if it depends on the original content, as it is for example the
case for informed (e.g. quantization based) schemes: in this case, in fact,
different assets watermarked with the same key have, in general, different
watermark signals added to them; thus the watermark signal estimated
from an asset is not useful for removing the watermark from another asset.
Another way for making the watermark signal dependent on the asset is to
generate it based (further than on a secret key) on the asset itself, e.g. by
using some hash value of the asset content: such an hash should be robust,
i.e. the same value should be obtained also after asset manipulations in
order to allow the correct recovery of the watermark.

8.3 The asymmetric case

We have seen in the previous section that symmetric schemes are not very
suitable for those applications requiring public watermark recovery. Even
if the recovery key could be kept secret, the sensitivity attack allows (in
general) to effectively estimate the detection boundary with a small complexity. Spurred by this problem, asymmetric techniques, that allow public
watermark recovery without the need to disclose enough information for
watermark removal, have been developed. In practice, asymmetric schemes
perform watermark embedding with a private key (that is kept secret), and
watermark recovery by means of a public key (available to anybody), in
such a way that:

1. It should be computationally impossible to estimate the private (embedding) key from the public (recovery) key,

2. The knowledge of the public key should not help the attacker to
   effectively remove (or make unreadable) the watermark.

Probably the first really asymmetric scheme that has been proposed is
the following: the feature vector $\mathbf{f}$ is split into two equal parts and to each
part the same pseudorandom signal is added:

$$
\begin{aligned}
f_{w,i} &= f_i + \gamma w_i, \\
f_{w,i+n/2} &= f_{i+n/2} + \gamma w_i,
\end{aligned}
\tag{8.6}
$$

for $1 \leq i \leq n/2$. The detector simply computes the correlation between
the first and the second part of the watermarked feature vector, i.e.:

$$
c = \frac{2}{n} \sum_{i=1}^{n/2} f_{w,i} f_{w,i+n/2}.
\tag{8.7}
$$

By assuming that the watermarked features have zero mean and are independent, it results that:

$$\mu_{c|H_0} = \frac{2}{n} \sum_{i=1}^{n/2} E[f_i] E[f_{i+n/2}] = 0, \tag{8.8}$$

while, under hypothesis H_1 we have:

$$\begin{aligned}
\mu_{c|H_1} &= \frac{2}{n} \sum_{i=1}^{n/2} E[f_i] E[f_{i+n/2}] + \frac{2}{n} \sum_{i=1}^{n/2} E[f_i] \gamma w_i + \\
&\quad \frac{2}{n} \sum_{i=1}^{n/2} E[f_{i+n/2}] \gamma w_i + \frac{2}{n} \sum_{i=1}^{n/2} \gamma^2 w_i^2 \\
&= \gamma^2 \overline{w^2} \approx \gamma^2 \sigma_w^2 = 0.
\end{aligned} \tag{8.9}$$

Yet from this last equation one of the drawbacks of asymmetric techniques is evident: $\mu_{c|H_1}$ depends on γ^2 instead of γ as it is for symmetric schemes (see for example equation (6.39)), and given that $\gamma < 1$ this implies that the robustness of asymmetric techniques is intrinsically lower than that of symmetric ones. To be more precise, it is useful to define an efficiency parameter that is often used to evaluate the degradation of performance of asymmetric schemes with respect to symmetric ones. For the simple case described above, the efficiency is defined as:

$$e = \sqrt{2} \frac{\mu_{c|H_1} - \mu_{c|H_0}}{\sqrt{\sigma_{c|H_1}^2 + \sigma_{c|H_0}^2}}; \tag{8.10}$$

the larger the efficiency the more robust is the technique. As an example, by referring to equations (6.33), (6.39), (6.34) and (6.40), and by assuming that the asset features are independent and have zero mean, and no attack is present, it results that for an additive detectable scheme:

$$e = \sqrt{n \frac{\gamma^2 \sigma_w^2}{\sigma_f^2}} = \frac{\sqrt{n}}{\sqrt{\mathrm{DWR}}}, \tag{8.11}$$

where DWR has been defined in equation (7.1). On the other side, For evaluating the efficiency of the just described asymmetric scheme, it needs to compute the variance of the detector response c under the usual two hypotheses H_0 and H_1. It can be easily verified that:

$$\sigma_{c|H_0}^2 = \frac{2}{n} \sigma_f^4, \tag{8.12}$$

and

$$\sigma_{c|H_1}^2 \approx \frac{2}{n}\sigma_f^4 + \frac{4}{n}\gamma^2\sigma_f^2\sigma_w^2, \tag{8.13}$$

we then have that:

$$e = \sqrt{2}\frac{\gamma^2\sigma_w^2}{\sqrt{\frac{4}{n}\sigma_f^2\left(\sigma_f^2 + \gamma^2\sigma_w^2\right)}} = \sqrt{\frac{n}{2}}\frac{1}{\sqrt{DWR(DWR+1)}} \approx \frac{\sqrt{n/2}}{DWR}, \tag{8.14}$$

where the approximation is valid because in general DWR $\gg$ 1. By comparing this result with equation (8.11), and by considering that DWR $\gg$ 1 it is confirmed that the robustness of this asymmetric scheme is lower than that exhibited by the corresponding symmetric techniques. This result is indeed general: it has been demonstrated, in fact, that all asymmetric techniques having a quadratic detection function, exhibit an efficiency which depends on the inverse of DWR, instead of the inverse of $\sqrt{DWR}$ as it is for symmetric schemes. On the other side, this just described algorithm really allows to detected the watermark presence without the need to disclose the secret embedding key: indeed, there is not a real public key, but, rather a public detection method. This characteristic also limits the validity of this method: although the knowledge of the detection parameters does not help to discover the secret embedding key, that would allow the easy removal of the watermark signal, it is always possible to perform the closest point attack.

Let us better investigate this issue. In general an asymmetric scheme based on a quadratic detection function can be defined by a matrix $A = \{a_{ij}\}_{i=1,n;j=1,n}$, and the detection rule written as:

$$c = \frac{1}{n}\sum_{i=1}^{n}\sum_{j=1}^{n}a_{ij}f_{w,i}f_{w,j} = \frac{1}{n}\mathbf{f}_w A\mathbf{f}_w^t. \tag{8.15}$$

For example, for the method described before we have:

$$A = 2\begin{bmatrix} 0_{n/2} & I_{n/2} \\ I_{n/2} & 0_{n/2} \end{bmatrix}, \tag{8.16}$$

where here $0_{n/2}$ stands for the all zero $n/2 \times n/2$ matrix, and $I_{n/2}$ for the $n/2 \times n/2$ identity matrix. The detection boundary is described by:

$$\frac{1}{n}\sum_{i=1}^{n}\sum_{j=1}^{n}a_{ij}f_if_j - T_c = 0 \tag{8.17}$$

where T_c is the detection threshold. The closest point attack consists of finding the point $\mathbf{f} = \{f_i\}_{i=1,n}$ belonging to the detection boundary that is

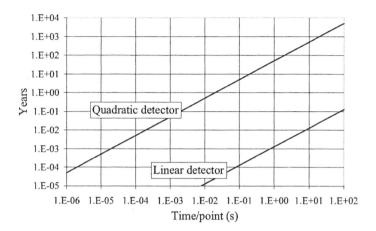

Figure 8.3: Plot of the time (years) needed for implementing the sensitivity attack vs the time (s) needed for finding a single boundary point. The dimension of the feature vector has been fixed to 40000 coefficients.

nearest to the watermarked feature vector $\mathbf{f}_w = \{f_{w,i}\}_{i=1,n}$: this problem can be solved by computing the absolute minimum of the Lagrangian

$$\arg\min_{f_i,\lambda} \left\{ \sum_{i=1}^{n} (f_i - f_{w,i})^2 + \lambda \left(\frac{1}{n} \sum_{i=1}^{n} \sum_{j=1}^{n} a_{ij} f_i f_j - T_c \right) \right\}. \qquad (8.18)$$

The minimization requires the resolution of the following set of non linear equations:

$$\begin{cases} \left[I_n + \dfrac{\lambda}{2n} \left(A + A^t \right) \right] \mathbf{f}^t = \mathbf{f}_w \\ \dfrac{1}{n} \mathbf{f} A \mathbf{f}^t = T_c \end{cases} \qquad (8.19)$$

For example, a possibility is to solve the first set of linear equations for different values of λ and test which of these solutions better satisfies the non linear equation at the bottom of (8.19). Another possibility is to use a gradient descent algorithm to directly minimize (8.18). The complexity of this type of attacks depends on n^2. To make this attacks more difficult, thus, it can be convenient to keep the matrix A secret, e.g. by embedding the detector inside an anti-forgery device (the scheme becomes then symmetric). In this case, the only possibility for an attacker is to estimate the matrix A (in practice the detection boundary), by means of the sensitivity attack described in section 8.2. The number of points of the detection boundary that need to be found in this case is of the order of n^2, that is a strong improvement with respect to the order n needed for linear (symmetric) detectors. As an example, in figure 8.3 the time (in years) needed to found the boundary points, if 40000 features are used, is plotted against the time (in seconds) needed to find 1 point: as an example, if the time for 1 point is 1 s, more than 50 years are needed to implement this attack on a quadratic (asymmetric) scheme, instead of the few ours required by a linear (symmetric) scheme. But the above described asymmetric scheme does not allow to choose a particular matrix A, which constitutes a big limitation. Furthermore this method does not allow to detect the presence of different watermarks, but only to test if an asset is watermarked with any watermark or not. Finally, the fact that the watermark signal is embedded twice in the feature vector, makes the method more sensitive to noise removal attacks (by taking the mean of the first part and of the second part of the vector, the variance of the original features is halved).

Other methods have been proposed that overcome the above limitations. In particular, based on the matrix form of the detector (equation (8.15)), it is possible to choose, as the additive watermarking vector $\mathbf{w}$, one of the eigenvectors of the matrix A, i.e. the embedding rule is given by

$$\mathbf{f}_w = \mathbf{f} + \gamma\mathbf{w}, \tag{8.20}$$

where $\mathbf{w}$ is the eigenvector of the matrix A corresponding to the eigenvalue λ_w. It can be demonstrated that[5]:

$$\mu_{c|H_0} = \sigma_f^2 \frac{1}{n}\mathrm{Tr}(A), \tag{8.22}$$

and

$$\mu_{c|H_1} = \sigma_f^2 \frac{1}{n}\mathrm{Tr}(A) + \gamma^2 \lambda_w \overline{w^2}, \tag{8.23}$$

where we have assumed that the features are independent identically distributed (which implies that $C_{\mathbf{f}_w|H_0} = C_{\mathbf{f}_w|H_1} = \sigma_f^2 I_n$) and have zero mean (which implies that $\mathbf{m}_{\mathbf{f}_w|H_0} = 0$ and $\mathbf{m}_{\mathbf{f}_w|H_1} = \gamma\mathbf{w}$), and we have remembered that $\mathbf{w}$ is an eigenvector of A (i.e. $A\mathbf{w} = \lambda_w \mathbf{w}$). It is worth noting that we can not use the approximation $\overline{w^2} \approx \sigma_w^2$ given that the watermark is not random.

---

[5]It is known that given a quadratic form $\mathbf{X}A\mathbf{X}^t$ of a random vector $\mathbf{X}$, its mean value is given by:

$$E[\mathbf{X}A\mathbf{X}^t] = \mathrm{Tr}(AC_X) + \mathbf{m}_X A\mathbf{m}_X^t \tag{8.21}$$

where $\mathrm{Tr}()$ is the trace of the matrix, i.e. the sum of its diagonal values, C_X is the covariance matrix of $\mathbf{X}$ and $\mathbf{m}_X$ is its mean vector.

Similarly, by further assuming that the features follow a normal distribution, it is possible to estimate the variance of the detector response under the two hypotheses[6]:

$$\sigma^2_{c|H_0} = 2\sigma^4_f \frac{1}{n^2} \text{Tr}(A^2), \tag{8.25}$$

and

$$\sigma^2_{c|H_1} = 2\sigma^4_f \frac{1}{n^2} \text{Tr}(A^2) + 4\frac{1}{n}\sigma^2_f \gamma^2 \lambda^2_w \overline{w^2}. \tag{8.26}$$

Given that the trace of a matrix is given by the sum of its eigenvalues, and that the eigenvalues of the matrix A^2 are obtained by those of matrix A by squaring them, it is possible to substitute:

$$\text{Tr}(A^2) = \sum_{k=1}^{n} \lambda^2_{w_k} = n\overline{\lambda^2_w} \tag{8.27}$$

where $\{\lambda_{w_k}\}_{k=1,n}$ are the n eigenvalues of matrix A, and where we let:

$$\overline{\lambda^2_w} = \frac{1}{n}\sum_{k=1}^{n} \lambda^2_{w_k}. \tag{8.28}$$

and finally obtain:

$$e = \frac{\sqrt{2}\gamma^2 \lambda_w \overline{w^2}}{\sqrt{4\sigma^4_f \frac{1}{n}\overline{\lambda^2_w} + 4\frac{1}{n}\sigma^2_f \gamma^2 \lambda^2_w \overline{w^2}}} = \sqrt{\frac{n}{2}} \frac{1}{\sqrt{\text{DWR}\left(\frac{\lambda^2_w}{\lambda^2_w}\text{DWR}+1\right)}} \propto \frac{\sqrt{n/2}}{\text{DWR}} \tag{8.29}$$

where $\text{DWR} = \sigma^2_f/\gamma^2\overline{w^2}$. Once again we obtain the same behavior, with respect to DWR, of the first asymmetric algorithm presented (see equation (8.14)), thus confirming the generality of this result. The possibility for a malicious attacker to get the embedding key, based on the knowledge of the detection key (i.e. of matrix A), is related to the difficulty to compute the eigenvectors of a large matrix: once these are known it is possible to try to subtract them, one at a time, until the one whose subtraction results in the worst (lower) detector answer is found. In practice it is not very difficult to obtain the secret (embedding) key from the public (detection) one.

In order to overcome this limit, another method has been proposed, which is based on a one way signal processing function: in particular it is

---

[6]On the other side the variance of the quadratic form, in the hypothesis that vector $\mathbf{X}$ follows a multivariate normal distribution, is given by:

$$\text{Var}[\mathbf{X}tA\mathbf{X}^t] = 2\text{Tr}(AC_X AC_X) + 4\mathbf{m}_X AC_X A\mathbf{m}^t_X \tag{8.24}$$

exploited the fact that it is impossible to recover a signal from its power spectrum. In practice, the additive watermark signal is obtained by filtering, with a kernel $\mathbf{h}$, a white Gaussian noise $\mathbf{v}$ with unitary variance:

$$\mathbf{f}_w = \mathbf{f} + \gamma \mathbf{h} \times \mathbf{v}, \tag{8.30}$$

Detection is performed by estimating the mean power spectral density (PSD) of the feature vector: by assuming again that the features are independent and identically distributed it results that if H_0 holds then the PSD is:

$$S_{H_0} = S_f(f) = \sigma_f^2, \tag{8.31}$$

i.e. a white PSD. On the other side, if H_1 is in force we get:

$$S_{H_1} = S_f(f) + \gamma^2 |H(f)|^2 S_v(f) = \sigma_f^2 + \gamma^2 |H(f)|^2. \tag{8.32}$$

Thus, to check for watermark presence, it only needs to compare the PSD with the function $|H(f)|^2$ (the comparison can be performed in an optimum way by resorting to a maximum likelihood formulation). It has been demonstrated that also in this case the detection function can be expressed as a quadratic function (see equation (8.15)), and that the efficiency depends on the inverse of DWR. From the public key ($|H(f)|^2$) it is impossible to obtain an estimate of the embedded signal (neither of filter $H(f)$, nor of v_i), thus satisfying the main requirements of asymmetric schemes. The main weakness of this method is that a clever attack can be performed by scaling the watermarked features in such a way that their PSD becomes more similar to a flat spectrum than to the spectrum defined by the public key: resistance to this attack is ultimately a matter of robustness, but this (as we have shown) is reduced for all those asymmetric methods based on quadratic detectors.

In general the fact that asymmetric schemes, as those described in this section, which are based on quadratic detectors, show a degradation of robustness (as measured by the efficiency parameter) with respect to symmetric techniques, makes them more sensitive to all the blind or informed attacks that we have analyzed previously: in practice the amount of degradation that has to be introduced for removing (or making unrecoverable) the watermark is much lower than for symmetric methods.

Another general consideration is worth to be drawn with reference to the two characteristics that should be exhibited by asymmetric algorithms, and that we listed at the beginning of the section. We have, in fact, seen that all the three algorithms analyzed satisfy the requirements that it is impossible (or very difficult in the case of the technique based on eigenvectors) to recover the embedding key from the recovery parameters. On the other side, for all three algorithms it is not too difficult to remove the watermark based solely on the recovery key, thanks to the closest point attack formalized in equation (8.18): that is, they fail to satisfy the second requirement. A feasible solution would be to design a detection function (and thus a detection boundary) so complicated to make the closest point attack computationally unviable: on the other side attention should be paid not to complicate embedding as well. Indeed asymmetric techniques are a step toward this direction, in that embedding is easily performed by addition of the watermark signal (private key), while the detection function is quite complicate (although still tractable).

8.4 Playing open cards

In the process of progressively reducing the amount of information to be kept secret, we now consider a situation in which everything (the algorithm, the embedding, and even the recovery parameters) is publicly available. This approach is based on the observation, presented at the end of section 8.3, that it should be highly difficult for an attacker to move a feature vector outside the detection region. To this regards, symmetric schemes can rely only on the secrecy of the embedding and detection keys, i.e. on the unavailability to the attacker of a description of the detection region. Asymmetric algorithms, on the other side, build a more complicated detection region, hoping to make difficult the closest point attack, and envisage the possibility for the embedder to rely on a further secret information (the embedding key) for efficiently moving non-watermarked feature vectors inside the detection region. The main problem of the asymmetric schemes presented in the literature so far, is that the detection region is not complicated enough to really make the closest point attack computationally not feasible. Furthermore in the asymmetric schemes we presented, the embedding key eases the embedding process, but, if known, also simplifies watermark removal.

On the basis of the previous observations it has been recently proposed a novel approach, which we call *playing open cards*. According to this approach, the detection region (boundary) should be designed in a very complicated way, so to make almost impossible the closest point attack. Simultaneously, a simpler region, completely lying inside the detection region, should be available to make embedding feasible. An example of how these two regions (and in particular their boundaries) could look like is given in figure 8.4: in this case, the embedder needs only to move the non-watermarked feature vector (the crossed dot) to the closest point (the watermarked feature vector represented by the filled dot) over the bold curve (which is completely enclosed by the detection region), while an attacker should try to move the watermarked feature vector on the closest

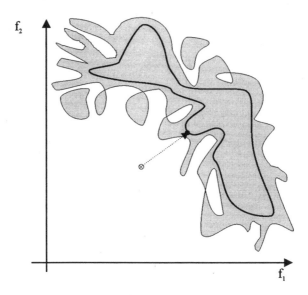

Figure 8.4: Example of a very complicated detection region (in grey), for which a simple curve completely included inside it is available (bold curve) to help the embedder. The embedder has simply to move the non-watermarked feature vector (white dot) to the closest point on the bold curve (black dot).

point over the boundary of the detection region, which can be a very complicated task, given the complex shape of the detection boundary. The knowledge of the embedding region boundary does not enable watermark removal and this permits to make this information publicly available. In practice, then, both the embedding key (the embedding boundary) and the recovery key (the detection boundary) are known to everybody, but this does not diminishes the security of the scheme.[7] Leaving everything publicly known makes unfair attacks meaningless. On the other side the richness of information available to the attacker can make fair attacks very effective (fair attacks are, on the other side, a matter of robustness more than security).

Another possibility for playing open cards would be to design a (publicly known) detection function making easy the embedding process (i.e. to enter the detection region), and computationally very difficult an attack (i.e. to exit from the detection region). As an example we could design a detection

---

[7]A quite similar approach has been presented at the end of section 8.2 where a fractal boundary is used for the detection region, and a straight line helps the embedding process. In that case, however, it is assumed that both the detection and the embedding boundaries are kept secret, i.e. the system does not play open cards.

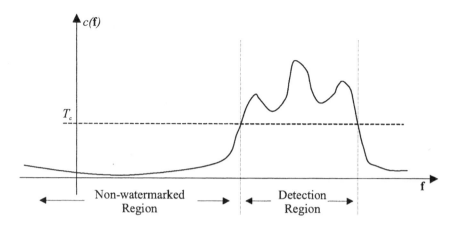

Figure 8.5: Example of a possible detection function $c(\mathbf{f})$ that although making quite easy the embedding process, is likely to highly complicate the task of the attacker.

function $c(\mathbf{f})$ with the following characteristics:

- $c(\mathbf{f})$ is regular enough (e.g. derivable) to allow the use of gradient descent techniques for finding its minima and maxima.

- Given a maximum distance d_{max}, the probability that there is at least a maximum $\mathbf{f}_m$ of the function $c(\mathbf{f})$ such that $||\mathbf{f}_m - \mathbf{f}|| < d_{max}$ should be approximately 1 (this assures that embedding is almost always possible).

- Given a real value c_0, and a feature vector $\mathbf{f}$, it is computationally difficult to find another feature vector $\mathbf{f}'$ such that $c(\mathbf{f}') = c_0$ and $||\mathbf{f}' - \mathbf{f}|| < d_{max}$. This is needed because otherwise the attacker could simply choose $c_0 < T_c$ (where T_c is the detection threshold) and find $\mathbf{f}'$ near to $\mathbf{f}$ such that $c(\mathbf{f}') = c_0$ to remove the watermark. This condition should hold in probability, i.e. it should be highly improbable (according to the distribution of $\mathbf{f}$) that a feature vector for which this does not hold is found.

- It exists a threshold T_c such that:
    - The $\mathrm{Prob}\,(c(\mathbf{f}) > T_c)$ is as small as we like (which grants the false alarm probability is small).
    - All local maxima of $c(\mathbf{f})$ are higher than the threshold T_c.
    - Most of the local minima of $c(\mathbf{f})$ are higher than the threshold T_c, but some are lower.

In figure 8.5 a sketch of how this function could look like is given. The embedding process could rely on the possibility of using gradient descent techniques for finding a local maximum of the function. The difficulty to move a feature vector outside the detection region is, on the other side, based on the fact that a gradient descent process starting from a local maximum is likely to be trapped in one of the surrounding local minima.

Up to today, no practical algorithm following this approach has been proposed, and it is thus not sure whether this path is feasible or not. Anyway this approach raises some questions about the real necessity of designing asymmetric watermarking schemes, given that, at least hypothetically, a secure watermarking technique could be devised that does not require any information to be kept secret.

8.5   Security based on protocol design

Up to now we only investigated the signal processing aspects of watermarking security, however a good help to improve the security of applications relying on watermarking technology could come from their integration, at the protocol level, with cryptography tools. As compared with watermarking, in fact, cryptography allows establish the security level of a technique more formally, and many secure tools have been developed and largely used today. It is the goal of this section to present an example of a protocol in which the integration between watermarking and cryptography can help to solve the security problems. This section, thus, does not pretend in any way to give a comprehensive description of the issues related to the integration between cryptography and watermarking, but only to show that great advantages can be obtained by moving in this direction.

As an example, we have already seen in section 2.1.2 the IBS protocol that makes secure a fingerprinting application by relying on a TTP that generate encrypted fingerprints, and on some nice properties of some asymmetric encryption algorithms (namely RSA) that allow to embed the watermark in the encrypted domain. Although this protocol succeeds in making safe a really untrusted environment, it does not allow to solve the problem of secure public watermark recovery, that we have demonstrated in this chapter to be one of the most challenging problems watermarking researchers must face with.

A possible way to attack this problem relies on the use of the zero knowledge proof paradigm. Without going in deep details, which is out of the scope of this book, a zero knowledge proof system is a protocol that is run between a prover P and a verifier V, and that allows P to convince V about a fact, without revealing any further property about the parameters involved in proofing the fact than the fact itself: as an example (which will be useful for watermarking) it is possible to prove that a number lies in a certain interval, without disclosing the number itself. Most zero knowledge systems relies on a cryptography primitive, named commitment scheme, that consists of a protocol Com to commit to a value m, and a protocol $Open$ that opens the commitment (i.e. it demonstrate that the value m was truly the one which generated the commitment $Com(m)$). Commitment schemes have to satisfy two properties:

Hiding property The commitment of a value ($Com(m)$) should not reveal any information about the committed value m, i.e. it should be (at least computationally) impossible to obtain m from $Com(m)$.

Binding property It should be (at least computationally) impossible for a dishonest committer to open a commitment to another value $m' \neq m$, i.e. the committer can not change his mind.

In particular there are commitment schemes that have some nice homomorphic properties, for example the committer can open the commitment $Com(m_1)Com(m_2)$ to the value $m_1 + m_2$: this particular type of commitment scheme will be used for the zero knowledge proof protocol that we are going to describe.

Let us assume that we are using a simple symmetric watermarking technique employing a correlation-based detector , in particular let us assume that detection is accomplished by comparing the correlation

$$\rho = \mathbf{f}' \cdot \mathbf{w} = \sum_{i=1}^{n} f_i' w_i, \tag{8.33}$$

between the possibly watermarked and attacked feature vector $\mathbf{f}'$ and the watermark signal $\mathbf{w}$, against a threshold T_ρ, that, for simplicity, we assume to be constant. Let us note that in this framework, in order to apply cryptographic primitives, both the host features and the watermark are assumed to take integer values. If this is not the case, $\mathbf{w}$ and $\mathbf{f}'$ are quantized prior to the applications of cryptographic primitives. First of all the commitments of the watermark signal $Com(\mathbf{w}) = (Com(w_1), Com(w_2), \ldots, Com(w_n))$[8] is made public: thanks to the hiding properties of the commitment scheme, it is impossible to obtain $\mathbf{w}$ from its commitment. Then P can prove to V that the correlation is higher than the threshold, without disclosing any information about the watermark $\mathbf{w}$, according to the following protocol:

1. P generates a commitment of T_ρ, namely $Com(T_\rho)$ and sends it to V.

---

[8]We assume that the commitment of a vector is the vector having as components the commitments of the coefficients of the original vector.

2. P immediately opens $Com(T_\rho)$ thus showing that it actually contains the threshold T_ρ.

3. Both P and V can compute $Com(\rho)$ solely based on the publicly available committed value of the watermark signal, and on the features extracted from the asset, thanks to the homomorphic properties of the commitment scheme, for which it results that:

$$Com(\rho) = \prod_{i=1}^{n} Com(w_i)^{f_i'}.$$

4. Now P, by exploiting a zero knowledge proof system available in the literature, proves in zero knowledge that the value committed by $Com(\rho)$ is higher than the threshold committed by $Com(T_\rho)$[9].

In practice then the verifier is convinced that the correlation is higher than the threshold, i.e. that the asset contains the watermark, without receiving any information about the watermark signal itself. The main drawback of this approach is that it is interactive, that is it requires a communication channel to be established between the verifier and the prover.

8.6 Further reading

The Secure Digital Music Initiative (SDMI) is an international consortium aimed at developing standard technologies for protecting the distribution of digital music. In September 2000 SDMI announced a three-week challenge inviting the public to try attack some watermarking schemes it has selected. The algorithms were not publicly available. For each watermarking algorithm SDMI provided three audio samples: one original non-watermarked audio, the same audio watermarked, a different watermarked audio fragment. Access was also granted to an *oracle* performing watermark detection and perceptual evaluation of the degradation introduced by the attacks. Although the setup of the challenge was very favorable for the challenger, given that the Kerckhoff's principle was violated, the challenge duration was very short, and the reply time of the oracle was relatively long, all the algorithms were successfully attacked by a team of researchers of the University of Princeton. This result caused a big reconsideration of the possibilities of watermarking technology to provide an effective solution to copy control, and, above all, pushed the watermarking community to more carefully consider the issues related to security. More details about this affair can be found in [61, 234].

---

[9]In practice it is shown that the value committed by $Com(\rho)$ lies in a certain interval defined by the threshold T_ρ and some very high value.

For the classification of the attacks that can be brought to a watermarking system, based on the type of data available to the attacker, we borrowed from the seminal paper by W. Diffie and M. Hellman [66], where a similar classification is proposed for cryptography tools. The Kerckhoff's principle, according to which the security of a system can not be based on the secrecy of the algorithm is known since 1883, when it was firstly presented in a paper about military cryptography [121].

Some attempts to develop a theoretical analysis of the security aspects of data hiding technology have been presented in the past, interested readers can refer to [34] and [156]. Some interesting considerations on watermarking security can also be found in [114] and [16] .

The sensitivity attack was first proposed with reference to the copy protection mechanism to be implemented in DVD recorders, more details about this attack can be found in [54, 116, 138]. The possibility to use a non-parametric fractal detection boundary to drastically reduce the effect of the sensitivity attack, has been proposed by A. Tewfik et al. [216, 145].

The first proposal for an asymmetric watermarking scheme consisted of a simple modification of a spread spectrum technique in which only part of the watermark signal was made publicly available: this part was sufficient for performing watermark detection, but not for removing the watermark [90]. More details about the more sophisticated asymmetric schemes presented in section 8.3 can be found in [73, 81, 82, 202, 220]. An excellent unified framework to model all the asymmetric watermarking techniques, and a detailed theoretical analysis of their performance is presented in [83]. A natural evolution of asymmetric schemes based on quadratic detection functions, is to implement higher order (more than quadratic) detectors, a proposal in this sense, along with an analysis of the achieved performance, is given in [97].

The open cards scenario was first introduce by M. L. Miller in [154] where it is argued that asymmetry is neither sufficient nor necessary for granting security. While it is not surprising that watermark asymmetry by itself is not sufficient to ensure security [10], since it also needs that both the conditions listed at the beginning of section 8.3 are satisfied (on the contrary, most of the presented watermarking schemes fail to satisfy the second requirement, if not both of them), the fact that asymmetry is not even necessary is a quite surprising result, and, although no proof is given that a system as that described in section 8.4 could be actually built, this approach can constitute an interesting research direction for the future.

The successful integration of cryptography protocols and watermarking technology for solving the problem of public watermark detection by means

---

[10]Also in the field of cryptography, asymmetric algorithms are safe only if very long keys are used

of a zero knowledge system is described in deeper detail in [2]. More details on commitment schemes and zero knowledge proof protocols can be found, for example, in the good book by N. Smart [201].

9

An information theoretic perspective

On several occasions throughout the book, we have drawn the attention of the reader to the analogy between digital watermarking and digital communication. In many cases such an analogy inspired us the development of new watermarking strategies both from the embedding and the detection/decoding points of view. In this chapter we move the analysis one step further, in that we will use a bunch of instruments borrowed from the mathematical theory of digital communications, namely information theory, to analyze the watermarking problem from as general as possible a perspective. More specifically, we will try to evaluate the ultimate limits of the performance achievable by any watermarking scheme subject to very general constraints, such as maximum allowed embedding and attack distortions .

As we will see in the following sections, some of the results obtained by looking at digital watermarking from an information theoretic perspective are rather surprising, in that they prove that some deep-seated opinions about watermarking are wrong (at least in principle). This is the case, for example, of the impact of detector/decoder blindness on watermarking reliability: however strange it may seem, under a proper set of hypotheses, it can be shown that detector/decoder blindness has no impact on the capacity of the watermarking channel[1].

Another benefit that is got by looking at digital watermarking from an information theoretic perspective, is that such an analysis provides a number of hints on optimal attacking and decoding/detection strategies. This

---

[1]No need saying that, in practice, having the possibility of accessing the original, non-marked asset, considerably simplifies the design of an effective detector/decoder.

is the case of the SCS (Scalar Costa's Scheme) watermarking algorithm, whose strategy is motivated by the analogy between digital watermarking and digital communication with side information at the encoder: a classical information theory problem whose solution has been known since the early 80's.

In addition to information theory concepts, a proper analysis of digital watermarking requires that some ideas borrowed from game theory are utilized. Digital watermarking, in fact, is a typical game, where two adversaries try to achieve two different, conflicting, goals. For this reason, in the following sections we will analyze the, so called, (Gaussian) watermarking game, whose solution will give us some fundamental insights about the critical balance between the rules of the game and the chances of the two actors of the game, i.e. the watermarker and the attacker[2].

Before starting our discussion, it is worth pointing out that information theoretic analysis of digital watermarking is a new research branch of data hiding, hence available results are liable to be refined, or even surpassed, in the next years. As a matter of fact, new results continue appearing in the related scientific literature, thus making it extremely difficult to give a complete picture of this field. For this reason the analysis contained in the next sections is an incomplete one. We tried to select only well-consolidated results who already had an important impact on algorithms development[3]. In addition, we will only present the main results without giving any proofs, since this would require much more than a brief chapter at the end of a book.

After some historical notes (section 9.1), we start by carefully defining the watermarking game 9.2, and by presenting the main results in a context which is as general as possible. Then, in section 9.3, we focus on a simplified version of the game, namely the additive attack watermarking game. In the same section we will present Costa's expression for the capacity of an additive Gaussian channel with side information at the encoder.

Finally, in the last section of the chapter (9.4), we describe a watermarking algorithm which derives directly from Costa's results. As we will see, such an algorithm has the potentiality of significantly outperforming the classical algorithms described so far.

---

[2]Indeed three actors could be individuated: the embedder, the attacker and the detector/decoder. However, by the light of the results we will discuss throughout the chapter we prefer to merge the embedder and the decoder/detector in a unique actor called the watermarker.

[3]For example, we will only focus on readable watermarking, since the extension to 1-bit watermarking is still a research subject.

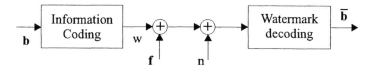

Figure 9.1: Watermark model used in the early days of watermarking research when the particular nature of the first form of noise affecting the watermark was not recognized.

9.1 Some historical notes

Though the close relationship between digital watermarking and digital communications was recognized since the very beginning of watermarking research, the real nature of the watermarking problem was not understood until late nineties. In the early days, in fact, the watermarking channel was modeled as in figure 9.1: the watermark signal was first mixed with the host features, then attack noise was added. Noticeably, mixing the watermark signal and the host features was looked at as a kind of noise addition, whereby the interference between the watermark and the host features was taken into account. Finally, watermark recovery was performed either with the aid of the original non-marked asset or blindly. This model, together with the observation that in the blind version of the channel, host features are not known to the detector/decoder, led to consider host features as an additional source of noise impairing the transmitted signal, i.e. the watermark. Of course, this was not the case if the detector/decoder could access the original asset, since in this case the first source of noise could be easily cancelled out. The main consequence of the above observations was that, due to the imperceptibility constraint which imposes that the watermark signal has a much lower strength than the host features, blind detection was possible only at the expense of a significant performance loss.

As it was realized lately, the above arguments failed to recognize that host features, even if unknown at the detector/decoder, are not a conventional source of noise, since they are known by the encoder. Communication through a classical channel, then, was not a proper model of the watermarking channel. Such an observation led to consider watermarking as power-constrained communication with side information at the encoder[4]. According to this model (see figure 9.2), the transmitted signal is impaired by two sources of noise, the first of which is known by the encoder. Such a channel had already been studied by Costa in 1983 with a somewhat surprising result: under certain hypotheses, the capacity of the channel does

---

[4]The power constraint directly derives from the imperceptibility constraint.

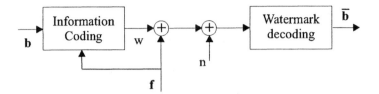

Figure 9.2: Watermarking seen as communication with side information. The only difference with respect to figure 9.1 consists in the fact that **f** is fed into the embedder.

not depend on the first source of noise. A more general analysis had been carried out by Gel'fand and Pinsker which analyzed digital communication over a channel whose state is known by the encoder but not by the decoder.

The analogy between communication with side information has led to considerable advances in watermarking theory, yet an additional step has to be done to properly model the watermarking problem. More specifically, it must be recognized the, so to say, active nature of the second source of noise. Such a noise, in fact, originates from the attacker's will to impair the watermark, and hence provision must be made for noise adaptivity, in that it is likely that the attacker will decide his strategy depending on the particular embedding rule used by the watermarker. This led researchers to cast digital watermarking in the framework of game theory, where the value of the game is defined in terms of achievable transmission rate, or, alternatively, mutual information.

This is exactly the viewpoint we will adopt in the next sections, where the watermarking game will be defined, and the value of the game evaluated. As it will be seen, such an exercise is extremely useful, since it permits to get some important insights into the ultimate achievable performance of any watermarking system[5].

9.2 The watermarking game

In this section we give an exact definition of the watermarking game, by carefully describing the players of the game, their conflicting goals, and the rules of the game. Then we present the main results, in terms of achievable transmission rates and watermarking channel capacity.

---

[5]Of course, theoretical models are never a true picture of reality, hence the results dictated by theory must be considered with great care, and can not be applied directly to practical situations. Nevertheless, the availability of a good theoretical framework is of outmost importance, since the insights it provides are an invaluable guide to the design of practical systems.

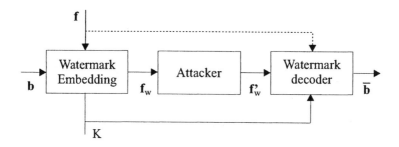

Figure 9.3: General watermarking game. As it can be seen this model is more general that those depicted in figures 9.1 and 9.2.

9.2.1 The rules of the game

Let us start by considering the general form of a watermarking system depicted in figure 1.1. In order to simplify the analysis, we restrict our discussion to readable watermarking schemes. In addition, we neglect the feature extraction process and its inverse (equations (1.3) through (1.5)), in such a way that the embedder, the attacker and the decoder operate directly on the host feature vector $\mathbf{f}$, rather then on the host asset A. Under these simplifying hypotheses, the general model of the watermarking game assumes the form shown in figure 9.3.

The first player we encounter is the embedder; his goal is to hide a message $\mathbf{b}$ within the host feature vector $\mathbf{f}$. In order to formalize such a process, let $\mathbf{b}$ be a generic message picked at random from the set $\mathbf{B}$ of possible messages, and let the cardinality of $\mathbf{B}$ be 2^{nR}, where by n we indicated the length of the feature vector $\mathbf{f}$, and R is called the transmission rate of the system. Data embedding is achieved by means of an embedding function $\mathcal{E}$, accepting at its input the to-be-hidden message $\mathbf{b}$, the host feature vector $\mathbf{f}$, and the value assumed by a random sequence K, usually referred to as the embedding key:

$$\mathbf{f}_w = \mathcal{E}(\mathbf{f}, \mathbf{b}, K), \qquad (9.1)$$

where $\mathbf{f}_w$ indicates the watermarked feature vector. Note that $\mathbf{f}$ itself is nothing but a sequence of random variables modelling the source which emits the to-be-marked assets. In order to satisfy the imperceptibility constraint, $\mathcal{E}$ must be designed so that the distance between $\mathbf{f}$ and $\mathbf{f}_w$ is lower than a given threshold. To be specific, let $d(\mathbf{f}, \mathbf{f}_w)$ be a generic non-negative function measuring the distortion introduced by the embedder. We require that:

$$d(\mathbf{f}, \mathbf{f}_w) \leq D_{\mathcal{E}}, \qquad (9.2)$$

where $D_{\mathcal{E}} > 0$ is a parameter of the game called embedding distortion. It is worth noting that $d(\mathbf{f}, \mathbf{f}_w)$ depends on the values assumed by $\mathbf{b}$, the feature sequence and K, hence the inequality in (9.2) has a probabilistic meaning. Though many interpretations are possible, we consider only the so called average distortion constraint, for which

$$E[d(\mathbf{f}, \mathbf{f}_w)] \leq D_{\mathcal{E}}, \qquad (9.3)$$

and the almost sure (a.s.) constraint, for which

$$P\{d(\mathbf{f}, \mathbf{f}_w) \leq D_{\mathcal{E}}\} = 1. \qquad (9.4)$$

As to the distortion function, it should be designed by relying on perceptual considerations, even in relation to the particular watermarking domain. The adoption of such a perceptual distortion measure, however, makes the watermarking game mathematically untractable, thus a simplified model is usually adopted. Here we deal exclusively with square error distortion, i.e. for any two vectors $\mathbf{x}$ and $\mathbf{y}$, we let:

$$d(\mathbf{x}, \mathbf{y}) = \frac{1}{n} \sum_{i=1}^{n} (x_i - y_i)^2. \qquad (9.5)$$

After the embedder, the next move is up to the attacker. The goal of the attacker is to map the marked feature vector $\mathbf{f}_w$ into an attacked vector $\mathbf{f}'_w$, in such a way that the extraction of $\mathbf{b}$ from $\mathbf{f}'_w$ is as difficult as possible. To be precise, the attacker first generates a random sequence $K_{\mathcal{A}}$, then applies to $\mathbf{f}_w$ an attack function $\mathcal{A}$, obtaining the attacked vector $\mathbf{f}'_w$:

$$\mathcal{A}(\mathbf{f}_w, K_{\mathcal{A}}) = \mathbf{f}'_{\mathbf{w}}. \qquad (9.6)$$

It is important to stress the importance of the random sequence $K_{\mathcal{A}}$, sometimes called the attack key, since the random nature of the attack only depends on it. For instance, if the attack consists of Gaussian noise addition, then $K_{\mathcal{A}}$ coincides with the noise sequence added to $\mathbf{f}_w$. As the embedder, the attacker must satisfy a constraint on the distortion he introduces, more specifically $\mathcal{A}$ must be designed in such a way that:

$$d(\mathbf{f}_w, \mathbf{f}'_w) \leq D_{\mathcal{A}}, \qquad (9.7)$$

where inequality must be matched almost surely or in the average. The positive constant $D_{\mathcal{A}}$ is called attack distortion.

The last player of the game is the decoder, whose goal is to recover the hidden message $\mathbf{b}$ from $\mathbf{f}'_w$. In order to perform his task, the decoder applies a decoding function $\mathcal{D}$ to $\mathbf{f}'_w$, i.e.

$$\mathcal{D}(\mathbf{f}'_w, K) = \hat{\mathbf{b}}, \qquad (9.8)$$

where $\hat{\mathbf{b}}$ is the estimated message, and where we have assumed that the embedder and the decoder share the same secret key K[6]. Equation (9.8) describes the blind version of the watermarking game, where the decoder does not know the non-marked feature vector $\mathbf{f}$. In the non-blind version of the game, such an equation must be replaced by:

$$\mathcal{D}(\mathbf{f}'_w, \mathbf{f}, K) = \hat{\mathbf{b}}. \tag{9.9}$$

We said that the goal of the attacker is to make the extraction of $\mathbf{b}$ from $\mathbf{f}'_w$ as difficult as possible. Such a difficulty is measured by the message error probability. To be specific, we first introduce an error function ε returning 1 if $\hat{\mathbf{b}} \neq \mathbf{b}$ and zero otherwise. As it can be readily seen ε is function of $\mathbf{f}$, $\mathbf{b}$, K, $K_\mathcal{A}$, $\mathcal{E}$, $\mathcal{A}$ and $\mathcal{D}$. We evaluate the reliability of the system by means of the average error probability, that is:

$$\bar{P}_e(\mathcal{E}, \mathcal{A}, \mathcal{D}) = E[\varepsilon(\mathbf{f}, \mathbf{b}, K, K_\mathcal{A}, \mathcal{E}, \mathcal{A}, \mathcal{D})], \tag{9.10}$$

where $\bar{P}_e(\mathcal{E}, \mathcal{A}, \mathcal{D})$ can be seen as a functional of $\mathcal{E}, \mathcal{A}, \mathcal{D}$ since expectation is taken with respect to $\mathbf{f}, \mathbf{b}, K$ and $K_\mathcal{A}$.

A very important aspect of any game, and the watermarking game in particular, is the sequence in which game players make their moves. In our case, it is obvious that the first move is up to the embedder. Next is the turn of the attacker, which we assume to have a full knowledge of the embedding function $\mathcal{E}$ but which does not know the secret key K, the original feature vector $\mathbf{f}$ and the hidden message $\mathbf{b}$. This assumption implies that the attacker can adaptively choose the attack function $\mathcal{A}$ according to the function $\mathcal{E}$ chosen by the embedder. The last move is for the decoder, however one may wonder if it is realistic to assume that the decoder perfectly knows $\mathcal{A}$. As a matter of fact, in most applications this is not the case, since the attacker will not publicize the attack strategy he chooses. Alternatively, the decoder may try to guess the particular $\mathcal{A}$ used by the attacker by analyzing the asset under inspection. Hereafter we will adopt a more conservative approach, in that we will assume that the decoder does not know anything about $\mathcal{A}$. Interestingly, and somewhat surprisingly, such an assumption does not have any impact on our analysis[7]. As a result we can merge the embedder and the decoder into a unique player

---

[6]In so doing we avoid dealing with asymmetric watermarking schemes.

[7]As it will be detailed in the following, the performance of the watermarking channel is usually defined in terms of capacity C. It turns our that a transmission rate R equal to C can be achieved even by assuming that the decoder does not know the attack rule. Conversely, even by assuming that the decoder has such a knowledge, the attacker may operate in such a way that no transmission rate $R > C$ can be used. These two observations together show that the ignorance of the decoder of the attack rule is not relevant to the final value of the watermarking game.

- the watermarker - acting before the attacker. A remarkable consequence of this assumptions is that the watermarker must design the embedder and the decoder without any knowledge of the attack, e.g. maximum likelihood decoding is not allowed.

In order to define the value of the game, we start by introducing the concept of achievable rate R. Given a message set $\mathbf{B}$ of cardinality 2^{nR}, we say that the rate R is achievable, if an embedding and a decoding rule $\mathcal{E}$ and $\mathcal{D}$ exist, such that for any possible choice of $\mathcal{A}$, the average error probability $\bar{P}_e(\mathcal{E}, \mathcal{A}, \mathcal{D})$ tends to zero, as n - the length of the host feature vector - tends to infinity. We define the coding capacity of the game as the supremum of all the achievable rates. We indicate such a capacity by C_b for the blind version of the game and by C_{nb} for the non-blind version. As it can be readily seen, C_b and C_{nb} depend on $D_{\mathcal{E}}$, $D_{\mathcal{A}}$ and the pdf of $\mathbf{f}$. In the sequel, when it is not necessary to distinguish between the blind and non-blind cases we will refer to the capacity of the watermarking game simply with C.

9.2.2 Some selected results

In this section we review some of the main results that have been obtained with respect to the watermarking channel described above. We give the results without demonstrating them, thus limiting our analysis to a discussion of the hypotheses behind the various theorems and the main consequences they bring. In some cases we sketch the layout of the proof, since in this way some useful hints about the actual design of a watermarking system can be obtained.

The first theorem we will consider is a very general one, since it upper bounds the capacity of the general watermarking game without putting any particular constraints on the form of the attack and the embedding rule. Before stating the theorem we need to define some auxiliary quantities. Let a be an auxiliary variable taking value in the interval[8]:

$$\mathcal{I}_a(D_{\mathcal{E}}, D_{\mathcal{A}}, \sigma_f^2) = \left\{ a : max\{D_{\mathcal{A}}, (\sigma_f - \sqrt{D_{\mathcal{E}}})^2\} \leq a \leq (\sigma_f + \sqrt{D_{\mathcal{E}}})^2 \right\}, \tag{9.11}$$

and let s be a function acting on a as follows:

$$s(a; D_{\mathcal{E}}, D_{\mathcal{A}}, \sigma_f^2) = \frac{D_{\mathcal{E}}}{D_{\mathcal{A}}} \left(1 - \frac{D_{\mathcal{A}}}{a} \right) \left(1 - \frac{(a - \sigma_f^2 - D_{\mathcal{E}})^2}{4 D_{\mathcal{E}} \sigma_f^2} \right). \tag{9.12}$$

The coding capacity of the watermarking game can be specified in terms

---

[8]In practice a corresponds to the power of the marked feature vector.

of a function C^* defined as:

$$C^*(D_\mathcal{E}, D_\mathcal{A}, \sigma_f^2) = \max_{a \in \mathcal{I}_a} \left[\frac{1}{2} \log_2(1 + s(a; D_\mathcal{E}, D_\mathcal{A}, \sigma_f^2)) \right], \qquad (9.13)$$

if $\mathcal{I}_a$ is not empty and $C^*(D_\mathcal{E}, D_\mathcal{A}, \sigma_f^2) = 0$ otherwise. We are now in the position of setting out the main results about the coding capacity of the watermarking game.

Theorem 9.1: *Let us consider a watermarking game with continuous real alphabets and squared error distortion, with distortion constraints to be met almost surely. Let $D_\mathcal{E}$ and $D_\mathcal{A}$ indicate the allowed distortion for the embedder and the attacker respectively. Let $p(\mathbf{f})$ be the pdf of the host features assumed to be ergodic. If:*

$$\begin{aligned} E[f_i^4] &< \infty, \\ E[f_i^2] &\le \sigma_f^2, \end{aligned} \qquad (9.14)$$

then:

$$C_b(D_\mathcal{E}, D_\mathcal{A}, p(\mathbf{f})) \le C_{nb}(D_\mathcal{E}, D_\mathcal{A}, p(\mathbf{f})) \le C^*(D_\mathcal{E}, D_\mathcal{A}, \sigma_f^2). \qquad (9.15)$$

Both the equalities are achieved if $\mathbf{f}$ *is an iid zero mean normally distributed sequence with variance σ_f^2.*

Discussion

The first observation we can make is that the above theorem is a very general one, since it does not put any limitations either on the form of the attack or on the embedding rule. In addition, letting the attacker play second, i.e. assuming that he knows both the embedding and the decoding strategy, is a very conservative assumption which is easily met in practice. Finally, having set the distortion constraints almost surely, guarantees that the probability that a host asset exists for which the channel capacity can not be reached is null. A limitation of the theorem, is that distortion is measured in terms of mean squared distance, thus disregarding perceptual considerations. Additionally, important attacks such as geometrical attacks or the gain attack can not be accommodated by this model, since, though perceptually irrelevant, they introduce a very high mean square error, thus making it very difficult to match the distortion constraint. A further limitation is that, with the noticeable exception of the Gaussian case, the theorem only gives an upper bound on the actual capacity of the channel.

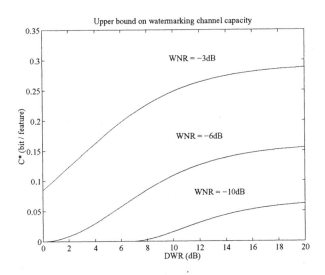

Figure 9.4: Capacity bound of the general watermarking game vs DWR.

In order to get more insight into the result of the theorem, it is instructive to plot $C^*(D_\mathcal{E}, D_\mathcal{A}, \sigma_f^2)$ as a function of $D_\mathcal{E}$, $D_\mathcal{A}$ and σ_f^2. In figure 9.4 the capacity bound given by theorem 9.1 is drawn as a function of DWR $(\sigma_f^2/D_\mathcal{E})$ for various values of WNR $(D_\mathcal{E}/D_\mathcal{A})$. It is at once evident that capacity increases with DWR. This may come as a surprise if one is used to think at host features as disturbing noise. Indeed this is not the case, on the contrary, large variance features are more easily marked than weak ones[9]. It is also instructive to plot $C^*(D_\mathcal{E}, D_\mathcal{A}, \sigma_f^2)$ against WNR (figure 9.5). We can see that a value of WNR exists below which channel capacity goes to zero. This happens whenever the attack strength enables the attacker to completely destroy the hidden information, e.g. by setting to zero the features hosting the watermark. From figure 9.5 it is also evident that such a threshold value of WNR depends on DWR, in accordance to the results given in figure 9.5.

Though from a general point of view theorem 9.1 only gives un upper bound on the achievable rate of the watermarking channel, the bound coincides with true channel capacity when the host features are independent and identically distributed Gaussian variables. This permits us to conclude that Gaussian features are the easiest to mark. This may be explained by observing that according to the optimum strategy, the embedder exploits

---

[9]This is an obvious result from a perceptual point of view: it is reasonable, in fact, that hiding a piece of data within a strong host signal is easier than hiding it in a weak signal.

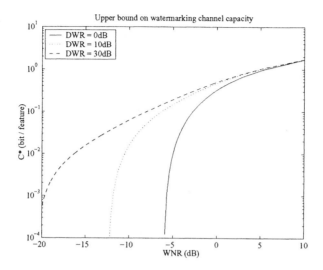

Figure 9.5: Capacity bound of the general watermarking game vs WNR.

the uncertainty of the host features to inject the hidden information within the host asset. It is natural, then, that Gaussian features achieve the maximum capacity since the Gaussian distribution is the one that maximizes uncertainty (source entropy) among all pdf with a given second order moment.

As a last result, theorem 9.1 permits to compare blind and non-blind decoding schemes. As it is expected, in general, decoder blindness results in a lower capacity, however this is not true if the host features form an iid normally distributed sequence. This is a very important result, since it demonstrates that, at least in the Gaussian case, no advantage in terms of achievable capacity has to be expected by granting the decoder the possibility to access the original, non-marked asset. Of course, this may not be true if a perspective other that channel capacity is used to judge system performance. For example, non-blind decoders are certainly simpler to implement than blind ones.

Outline of proof

We now give an outline of the proof of theorem 9.1. The outline is not intended to substitute the true proof, anyway it is sufficient to give some hints about the possible implementation of an optimum watermarking system. We start by considering the achievability of the capacity bound, which is proved by assuming that the host features form an iid zero mean Gaussian sequence. The proof is based on a particular embedding and decoding strategy (sometimes called random binning). The embedder generates a random codebook $\mathcal{U}$ consisting of 2^{nR_t} iid entries. The codebook is then partitioned into 2^{nR} bins (subcodebooks) each containing $2^{n(R_t-R)}$ codewords. The codewords are generated so that they are uniformly distributed over a sphere whose radius depends on a (see equation(9.11)). All the codewords in the same bin are associated to the same message in $\mathbf{B}$. Given the to-be-marked feature vector $\mathbf{f}$ and the to-be-hidden message $\mathbf{b}$, the embedder chooses the closest codeword in the bin indexed by $\mathbf{b}$. Let us indicate such a codeword by $\mathbf{u}(\mathbf{f})$. The marked feature vector is defined as:

$$\mathbf{f}_w = \mathbf{u}(\mathbf{f}) + (1 - \alpha)\mathbf{f}, \tag{9.16}$$

where α is a constant depending on a. The rates R and R_t are chosen so to ensure that the embedding distortion constraint is met with high probability. The decoder looks at all the codewords in $\mathcal{U}$ and picks up the one which is closest to the received feature vector $\mathbf{f}'_w$. Then it outputs the message $\hat{\mathbf{b}}$ corresponding to the bin the selected codeword belongs to. As it can be seen the decoder does not need to know the attack strategy. The actual proof of the theorem demonstrates that for any R lower than C^*, the error probability $\overline{P}_\varepsilon(\mathcal{E}, \mathcal{A}, \mathcal{D})$ tends to zero for $n \to \infty$.

The proof of the converse part of theorem 9.1 is more cumbersome. Basically, it can be shown that in order to prevent any rate larger than $C^*(D_\mathcal{E}, D_\mathcal{A}, \sigma_f^2)$ to be achieved, the attacker first estimates the value of a used by the embedder, then it chooses $\mathbf{f}'_w$ so to minimize the mutual information between $\mathbf{f}_w$ and $\mathbf{f}'_w$. Note that is so doing the attacker does not use any knowledge about the decoder strategy, nevertheless he succeeds in keeping the transmission rate below C^*. Stated in another way, even if we assume that the decoder knows the attacker strategy he can not do anything to increase the transmission rate, thus justifying our assumption that the decoder is ignorant of attacker's operations.

9.2.3 Capacity under average distortion constraints

The assumption that the distortion constraints must be meet almost surely is a critical one. As a matter of fact, the following theorem shows that if average distortion constraints are adopted, then the watermarking channel capacity is zero.

Theorem 9.2: *Let us consider a watermarking game with continuous real alphabets and squared error distortion, with average distortion constraints $D_\mathcal{E}$ and $D_\mathcal{A}$. Let $p(\mathbf{f})$ be the pdf of the host features, and assume* *that:*

$$\liminf_{n \to \infty} E\left[\frac{1}{n}\|\mathbf{f}\|^2\right] < \infty, \tag{9.17}$$

then

$$C_b(D_\mathcal{E}, D_\mathcal{A}, p(\mathbf{f})) = C_{nb}(D_\mathcal{E}, D_\mathcal{A}, p(\mathbf{f})) = 0, \tag{9.18}$$

for any value of $D_\mathcal{E}$ and $D_\mathcal{A}$ ($D_\mathcal{A} \neq 0$).

The basic idea behind the proof of Theorem 9.2 is that if the power of marked features is limited, which is surely the case due to the embedding distortion constraint and (9.17), then the attacker can set $\mathbf{f}_w$ to zero with some fixed probability p. By letting such a probability be small enough, yet strictly larger than zero, the attack distortion constraint can always be met. However, the error probability will never be zero, since it will at least be equal to p.

It is worth noting that for such a strategy to be effective the attack must depend on the marked features. If this is not possible, as in some versions of the additive attack watermarking game described in the next section, a non null capacity may still be possible even if average distortion constraints are in effect.

9.3   The additive attack watermarking game

In this section we analyze a simplified version of the general game described so far, namely the additive attack watermarking game, in which possible attacks are limited to noise addition.

9.3.1   Game definition and main results

The general watermarking game described in the previous section may be reformulated is such a way that both embedding and attack are seen as signal/noise addition. Such a point of view is exemplified in figure 9.2. Here the watermarked features are obtained by adding to $\mathbf{f}$ a watermark signal $\mathbf{w}$:

$$\mathbf{f}_w = \mathbf{f} + \mathbf{w}. \tag{9.19}$$

A similar approach is used to describe attacks: the attacked feature vector is obtained by adding a noise vector $\mathbf{n}$ to $\mathbf{f}$:

$$\mathbf{f}'_w = \mathbf{f}_w + \mathbf{n}. \tag{9.20}$$

The model depicted in figure 9.2 clearly coincides with the watermarking game illustrated in figure 9.3, if we allow that the watermark signal and the attack noise depend on $\mathbf{f}$ and $\mathbf{f}_w$ respectively. The model revised according to the additive noise perspective also explains the term informed embedding which is commonly used to indicate an embedding strategy in which $\mathbf{w}$ depends on $\mathbf{f}$.

Motivated by the above analysis, we now introduce a simplified version of the general watermarking game, in which the attacker is only allowed to add a noise vector which is independent of the to-be-attacked features $\mathbf{f}_w$. Stated in another way we let:

$$\mathcal{A}(\mathbf{f}_w, K_A) = \mathbf{f}'_{\mathbf{w}} = \mathbf{f}_w + \mathbf{n}, \qquad (9.21)$$

where $\mathbf{n}$ is generated independently of $\mathbf{f}_w$. We call this version of the game the additive attack watermarking game. The following theorem upper bounds the capacity of this game.

Theorem 9.3: *Let us consider an additive attack watermarking game with continuous real alphabets and squared error distortion, with distortion constraints to be met almost surely. Let $D_\mathcal{E}$ and D_A indicate the allowed distortion for the embedder and the attacker respectively. Then, for any host feature distribution $p(\mathbf{f})$:*

$$C_b^a(D_\mathcal{E}, D_A, p(\mathbf{f})) \le C_{nb}^a(D_\mathcal{E}, D_A, p(\mathbf{f})) = \frac{1}{2}\log_2\left(1 + \frac{D_\mathcal{E}}{D_A}\right). \qquad (9.22)$$

Equality is achieved if $\mathbf{f}$ is an iid normally distributed sequence of any variance.

Discussion

Theorem 9.3 is important since it permits to evaluate to effectiveness of an additive attack with respect to the more general class of distortion-limited attacks. To this aim it is instructive to compare the bounds given by theorem 9.1 and 9.3 Such a comparison is depicted in figure 9.6. As it can be seen, a major difference between the two bounds is that the one valid for the additive attack case does not depend on host feature variance σ_f^2. In addition, the additive bound is a loose one for low values of DWR or high noise levels. Then, we can conclude that it is suboptimal (highly suboptimal in certain cases) for the attacker to restrict itself to an additive attack.

As an additional remark, we observe that in this case the theorem upper bounds only the capacity of the blind version of the game, whereas for non-blind decoding, equation (9.22) gives the actual capacity of the channel. Finally it is confirmed that the Gaussian iid features are the easiest to watermark.

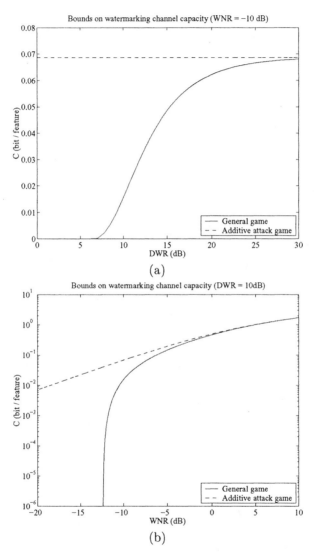

Figure 9.6: Comparison between the Capacity bounds of the general and the additive attack watermarking game as a function of DWR (a) ab WNR (b). As it can be seen the additive attack allows a higher capacity this revealing its highly non-optimal nature.

9.3.2 Costa's writing on dirty paper

The most popular, though limited in scope, result about communication with side information at the encoder is due to M. Costa [50]. He addresses a special case of communication with side information where two independent sources of iid Gaussian noise impair the transmission. The former noise is non-causally known at the encoder (but not to the decoder), whereas the second source of noise is a classical one, and it is not known either to the encoder or the decoder. Costa's model, then, closely resembles the additive attack watermarking game depicted in figure 9.2, with the only difference that both $\mathbf{f}$ and $\mathbf{n}$ are iid sequences following a normal pdf. Additionally, Costa follows a classical communication perspective, hence the characteristics of $\mathbf{n}$ are fixed, since the presence of an active attacker explicitly aiming at impairing transmission is not considered. The main result found by Costa is that channel capacity is not affected by the presence of the first source of noise, namely $\mathbf{f}$. This result is more clearly stated in the following theorem.

Theorem 9.4 (M. Costa): *Let us consider the communication channel with side information at the encoder depicted in figure 9.2. Let us assume that $\mathbf{f}$ and $\mathbf{n}$ are independent zero mean normally distributed iid sequences having variance σ_f^2 and σ_n^2. If encoding is subject to the following power constraint:*

$$\frac{1}{n}\|\mathbf{w}\|^2 \leq P, \qquad (9.23)$$

then the capacity of the channel is

$$C = \frac{1}{2}\log_2\left(1 + \frac{P}{\sigma_n^2}\right), \qquad (9.24)$$

i.e. the same capacity of a conventional AWGN channel where only the second source of noise is present.

Discussion

Theorem 9.4 has played a fundamental role in the development of watermarking theory, since it was by looking at Costa's work that watermarking researchers realized that decoder blindness may not have a dramatic impact on system performance as it was believed before. Indeed, at least in the ideal conditions of Costa's theorem, the negative impact of decoder blindness may be reduced to zero. It is also instructive to compare the results expressed by theorems 9.3 and 9.4. The first one is by far more general.

First in theorem 9.3 the problem is looked from a game theory perspective, thus requiring that channel capacity is determined by considering all possible kinds of (additive) attacks. Secondly, in 9.3 **n** neither needs to follow a Gaussian pdf nor to be an iid sequence (it only needs not to depend on $\mathbf{f}_w$). Additionally, theorem 9.3 gives an upper bound of the watermarking capacity even when the host features are not normally distributed (even if in this case the capacity does not necessarily coincides with that found by Costa), whereas Costa's theorem only addresses the everything-is-Gaussian case.

A further comparison between the two theorems also reveals that when the host features form an iid normally distributed sequence, the addition of Gaussian noise represents the worst attack among additive ones. This is readily seen by noting that restricting the attack to Gaussian noise addition does not reduce the capacity of the watermarking channel.

Proof outline

Looking at the proof of Theorem 9.4 is an instructive exercise since in this way it is possible to get some useful insights about the way channel capacity can be achieved. As a matter of fact, several effective watermarking algorithms have been proposed inspired by the proof of theorem 9.4. For this reason, in the sequel we give a precise outline of such a proof. To do so we need to adopt an information theoretic formalism. Let us start by assuming that the host feature sequence **f** is the output of a memoryless Gaussian source F. In the same way, the noise sequence **n** is seen as the output of a second memoryless Gaussian source N. We further assume that F and N are independent of each other. The evaluation of the channel capacity for this communication model follows the same path traced by a number of works analyzing the capacity of a channel with random state, when the random state is non-causally available at the encoder [84, 93]. More specifically, by following Gel'fand and Pinsker [84] it can be shown that the capacity of the channel studied by Costa has the form:

$$C = \max_{p(U,W|F)} (I(U;F') - I(U;F)), \qquad (9.25)$$

where I denotes mutual information. Note that the same symbol is used both to indicate the information source and the random variable describing the output of the source. Thus, F indicates both the memoryless source emitting the host feature sequence and the random output of the source at any given instant. The reason for the presence in equation (9.25) of the auxiliary source of randomness U will be clear from the subsequent discussion.

Before going on with the outline of the proof, it is necessary to introduce the concept of typical sequences. Without pretending to be rigorous, suppose we are given a memoryless discrete random source X, drawing from a finite alphabet $\mathcal{A}_X$. Any output sequence $\mathbf{x}$ can be characterized by the relative frequency with which the various elements of $\mathcal{A}_X$ are present in it. If these frequencies agree with the a priori probabilities of the symbols of $\mathcal{A}_W$, then the sequence $\mathbf{x}$ is said to be a typical sequence. It can be shown that when n tends to infinity, the probability that a non-typical sequence is emitted tends to 0. The above concept can be easily extended to continuous random variables and to sources with memories. In addition, given two sources X and Y the same arguments can be used to introduce the notion of jointly typical sequences[10].

The proof of equation (9.25) goes through a random binning argument similar to that described in the outline of the proof of theorem 9.1. To be specific, capacity is achieved as follows. Let us assume, for simplicity, that input alphabets are finite, the extension to the continuous case being easy. We first generate a codebook $\mathcal{U}$ consisting of 2^{nR_t} entries (the sequences $\mathbf{u}$) which are randomly generated so to span uniformly the set of typical sequences of U. Then $\mathcal{U}$ is randomly (and uniformly) split into 2^{nR} bins (sub-codebooks) each containing $2^{n(R_t-R)}$ codewords. By observing that the message set $\mathbf{B}$ contains exactly 2^{nR} messages, it is possible to associate each message $\mathbf{b}$ to a bin of $\mathcal{U}$. In order to transmit a message $\mathbf{b}$, the value of $\mathbf{f}$ is analyzed, then an entry in the bin indexed by $\mathbf{b}$ is looked for which is jointly typical with $\mathbf{f}$[11]. Next we transmit a sequence $\mathbf{w}$ which is jointly typical with $\mathbf{u}$ and $\mathbf{f}$. At the other side of the transmission channel, the decoder receives a sequence $\mathbf{f}'_w$. In order to estimate the transmitted message $\mathbf{b}$, the decoder looks for a unique sequence $\mathbf{u}^*$ in $\mathcal{U}$ which is jointly typical with $\mathbf{f}'_w$ and outputs the message corresponding to the bin $\mathbf{u}^*$ belongs to. The decoder declares an error if more than one, or no such typical sequence exists. If $R < C$ then the error probability averaged over al possible codes $\mathcal{U}$ tends to 0 as the length n of the transmitted sequence tends to infinity.

The further result provided by Costa is that, in the hypothesis of theorem 9.4, the capacity coincides with that of an AWGN channel in which the first source of noise is absent. In addition, a practical way to choose a capacity-achieving auxiliary variable U is provided.

---

[10]Actually this is a rather heuristic definition of typical sequences, however for the demonstration of Costa's theorem it only needs to resort to a weaker form of typicality. More specifically, a sequence $\mathbf{x}$ is said to be typical (at level ε) if $2^{-n(H(X)+\varepsilon)} \leq p(x_1, x_2 \ldots x_n) \leq 2^{-n(H(X)-\varepsilon)}$, where $H(X)$ is the entropy of source X. For a more detailed, yet tutorial, introduction to typical sequences readers may refer to [51, 75].

[11]An error is output if any such a typical sequence is not found.

In order to prove theorem 9.4, Costa argues that letting

$$\mathbf{u} = \mathbf{w} + \alpha \mathbf{f}, \tag{9.26}$$

permits to achieve a transmission rate equal to the channel capacity as defined by Gel'fand and Pinsker. As it is proved in [50] this is indeed the case if we let:

$$\alpha^* = \frac{P}{P + \sigma_n^2}. \tag{9.27}$$

Furthermore, with this choice we have that C is expressed by equation (9.24).

To summarize, the embedding/decoding strategy which permits to achieve Costa's capacity works as follows. The embedder generates 2^{nR_t} sequences $\{\mathbf{u}_i\}_{i=1}^{2^{nR_t}}$ according to a Gaussian distribution having zero mean and variance $P + (\alpha^*)^2 \sigma_f^2$. Such sequences are then split into 2^{nR} bins each associated to a message in $\mathbf{B}$. The codebook $\mathcal{U}$ with all sequences $\mathbf{u}_i$ and its subdivision into bins is known to both the embedder and the decoder. Given a host feature sequence $\mathbf{f}$ and the message $\mathbf{b}$ to be transmitted, the embedder looks for a sequence $\mathbf{u}$ in the bin indexed by $\mathbf{b}$ such that:

$$|(\mathbf{u} - \alpha^* \mathbf{f}) \cdot \mathbf{f}| \leq \varepsilon, \tag{9.28}$$

for a small ε. Then he transmits $\mathbf{w} = \mathbf{u} - \alpha^* \mathbf{f}$. Equation (9.28) ensures that $\mathbf{w}$ and $\mathbf{f}$ are nearly orthogonal, a condition which is equivalent to ensuring that $\mathbf{u}, \mathbf{f}$ and $\mathbf{w}$ are jointly typical. Upon receiving a sequence $\mathbf{f}'_w$ the decoder looks for a sequence in $\mathcal{U}$ which is jointly typical with $\mathbf{f}'_w$ and outputs the message associated to the bin the decoded sequence belongs to.

In the next section we see how by replacing the randomly generated codebook $\mathcal{U}$ with a structured codebook permits to design a capacity achieving (at least theoretically) watermarking algorithm.

9.4 Lattice-based capacity-achieving watermarking

As we have seen in the previous section, for achieving the capacity limit in presence of Gaussian independent identically distributed features, and of an AWGN attack, it needs to build a very large random multidimensional codebook that is partitioned into a number of bins, each bin being associated with a message. Managing such a large random multidimensional codebook for coding and decoding is not feasible, research has then focused on looking for sub-optimal structured codebooks, allowing to at least approach the theoretical performance suggested by Costa. An attempt in this direction is constituted by the so called Scalar Costa Scheme (SCS) that we are going to describe in this section.

The SCS is based on the use of an n-dimensional lattice codebook $\mathcal{U}$ obtained by the Cartesian product of n, 1-dimensional, identical codebooks $\mathcal{U} = \mathcal{U}^1 \times \mathcal{U}^1 \ldots \mathcal{U}^1$. To be more precise, let us assume that the message to be sent is encoded in a sequence of n bits $b_i \in \{0,1\}^{12}$, which can also be obtained by error correction coding the original message bits, each 1-dimensional codebook can be written as:

$$\mathcal{U}^1 = \mathcal{U}_0^1 \cup \mathcal{U}_1^1, \tag{9.29}$$

where each sub-codebook is given by:

$$\mathcal{U}_b^1 = \left\{ u = lD + b\frac{D}{2} : l \in \mathbb{Z} \right\}, \tag{9.30}$$

with $b \in \{0,1\}$. The codebook $\mathcal{U}$ can be partitioned into 2^n bins, each one associated with a combination of n bits, by suitably combining the sub-codebooks, i.e. the bin associated with the bit sequence $\mathbf{b} = \{b_1, b_2, \ldots, b_n\}$ is:

$$\mathcal{U}_\mathbf{b} = \mathcal{U}_{b_1}^1 \times \mathcal{U}_{b_2}^1 \ldots \mathcal{U}_{b_n}^1. \tag{9.31}$$

An example of how a 2-dimensional codebook looks like is given in figure 4.26, where each bin is marked by a different symbol. For security reasons (and for other purposes that will be more clear later) it is convenient to randomly translate each of the 1-dimensional codebooks by a value kD, where k is uniformly distributed in $[0,1)$, thus yielding:

$$\mathcal{U}_\mathbf{b} = \mathcal{U}_{b_1}^1(k_1) \times \mathcal{U}_{b_2}^1(k_2) \ldots \mathcal{U}_{b_n}^1(k_n), \tag{9.32}$$

where

$$\mathcal{U}_b^1(k) = \left\{ u = (l+k)D + b\frac{D}{2} : l \in \mathbb{Z} \right\}, \tag{9.33}$$

in such a way that anybody who does not know the sequence $\{k_1, k_2, \ldots, k_n\}$ can not recover the codebook and read the embedded message.

Once the codebook has been built, it needs to find, according to Costa's approach, the sequence (codebook entry) $\mathbf{u}$, belonging to the bin associated with the message $\mathbf{b}$, such that $\mathbf{u} - \alpha\mathbf{f}$ is nearly orthogonal to $\mathbf{f}$, as from equation (9.28). This is equivalent to find the sequence $\mathbf{q} = \mathbf{u}/\alpha - \mathbf{f}$ which is nearly orthogonal to $\mathbf{f}$. A sequence $\mathbf{q}$ satisfying this constraint can be obtained as the quantization error resulting from the quantization of the

---

[12]The system is easily generalizable to multilevel signaling.

features f_i with the quantizer obtained by scaling by $1/\alpha$ the entries of the codebook $\mathcal{U}$, i.e.:

$$q_i = \mathcal{Q}_\Delta \left\{ f_i - \Delta \left[\frac{b}{2} + k_i \right] \right\} - \left\{ f_i - \Delta \left[\frac{b}{2} + k_i \right] \right\}, \qquad (9.34)$$

where $\Delta = D/\alpha$. In fact we have that:

$$\sum_{i=1}^{n} f_i q_i \approx nE[fq] = E[f]E[q] = 0 \qquad (9.35)$$

where the second and third equalities follow from the fact that, before quantization, the features f_i are translated by a random value uniformly distributed in the interval $[0, \Delta)$ (see equation (9.34)): it is well known from the theory of dither modulation[13] that in this case the quantization noise does not depend on the original features, its mean is zero, and its variance given by $\Delta^2/12$. Still following Costa's approach the sequence $\mathbf{u} - \alpha\mathbf{f} = \alpha\mathbf{q}$ is sent over the channel, where the first step is the addition of the host feature vector $\mathbf{f}$, i.e., the watermarked feature vector is obtained as:

$$f_{w,i} = f_i + \alpha q_i. \qquad (9.36)$$

It is, then, immediate to verify that the watermark energy is:

$$\sigma_w^2 = \alpha^2 \frac{\Delta^2}{12}, \qquad (9.37)$$

which, once fixed, imposes a relation between the parameter α, and the quantization step Δ. It is also interesting to have a look at the input output characteristic of this watermark embedding scheme: to this aim let us suppose, for example, that $b_i = k_i = 0$, it is then easy to get:

$$f_{w,i} = \mathcal{Q}_\Delta\{f_i\} + (1 - \alpha)(f_i - \mathcal{Q}_\Delta\{f_i\}), \qquad (9.38)$$

whose trend is sketched in figure 9.7 for $\alpha > 0.5$. Finally let us note that the SQIM(or DM) technique described at page 148, corresponds to the SCS when $\alpha = 1$.

The decoder will quantize the received (possibly attacked features) with the same quantizers, and compute the decision variables:

$$r_{b,i} = \mathcal{Q}_\Delta \left\{ f_{w,i} - \Delta \left[\frac{b}{2} + k_i \right] \right\} - \left\{ f_{w,i} - \Delta \left[\frac{b}{2} + k_i \right] \right\}, \qquad (9.39)$$

---

[13]Interested readers can find the main results on dither quantization theory in [196].

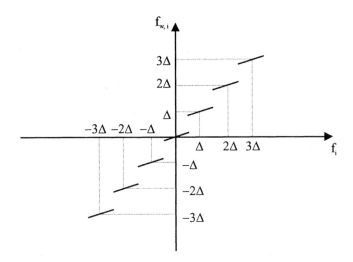

Figure 9.7: Input-output relation for SCS: it has been assumed that $b_i = k_i = 0$, and $\alpha > 0.5$. It is evident that the relation is reversible.

which, if error correction coding is not applied, can be directly used for deciding, by thresholding, on the i-th bit. On the contrary, if some sort of error correction has been applied, it can be fed to a soft decoder performing Maximum Likelihood decoding.

Up to now we have said nothing about the choice of the parameter α: Costa chooses it by maximizing the channel mutual information, thus obtaining:

$$\alpha = \frac{\sigma_w^2}{\sigma_w^2 + \sigma_n^2}, \tag{9.40}$$

where σ_n^2 is the variance of the additive Gaussian noise (the attack). Given that in the SCS a sub-optimum codebook is used, this choice of α could be no longer optimum. Let us see here, then, how this parameter can be selected for this particular watermarking scheme. The criterion is similar to that of Costa, in that the mutual information of the scheme is maximized[14]. In particular it is first assumed that the attack consists of the addition of white Gaussian Noise, i.e. that $f'_{w,i} = f_{w,i} + n_i$, where $p(n_i) = \mathcal{N}(0, \sigma_n^2)$: given that also the host features are considered to be independent and identically distributed, and that the embedding is performed component-wise, it is possible to consider the transmission of only 1 sample (we will thus omit in the following the index i), thus yielding the following maximization

---

[14]See in section 1.2.1 some words about the difference between the general watermarking capacity and the capacity of a given technique.

criterion:

$$\alpha_{\text{SCS}}^* = \arg\max_{\alpha} I(F_w'; B), \qquad (9.41)$$

where

$$I(F_w; B) = -\int p(f_w') \log_2 p(f_w') df_w' + \frac{1}{2} \sum_{b \in \{0,1\}} \int p(f_w'|b) \log_2 p(f_w'|b) df_w'$$

$$(9.42)$$

where we have assumed that the two symbols (bits) are equiprobable, and we have omitted the dependence on the key. To estimate the mutual information it needs to compute the pdf of the attacked feature f_w': given that the additive noise is independent on the watermarked features, this can be done as:

$$p(f_w'|b) = p(f_w|b) \otimes \mathcal{N}(0, \sigma_n^2), \qquad (9.43)$$

and

$$p(f_w') = \frac{1}{2} \sum_{b \in \{0,1\}} p(f_w'|b). \qquad (9.44)$$

In turn, the pdf of the watermarked feature $p(f_w|b)$ can be obtained from that of the original feature $p(f)$, and by considering the input output relation of the embedding scheme (as for example the one reported in figure 9.7 for the bit 0). Analytical derivation of these pdf's is not possible, thus only a numerical solution is viable. In this way it can be shown that the optimum value for the parameter α can be approximated by:

$$\alpha_{\text{SCS}}^* = \sqrt{\frac{\sigma_w^2}{\sigma_w^2 + 2.71\sigma_n^2}} \qquad (9.45)$$

and results to depend on the WNR: this implies that for precisely setting α, the embedder should exactly know the strength of the attack (i.e. σ_n^2), which usually is not possible.

In figure 9.8 the capacity achievable with SCS[15] as a function of WNR is compared with ideal Costa's capacity:

$$C_{\text{Costa}} = \frac{1}{2} \log_2 \left(1 + \frac{\sigma_w^2}{\sigma_n^2}\right) = \frac{1}{2} \log_2 (1 + \text{WNR}) \qquad (9.46)$$

and with the capacity achievable with additive techniques (e.g. the theoretical capacity for an additive spread spectrum watermark):

$$C_{\text{SS}} = \frac{1}{2} \log_2 \left(1 + \frac{\sigma_w^2}{\sigma_f^2 + \sigma_n^2}\right) = \frac{1}{2} \log_2 \left(1 + \frac{\text{WNR}}{\text{WNR} \cdot \text{DWR} + 1}\right). \qquad (9.47)$$

---

[15]We took this result from the paper by Eggers et al. [70].

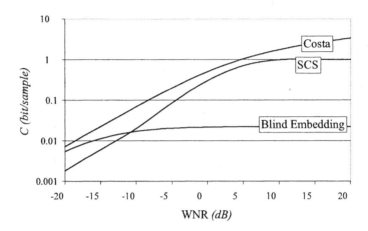

Figure 9.8: Plot of the capacity (in *bit/sample*) achievable by SCS against WNR, compared with ideal Costa's capacity, and with the capacity achievable classically with additive watermarking.

The DWR has been fixed to 15 *dB* for the additive case, and the optimum $\alpha*_{SCS}$ has been used for each WNR value for the SCS (this is implicit for the Costa's capacity). From the figure it is evident that SCS approaches the theoretical Costa's capacity much better than the blind embedding additive scheme over a wide range of WNR values. On the other side, at very low WNR (i.e. where the effect of the non rejected host becomes negligible with respect to channel noise) the blind additive approach works better. Given that we have considered only a binary SCS, capacity does not raise above 1 *bit/sample*, multilevel implementations of SCS are anyway also possible. On the contrary, for very large values of WNR the capacity of the blind scheme is limited by the DWR value (the asymptote is at $1/2 \log_2(1 + 1/\text{DWR})$).

9.5   Equi-energetic structured code-books

In the previous section we saw that lattice-based watermarking, namely SCS watermarking, is a convenient way to turn Costa's principles into practice avoiding the problems associated to the large codebooks necessary to implement the random binning algorithm.

A major problem with lattice-based codebooks, is that they are vulnerable against value-metric scaling of the host features (gain attack), a very common operation which consists in multiplying the host feature sequence by a constant gain factor g which is unknown to the decoder. This weakness derives from the choice of using the host features amplitude to code the hidden signal, and it is shared by all the techniques that use a codebook $\mathcal{U}$ with entries having different energies.

As a matter of fact, in its original form, random binning watermarking does not imply any weakness against value-metric scaling. For n sufficiently large, in fact, all the typical sequences have approximately the same energy, since they are uniformly distributed over an n-dimensional sphere with radius $\sqrt{\sigma_n^2 + (\alpha^*)^2 \sigma_f^2}$. On the contrary, the problems with value-metric scaling derive from the use of lattice-based codebooks instead of equi-energetic codes. Unfortunately, the values of n adopted in practice are not large enough to consider only equi-energetic codewords, since for these n typical sequences with significantly different energies are likely to occur. Of course, one may still force the codebook entries to have the same energy, however, in this way the simple embedding strategy described in equation (9.26) can no more be used since the host feature sequence is likely to be very distant from the closest codebook entry associated to the to-be-hidden message. In this case, a more sophisticated embedding strategy is needed to ensure that the watermarked feature sequence falls inside the correct decoding region. An example of the above approach was described in section 4.3.2, where a general optimum informed embedding strategy under a fixed robustness constraint was discussed.

9.6 Further reading

This chapter heavily relies on information theory concepts. For a good and deep introduction to this topic readers may refer to a number of good books that have been published in the last decades. Among them the excellent book by T. M. Cover and J. A. Thomas [51] is very closed to the perspective used in this book. Alternatively, readers may refer to [62].

Something similar can be said about game theory, since, as we have seen, digital watermarking is nothing but a game between the watermarker and the attacker. Even in this case, readers may choose among several good introductory books, for example the text by R. Gibbons [85].

Some specific works, originally non intended to deal with data hiding, need to be considered as background for the correct application of information theory concepts to digital watermarking. These surely include the work by S. I. Gel'fand and M. S. Pinsker on coding with channels with random states [84] and the seminal work by M. H. M. Costa [50], in which it is argued that for the Gaussian case ignorance of the side information by the decoder does not have any impact on channel capacity.

The importance of modelling the watermarking channel as a communication channel with side information at the encoder was first pointed out by I. J. Cox, M. L. Miller and A. L. McKellips in [57], and independently by B. Chen and G. Wornell in [37] and later on, and more comprehensively in [39].

The development of a general data hiding theory based on information theory concepts is still in the phase of being developed, hence the brief analysis we gave in this chapter is a limited one. Moreover, new results continuously appear extending our understanding of the watermarking problem. For the same reason, it is impossible to give a list of definitive readings covering this rapidly evolving field. A list of the most important and comprehensive works among those that have been published up to date, certainly includes the works by P. Moulin [164, 157, 159, 163], those by B. Chen and G. Wornell [36, 39], those by A. S. Cohen and A. Lapidoth [45, 46] and those by N. Merhav [150, 205].

The information theoretic analysis of digital watermarking has spurred the interest towards dirty paper coding. An excellent introduction to this topic at a heuristic level is given by M. L. Miller, G. J. Doerr and I. J. Cox in [155], whereas for a more theoretical analysis, readers may refer to [42, 41, 43, 183, 184]

The Scalar Costa Scheme has been proposed by J. J. Eggers, J. K. Su and B. Girod [70, 71, 72, 74] who developed a deep theoretical analysis of this method, however the approach was previously proposed by B. Chen and G. Wornell in [38, 39], who proposed a different criterion for setting the parameter α, minimizing an approximation of the error probability.

For an introduction to dither quantization , which is extensively used in the SCS approach, readers may refer to the classical work by L. Schuchman [196].

Equi-energetic dirty paper coding, is a promising research field, since it permits to achieve the excellent performance of lattice-based watermarking, without running into the problems set by the gain attack. Interested readers may find two different perspectives on equi-energetic dirty paper coding in the works by M. L. Miller et al. [151, 155], and A. Abrardo and M. Barni [1].

Bibliography

1. A. Abrardo and M. Barni, *Orthogonal dirty paper coding for informed watermarking*, in Security, Steganography, and Watermarking of Multimedia Contents VI, Proc. SPIE Vol. 5306, P. W. Wong and E. J. Delp, eds., San Jose, CA, USA, January 2004.

2. A. Adelsbach, S. Katzenbeisser, and A.-R. Sadeghi, *Crytpography meets watermarking: detecting watermarks with minimal or zero knowledge disclosure*, in Proc. XI Europ. Signal Processing Conf., EUSIPCO'02, vol. I, Toulouse, France, September 2002, pp. 446–449.

3. O. K. Al-Shaykh, E. Miloslavsky, T. Nomura, R. Neff, and A. Zakhor, *Video compression using matching pursuits*, IEEE Trans. on Circuits and Systems for Video Technology, 9 (1999), no. 1, pp. 123 –143.

4. M. Alghoniemy and A. H. Tewfik, *Geometric distortion correction in image watermarking*, in Security and Watermarking of Multimedia Contents II, Proc. SPIE Vol. 3971, P. W. Wong and E. J. Delp, eds., San Jose, CA, USA, January 2000, pp. 82–89.

5. R. Baitello, M. Barni, F. Bartolini, and V. Cappellini, *From watermark detection to watermark decoding: a PPM approach*, Signal Processing, 81 (2001), no. 6, pp. 1261–1271.

6. B. Barnett and D. E. Pearson, *Attack operators for digitally watermarked images*, IEE Proceedings on Vision Image Signal Processing, 145 (1998), no. 4, pp. 271–279.

7. R. Barnett and D. E. Pearson, *Frequency mode LR attack operator for digitally watermarked images*, Electronic Letters, 34 (1998), no. 19, pp. 1837–1839.

8. M. Barni, F. Bartolini, V. Cappellini, and N. Checcacci, *Object watermarking for MPEG4 video streams copy protection*, in Security and Watermarking of Multimedia Contents II, Proc. SPIE Vol. 3971, P. W.

Wong and E. J. Delp, eds., San Jose, CA, USA, January 2000, pp. 465–476.

9. M. Barni, F. Bartolini, V. Cappellini, and A. Piva, *A DCT-domain system for robust image watermarking*, Signal Processing, 66 (1998), no. 3, pp. 357–372.

10. M. Barni, F. Bartolini, V. Cappellini, A. Piva, and F. Rigacci, *A MAP identification criterion for DCT-based watermarking*, in Proc. IX Europ. Signal Processing Conf., EUSIPCO'98, Rhodos, Greece, Sept. 1998, pp. 17–20.

11. M. Barni, F. Bartolini, and A. De Rosa, *On the performance of multiplicative spread spectrum watermarking*, in Proc. IEEE Work. on Multimedia signal Processing, MMSP'02, San Thomas, Virgin Islands, USA, December 2002.

12. M. Barni, F. Bartolini, A. De Rosa, and A. Piva, *A new decoder for the optimum recovery of non-additive watermarks*, IEEE Trans. on Image Processing, 10 (2001), no. 5, pp. 755–766.

13. M. Barni, F. Bartolini, A. De Rosa, and A. Piva, *Color image watermarking in the Karhunen-Loeve transform domain*, Journal of Electronic Imaging, 11 (2002), no. 1, pp. 87–95.

14. M. Barni, F. Bartolini, A. De Rosa, and A. Piva, *Optimal decoding and detection of multiplicative watermarks*, IEEE Trans. on Signal Processing, 4 (2003), no. 51.

15. M. Barni, F. Bartolini, E.Magli, and G. Olmo, *Watermarking techniques for electronic delivery of remote sensing images*, Optical Engineering, 41 (2002), no. 9, pp. 2111–2119.

16. M. Barni, F. Bartolini, and T. Furon, *A general framework for robust watermarking security*, Signal Processing, 83 (2003), no. 10, pp. 2069–2084.

17. M. Barni, F. Bartolini, A. Manetti, and A. Piva, *A data hiding approach for correcting errors in H.263 video transmitted over a noisy channel*, in Proc. IEEE Work. on Multimedia signal Processing, MMSP'01, Cannes, France, October 2001, pp. 65–70.

18. M. Barni, F. Bartolini, and A. Piva, *Improved wavelet-based watermarking through pixel-wise masking*, IEEE Trans. on Image Processing, 10 (2001), no. 5, pp. 783–791.

19. M. Barni, F. Bartolini, and A. Piva, *Multichannel watermarking of color images*, IEEE Trans. on Circuits and Systems for Video Technology, 12 (2002), no. 3, pp. 142–156.

20. M. Barni, F. Bartolini, A. Piva, and F. Salucco, *Robust watermarking of cartographic images*, EURASIP Journal on Applied Signal Processing, 2002 (2002), no. 2, pp. 197–208.

21. P. G. J. Barten, *Evaluation of subjective image quality with the square-root intergral method*, Journal of Optical Society of America A, 7 (1990), no. 10, pp. 2024–2031.

22. F. Bartolini, M. Barni, V. Cappellini, and A. Piva, *Mask building for perceptually hiding frequency embedded watermarks*, in Proc. 5th IEEE Int. Conf. on Image Processing, ICIP'98, vol. I, Chicago, IL, USA, October 1998, pp. 450–454.

23. F. Bartolini, I. Tefas, M. Barni, and I. Pitas, *Image authentication techniques for surveillance applications*, Proceedings of the IEEE, (2001), no. 10, pp. 1403–1418.

24. P. Bas, J. Chassery, and F. Davoine, *Using the fractal code to watermark images*, in Proc. 5th IEEE Int. Conf. on Image Processing, ICIP'98, vol. I, Chicago, IL, USA, October 1998, pp. 469 –473.

25. P. Bas, J. Chassery, and B. Macq, *Robust watermarking based on the warping of pre-defined triangular patterns*, in Security and Watermarking of Multimedia Contents II, Proc. SPIE Vol. 3971, P. W. Wong and E. J. Delp, eds., San Jose, CA, USA, January 2000, pp. 99–109.

26. P. Bas and B. Macq, *A new video-object watermarking scheme robust to object manipulation*, in Proc. 8th IEEE Int. Conf. on Image Processing, ICIP'01, vol. II, Thessaloniki, Greece, October 2001, pp. 526 –529.

27. P. Bassia, I. Pitas, and N. Nikolaidis, *Robust audio watermarking in the time domain*, IEEE Trans. on Multimedia, 3 (2001), no. 2, pp. 232–241.

28. R. Bäuml, J. J. Eggers, R. Tzschoppe, and J. Huber, *A channel model for watermarks subject to desynchronization attacks*, in Security and Watermarking of Multimedia Contents IV, Proc. SPIE Vol. 4675, P. W. Wong and E. J. Delp, eds., San Jose, CA, USA, January 2002, pp. 281–292.

29. W. Bender, D. Gruhl, and A. Lu, *Echo hiding*, in Proc. 1st Int. Work. on Information Hiding, IH'96, R. Anderson, ed., vol. 1174 of Lecture Notes in Computer Science, Cambridge, UK, May/June 1996, Springer Verlag, pp. 295–315.

30. W. R. Bender, D. Gruhl, and N. Morimoto, *Techniques for data hiding*, in Storage and Retrieval of Image and Video Databases III, Proc. SPIE Vol. 2420, W. Niblack and R. C. Jain, eds., San Jose, CA, USA, February 1995, pp. 164–173.

31. K. A. Birney and T. R. Fischer, *On the modeling of DCT and subband image data for compression*, IEEE Trans. on Image Processing, 4 (1995), no. 2, pp. 186 –193.

32. J. A. Bloom, I. J. Cox, T. Kalker, J.-P. Linnartz, M. L. Miller, and C. B. S. Traw, *Copy protection for DVD video*, Proceedings of the IEEE, 87 (1999), no. 7, pp. 1267–1276.

33. D. Boneh and J. Shaw, *Collusion-secure fingerprinting for digital data*, IEEE Trans. on Information Theory, 44 (1998), no. 5, pp. 1897–1905.

34. C. Cachin, *An information-theoretic model for steganography*, in Proc. 2nd Int. Work. on Information Hiding, IH'98, D. Aucsmith, ed., vol. 1525 of Lecture Notes in Computer Science, Portland, OR, USA, April 1998, Springer Verlag, pp. 306–318.

35. P. Campisi, D. Kundur, D. Hatzinakos, and A.Neri, *Compressive data hiding: An unconventional approach for improved color image coding*, EURASIP Journal on Applied Signal Processing, 2002 (2002), no. 2, pp. 152–163.

36. B. Chen, *Design and Analysis of Digital Watermarking, Information Embedding, and Data Hiding Systems*, Ph.D Thesis, MIT - Cambridge, 2000.

37. B. Chen and G. Wornell, *Achievable performance of digital watermarking schemes*, in Proc. IEEE Int. Conf. on Multimedia Computing and Systems, ICMCS '99, vol. 1, Florence, Italy, June 1999, pp. 13–18.

38. B. Chen and G. Wornell, *Preprocessed and postprocessed quantization index modulation methods for digital watermarking*, in Security and Watermarking of Multimedia Contents II, Proc. SPIE Vol. 3971, P. W. Wong and E. J. Delp, eds., San Jose, CA, USA, January 2000, pp. 48–59.

39. B. Chen and G. Wornell, *Quantization index modulation: a class of provably good methods for digital watermarking and information embedding*, IEEE Trans. on Information Theory, 47 (2001), no. 4, pp. 1423–1443.

40. Q. Cheng and T. S. Huang, *Optimum detection and decoding of multiplicative watermarks in DFT domain*, in Proc. IEEE Int. Conf. on Acoustic Speech and Signal Processing, ICASSP'02, vol. 4, Orlando, FL, USA, May 2002, pp. 3477–3480.

41. J. Chou, S. S. Pradhan, L. El Ghaoui, and K. Ramchandran, *Watermarking based on duality with distributed source coding and robust optimization principles*, in Proc. 7th IEEE Int. Conf. on Image Processing, ICIP'00, vol. 1, Vancouver, Canada, September 2000, pp. 585–588.

42. J. Chou, S. S. Pradhan, and K. Ramchandran, *On the duality between distributed source coding and data hiding*, in Proc. 33rd Asilomar Conf. on Signals, Systems, and Computers, vol. II, Pacific Grove, CA, USA, October 1999, pp. 1503–1507.

43. J. Chou and K. Ramchandran, *Robust turbo-based data hiding for image and video sources*, in Proc. 9th IEEE Int. Conf. on Image Processing, ICIP'02, vol. 2, Rochester, NY, USA, September 2002, pp. 133–136.

44. R. J. Clarke, *Transform Coding of Images*, Academic Press, New York, 1985.

45. A. S. Cohen, *Information Theoretical Analysis of Watermarmking Systems*, Ph.D Thesis, MIT - Cambridge, 2001.

46. A. S. Cohen and A. Lapidoth, *The gaussian watermarking game*, IEEE Trans. on Information Theory, 48 (2002), no. 6, pp. 1639–1667.

47. D. Coltuc and P. Bolon, *Robust watermarking by histogram specification*, in Proc. 6th IEEE Int. Conf. on Image Processing, ICIP'99, vol. II, Kobe, Japan, October 1999, pp. 236 –239.

48. D. Coltuc and P. Bolon, *Watermarking by histogram specification*, in Security and Watermarking of Multimedia Contents, Proc. SPIE Vol. 3657, P. W. Wong and E. J. Delp, eds., San Jose, CA, USA, January 1999, pp. 252–263.

49. S. Comes and B. Macq, *Human visual quality criterion*, in Visual Communications and Image Processing, Proc. SPIE Vol. 1360, M. Kunt, ed., vol. I, Lausanne, Switzerland, October 1990, pp. 2–13.

50. M. H. M. Costa, *Writing on dirty paper*, IEEE Trans. on Information Theory, 29 (1983), no. 3, pp. 439–441.

51. T. M. Cover and J. A. Thomas, *Elements of Information Theory*, Wiley, New York, 1991.

52. C. P. Cox, *A Handbook of Introductory Statistical Methods*, Wiley, New York, 1991.

53. I. J. Cox, J. Kilian, T. Leighton, and T. Shamoon, *Secure spread spectrum watermarking for multimedia*, IEEE Trans. on Image Processing, 6 (1997), no. 12, pp. 1673–1687.

54. I. J. Cox and J. P. M. G. Linnartz, *Some general methods for tampering with watermarks*, IEEE Journal of Selected Areas in Communications, 16 (1998), no. 4, pp. 587–593.

55. I. J. Cox and M. L. Miller, *The first 50 years of electronic watermarking*, EURASIP Journal on Applied Signal Processing, 2002 (2002), no. 2, pp. 126–132.

56. I. J. Cox, M. L. Miller, and J. A. Bloom, *Digital Watermarking*, Morgan Kaufmann, 2001.

57. I. J. Cox, M. L. Miller, and A. L. McKellips, *Watermarking as communications with side information*, Proceedings of the IEEE, 87 (1999), no. 7, pp. 1127–1141.

58. S. Craver and S. Katzenbeisser, *Security analysis of public-key watermarking schemes*, in Mathematics of Data/Image Coding, Compression, and Encryption IV, Proc. SPIE Vol. 4475, M. S. Schmalz, ed., San Diego, CA, USA, July 2001, pp. 172–182.

59. S. Craver, N. Memon, B. L. Yeo, and M. M. Yeung, *On the invertibility of invisible watermarking techniques*, in Proc. 4th IEEE Int. Conf. on Image Processing, ICIP'97, vol. I, Santa Barbara, CA, USA, October 1997, pp. 540–543.

60. S. Craver, N. Memon, B. L. Yeo, and M. M. Yeung, *Resolving rightful ownership with invisible watermarking techniques: limitations, attacks and implications*, IEEE Journal of Selected Areas in Communications, 4 (1998), no. 16, pp. 573–586.

61. S. Craver and J. P. Stern, *Lessons learned from SDMI*, in Proc. IEEE Work. on Multimedia signal Processing, MMSP'01, Cannes, France, October 2001, pp. 213–218.

62. I. Csiszar and J. Korner, *Information Theory: Coding Theory for Discrete Memoryless Systems*, Academic Press, New York, 1981.

63. S. Daly, *Engineering observations from spatiovelocity and spatiotemporal visual models*, in Vision Models and Applications to Image and Video Processing, C. van den Branden Lambrecht, ed., Kluwer Academic Publishers, Dordrecht, The Netherlands, 2001, pp. 179–200.

64. A. De Rosa, M. Barni, F. Bartolini, and A. Piva, *A watermark capacity measure incorporating a model of the human visual system*, in Security and Watermarking of Multimedia Contents III, Proc. SPIE Vol. 4314, P. W. Wong and E. J. Delp, eds., San Jose, CA, USA, January 2001, pp. 483–94.

65. S. Decker, *Engineering considerations in commercial watermarking*, IEEE Communications Magazine, 39 (2001), no. 8, pp. 128–133.

66. W. Diffie and M. Hellman, *New directions in cryptography*, IEEE Trans. on Information Theory, 22 (1976), no. 6, pp. 644–54.

67. J. Dittmann, P. Schmitt, E. Saar, J. Schwenk, and J. Ueberberg, *Combining digital watermarks and collusion secure fingerprints for digital images*, SPIE Journal of Electronic Imaging, 9 (2000), no. 4, pp. 456–467.

68. J. Dittmann, M. Steinebach, P. Wohlmacher, and R. Ackermann, *Digital watermarking enabling e-commerce strategies: conditional and user specific access to services and resources*, EURASIP Journal on Applied Signal Processing, 2002 (2002), no. 2, pp. 174–184.

69. J. J. Eggers, R. Bäuml, and B. Girod, *Estimation of amplitude modifications before SCS watermark detection*, in Security and Watermarking of Multimedia Contents IV, Proc. SPIE Vol. 4675, P. W. Wong and E. J. Delp, eds., San Jose, CA, January 2002, pp. 387–398.

70. J. J. Eggers, R. Bäuml, R. Tzschoppe, and B. Girod, *Scalar Costa scheme for information embedding*, IEEE Trans. on Signal Processing, 4 (2003), no. 51.

71. J. J. Eggers and B. Girod, *Informed Watermarking*, Kluwer Academic Publishers, 2002.

72. J. J. Eggers, J. K. Su, and B. Girod, *A blind watermarking scheme based on structured codebooks*, in Proc. IEE Seminar on Secure Images and Image Authentication, London, UK, April 2000, pp. 4/1–4/21.

73. J. J. Eggers, J. K. Su, and B. Girod, *Public key watermarking by eigenvectors of linear transforms*, in Proc. X Europ. Signal Processing Conf., EUSIPCO'00, vol. III, Tampere, Finland, September 2000, pp. 1685–1688.

74. J. J. Eggers, J. K. Su, and B. Girod, *Performance of a practical blind watermarking scheme*, in Security and Watermarking of Multimedia Contents III, Proc. SPIE Vol. 4314, P. W. Wong and E. J. Delp, eds., San Jose, CA, USA, January 2001, pp. 594–605.

75. A. El Gamal and T. M. Cover, *Multiple user information theory*, Proceedings of the IEEE, 68 (1980), no. 12, pp. 1466–1485.

76. G. T. Fechner, *Elemente der psychophysiscs*, Breitkopf, Leipzing, 1860.

77. C. Fei, D. Kundur, and R. H. Kwong, *Transform-based hybrid data hiding for improved robustness in the presence of perceptual coding*, in Mathematics of Data/Image Coding, Compression and Encryption IV, with Applications, Proc. SPIE Vol. 4475, M. S. Schmalz, ed., San Diego, CA, USA, July 2001, pp. 203–212.

78. J. Fridrich, *Key-dependent random image transforms and their applications in image watermarking*, in Proc. Int. Conf. on Imaging Science, Systems and Technology, CISST '99, Las Vegas, NV, USA, June/July 1999, pp. 237–243.

79. J. Fridrich, *Security of fragile authentication watermarks with localization*, in Security and Watermarking of Multimedia Contents IV, Proc. SPIE Vol. 4675, P. W. Wong and E. J. Delp, eds., San Jose, CA, USA, January 2002, pp. 691–700.

80. J. Fridrich and M. Goljan, *Images with self-correcting capabilities*, in Proc. 6th IEEE Int. Conf. on Image Processing, ICIP'99, vol. III, Kobe, Japan, October 1999, pp. 792–796.

81. T. Furon and P. Duhamel, *An asymmetric public detection watermarking technique*, in Proc. 3rd Int. Work. on Information Hiding, IH'99, A. Pfitzmann, ed., vol. 1768 of Lecture Notes in Computer Science, Dresden - Germany, September 1999, Springer Verlag, pp. 88–100.

82. T. Furon and P. Duhamel, *Robustness of an asymmetric watermarking technique*, in Proc. 7th IEEE Int. Conf. on Image Processing, ICIP'00, vol. III, Vancouver, Canada, September 2000, pp. 21–24.

83. T. Furon, I. Venturini, and P. Duhamel, *An unified approach of asymmetric watermarking schemes*, in Security and Watermarking of Multimedia Contents III, Proc. SPIE Vol. 4314, E. J. D. P.W. Wong, ed., San Jose, CA, January 2001, pp. 269–279.

84. S. I. Gelf'and and M. S. Pinsker, *Coding for channel with random parameters*, Problems of Control and Information Theory, 9 (1980), no. 1, pp. 19–31.

85. R. Gibbons, *Game Theory for Applied Economists*, Princeton Unievrsity Press, Princeton, NJ, 1992.

86. R. Gold, *Maximal recursive sequences with 3-valued recursive cross-correlation functions*, IEEE Trans. on Information Theory, 14 (1968), no. 1, pp. 154–156.

87. S. W. Golomb, *Shift Register Sequences*, Holden Day, San Francisco, CA, 1967.

88. J. Haitsma and T. Kalker, *A watermarking scheme for digital cinema*, in Proc. 8th IEEE Int. Conf. on Image Processing, ICIP'01, vol. II, Thessaloniki, Greece, October 2001, pp. 487–489.

89. S. Harrington, *Computer graphics: a programming approach*, McGraw-Hill, 1987.

90. F. Hartung and B. Girod, *Fast public-key watermarking of compressed video*, in Proc. 4th IEEE Int. Conf. on Image Processing, ICIP'97, vol. I, Santa Barbara, CA, USA, October 1997, pp. 528–531.

91. F. Hartung and B. Girod, *Watermarking of uncompressed and compressed video*, Signal Processing, 66 (1998), no. 3, pp. 283–301.

92. F. Hartung and M. Kutter, *Multimedia watermarking techniques*, Proceedings of the IEEE, 87 (1999), no. 7, pp. 1079–1107.

93. C. Heegard and A. A. El Gamal, *On the capacity of computer memory with defects*, IEEE Trans. on Information Theory, 29 (1983), pp. 731–739.

94. J. R. Hernandez, M. Amado, and F. Perez-Gonzales, *DCT-domain watermarking techniques for still images: detector performance analysis and a new structure*, IEEE Trans. on Image Processing, 9 (2000), no. 1, pp. 55–68.

95. J. R. Hernandez and F. Perez-Gonzales, *Statistical analysis of watermarking schemes for copyright protection of images*, Proceedings of the IEEE, 87 (1999), no. 7, pp. 1142–1166.

96. J. R. Hernandez, J. M. Rodriguez, and F. Perez-Gonzales, *Improving the performance of spatial watermarking of images using channel coding*, Signal Processing, 80 (2000), no. 7, pp. 1261–1279.

97. N. Hurley and G. C. M. Silvestre, *Nth-order audio watermarking*, in Security and Watermarking of Multimedia Contents IV, Proc. SPIE Vol. 4675, P. W. Wong and E. J. Delp, eds., San Jose, CA, USA, January 2002, pp. 102–109.

98. International Commission on Illumination (CIE), *Colorimetry, CIE Publication no. 15 (E-1.3.1)*, Bureau Central de la CIE, Wien, Austria, 1971.

99. International Commission on Illumination (CIE), *Recommendations on uniform color spaces, color difference equation, psychometric color terms, Supplement to CIE Publication No. 15 (E.-1.3.1)*, Bureau Central de la CIE, Wien, Austria, 1971.

100. International Commission on Illumination (CIE), *Industrial color-difference evaluation, CIE Publication no. 116-95*, Bureau Central de la CIE, Wien, Austria, 1995.

101. S. Isabelle and G. Wornell, *Statistical analysis and spectral estimation techniques for one-dimensional chaotic signals*, IEEE Trans. on Signal Processing, 45 (1997), no. 6, pp. 1495–1506.

102. ISO/IEC, JTC1/SC29/WG11, *IS11172-3 - Information Technology - Coding of moving pictures and associated audio for digital storage media at up to about 1.5 Mbit/s - Part 3: Audio*, ISO, 1992. (MPEG-1 - Audio).

103. ISO/IEC JTC1/SC29/WG11, *IS11172-2 - Information technology – Coding of moving pictures and associated audio for digital storage media at up to about 1,5 Mbit/s – Part 2: Video*, ISO, 1993. (MPEG-1 - Video).

104. ISO/IEC, JTC1/SC29/WG11, *IS13818-1 - Information Technology - Generic coding of moving pictures and associated audio information - Part 3: Audio*, ISO, 1994. (MPEG-2 - Audio).

105. ISO/IEC, JTC1/SC29/WG11, *IS13818-2 - Information Technology - Generic coding of moving pictures and associated audio information - Part 2: Video*, ISO, 1996. (MPEG-2 - Video).

106. ISO/IEC JTC1/SC2/WG11, *IS14496-2 - Information technology - Coding of audio-visual objects - Part 2: Visual*, ISO, 2001. (MPEG4).

107. ISO/IEC, JTC1/SC2/WG8, *IS10918-1 - Information Technology - Digital compression and coding of continuous-tone still images*, ISO, 1992. (JPEG).

108. ISO/IEC JTC1/SC2/WG8, *IS15444-1 - Information technology - JPEG 2000 image coding system - Part 1: Core coding system*, ISO, 2000. (JPEG2000).

109. ITU-R, *Rec. 601 - Encoding parameters of digital television for studios*, ITU, 1990.

110. ITU-R, *Rec. BT.709 - Basic parameters value for the HDTV standard for the studio and for international programme exchange*, ITU, 1990.

111. A. E. Jacquin, *Image coding based on a fractal theory of iterated contractive image transformations*, IEEE Trans. on Image Processing, 1 (1992), no. 1, pp. 18–30.

112. A. K. Jain, *Fundamentals of digital images processing*, Prentice Hall, Englewood Cliffs, NJ, USA, 1989.

113. N. F. Johnson and S. Katzenbeisser, *A survey of steganographic techniques*, in Information Hiding, S. Katzenbeisser and F. Petitcolas, eds., Artech House, Norwood, MA, 2000, pp. 43–78.

114. T. Kalker, *Considerations on watermarking security*, in Proc. IEEE Work. on Multimedia signal Processing, MMSP'01, Cannes, France, October 2001, pp. 201–206.

115. T. Kalker, G. Depovere, J. Haitsma, and M. Maes, *A video watermarking system for broadcast monitoring*, in Security and Watermarking of Multimedia Contents, Proc. SPIE Vol. 3657, P. W. Wong and E. J. Delp, eds., San Jose, CA, January 1999, pp. 103–112.

116. T. Kalker, J. P. Linnartz, and M. van Dijk, *Watermark estimation thorugh detector analysis*, in Proc. 5th IEEE Int. Conf. on Image Processing, ICIP'98, vol. I, Chicago, IL, USA, October 1998, pp. 425–429.

117. T. Kasami, *Weight distribution formula for some class of cyclic codes*, Coordinated Science Laboratory, University of Illinois, Urbana, IL, Tech. Report, (1966), no. R-285.

118. S. M. Kay, *Fundamentals of Statistical Signal Processing: Detection Theory*, vol. II, Prentice Hall, 1998.

119. D. Kelly, *Motion and vision. II. stabilized spatio-temporal threshold surface*, Journal of Optical Society of America, 69 (1979), no. 10, pp. 1340–1349.

120. M. G. Kendall and A. Stuart, *The Advanced Theory of Statistics*, vol. I, Charles Griffin and Company Limited, London, 1983.

121. A. Kerckhoffs, *La cryptografie militaire*, Journal des Sciences Militaire, 9 (1883), pp. 5–38.

122. D. Kirovski and H. Malvar, *Spread-spectrum audio watermarking: requirements, applications, and limitations*, in Proc. IEEE Work. on Multimedia signal Processing, MMSP'01, Cannes, France, October 2001, pp. 219 –224.

123. D. E. Knuth, *The Art of Computer Programming, 3rd Edition*, Addison Wesley, Reading, 1997-98.

124. D. Kundur, *Energy allocation principles for high capacity data hiding*, in Proc. 7th IEEE Int. Conf. on Image Processing, ICIP'00, vol. 1, Vancouver, Canada, September 2000, pp. 423–426.

125. D. Kundur, *Water-filling for watermarking?*, in Proc. IEEE Int. Conf. On Multimedia and Expo, ICME'00, vol. III, New York, NY, USA, August 2000, pp. 1287–1290.

126. D. Kundur and D. Hatzinakos, *Digital watermarking for telltale tamper proofing and authentication*, Proceedings of the IEEE, 87 (1999), no. 7, pp. 1167–1180.

127. D. Kundur and D. Hatzinakos, *Diversity and attack characterization for improved robust watermarking*, IEEE Trans. on Signal Processing, 49 (2001), no. 10, pp. 2383–2396.

128. M. Kutter, *Digital Image Watermarking: Hiding Information in Images*, Ph.D Thesis, EPFL - Lausanne, 1999.

129. M. Kutter, F. Jordan, and F. Bossen, *Digital signature of color images using amplitude modulation*, in Storage and Retrieval of Image and Video Databases V, Proc. SPIE Vol. 3022, I. K. Sethi and R. C. Jain, eds., San Jose, CA, February 1997, pp. 518–526.

130. M. Kutter and F. A. P. Petiticolas, *Fair benchmark for image watermarking systems*, in Security and Watermarking of Multimedia Contents, Proc. SPIE Vol. 3657, P. W. Wong and E. J. Delp, eds., San Jose, CA, USA, January 1999, pp. 226–239.

131. G. C. Langelaar, I. Setyawan, and R. L. Lagendijk, *Watermarking digital image and video data. a state-of-the-art overview*, IEEE Signal Processing Magazine, 17 (2000), no. 5, pp. 20–46.

132. G. E. Legge, *Spatial frequency in human vision: binocular interaction*, Journal of Optical Society of America, 69 (1979), no. 6, pp. 838–847.

133. G. E. Legge and J. M. Foley, *Contrast masking in human vision*, Journal of Optical Society of America, 70 (1980), no. 12, pp. 1458–1471.

134. A. S. Lewis and G. Knowles, *Image compression using the 2-D wavelet transform*, IEEE Trans. on Image Processing, 1 (1992), no. 2, pp. 244–250.

135. C. Y. Lin, M. Wu, J. A. Bloom, I. J. Cox, M. L. Miller, and Y. M. Lui, *Rotation, scale, and translation resilient watermarking for images*, IEEE Trans. on Image Processing, 10 (2001), no. 5, pp. 767–782.

136. S. Lin and D. J. Costello, *Error Control Coding: Fundamentals and Applications*, Prentice-Hall, Englewood Cliffs, NJ, 1983.

137. J. P. M. G. Linnartz and J. Talstra, *MPEG PTY-marks: Cheap detection of embedded copyright data in DVD-video*, in Proc. 5th European Symposium on Research in Computer Security ESORICS, J. Quisquater, Y. Deswarte, C. Meadows, and D. Gollmann, eds., Louvain-la-Neuve, Belgium, September 1998, Springer Verlag, pp. 221–240.

138. J. P. M. G. Linnartz and M. van Dijk, *Analysis of the sensitivity attack against electronic watermarks in images*, in Proc. 2nd Int. Work. on Information Hiding, IH'98, D. Aucsmith, ed., vol. 1525 of Lecture Notes in Computer Science, Portland, OR, USA, April 1998, Springer Verlag, pp. 258–272.

139. P. Loo and N. Kingsbury, *Digital watermarking using complex wavelets*, in Proc. 7th IEEE Int. Conf. on Image Processing, ICIP'00, vol. 3, Vancouver, Canada, September 2000, pp. 29–32.

140. M. Maes, T. Kalker, J. Haitsma, and G. Depovere, *Exploiting shift invariance to obtain a high payload in digital image watermarking*, in Proc. IEEE Int. Conf. on Multimedia Computing and Systems, ICMCS '99, vol. I, Florence, Italy, June 1999, pp. 7–12.

141. M. Maes, T. Kalker, J.-P. Linnartz, J. T. F. G. Depovere, and J. Haitsma, *Digital watermarking for DVD video copy protection*, IEEE Signal Processing Magazine, 17 (2000), no. 5, pp. 47–57.

142. M. J. J. J. B. Maes and C. W. A. M. van Overheld, *Digital water-marking by geometric warping*, in Proc. 5th IEEE Int. Conf. on Image Processing, ICIP'98, vol. 2, Chicago, IL, USA, October 1998, pp. 424–426.

143. S. Mallat, *A theory for multiresolution signal decomposition: the wavelet representation*, IEEE Trans. on Pattern Analysis and Machine Intelligence, 11 (1989), no. 7, pp. 674–693.

144. S. Mallat, *A Wavelet Tour of Signal Processing*, Academic Press, 1998.

145. M. F. Mansour and A. H. Tewfik, *Secure detection of public water-marks with fractal decision boundary*, in Proc. XI Europ. Signal Processing Conf., EUSIPCO'02, Toulouse, France, September 2002.

146. S. Martucci, *Symmetric convolution and the discrete sine and co-sine transforms*, IEEE Trans. on Signal Processing, 42 (1994), no. 5, pp. 1038–1051.

147. G. Mazzini, G. Setti, and R. Rovatti, *Chaotic complex spreading se-quences for asynchronous DS-CDMA - Part I: system modeling and results*, IEEE Trans. on Circuits and Systems I, 44 (1997), no. 10, pp. 937–947.

148. P. Meerwald and A. Uhl, *Watermark security via wavelet filter parametrization*, in Proc. 8th IEEE Int. Conf. on Image Processing, ICIP'01, vol. III, Thessaloniki, Greece, October 2001, pp. 1027–1030.

149. N. Memon and P. W. Wong, *A buyer-seller watermarking protocol*, IEEE Trans. on Image Processing, 10 (2001), no. 4, pp. 643–649.

150. N. Merhav, *On random coding error exponents of watermarking sys-tems*, IEEE Trans. on Information Theory, 46 (2000), no. 2, pp. 420–430.

151. M. Miller, *Watermarking with dirty-paper codes*, in Proc. 8th IEEE Int. Conf. on Image Processing, ICIP'01, vol. II, Thessaloniki, Greece, October 2001, pp. 538–541.

152. M. Miller and J. A. Bloom, *Computing the probability of false water-mark detection*, in Proc. 3rd Int. Work. on Information Hiding, IH'99, A. Pfitzmann, ed., vol. 1768 of Lecture Notes in Computer Science, Dresden - Germany, September 1999, Springer Verlag, pp. 146–158.

153. M. Miller, I. Cox, and J. Bloom, *Informed embedding: exploiting im-age and detector information during watermark insertion*, in Proc.

7th IEEE Int. Conf. on Image Processing, ICIP'00, vol. I, Vancouver, Canada, September 2000, pp. 1–4.

154. M. L. Miller, *Is asymmetric watermarking necessary or sufficient?*, in Proc. XI Europ. Signal Processing Conf., EUSIPCO'02, vol. I, Toulose, France, September 2002, pp. 291–294.

155. M. L. Miller, G. J. Doerr, and I. J. Cox, *Applying informed coding and embedding to design a robust, high capacity, watermark*, IEEE Trans. on Image Processing. Submitted paper.

156. T. Mittelholzer, *An infomation-theoritic approach to steganography and watermarking*, in Proc. 3rd Int. Work. on Information Hiding, IH'99, A. Pfitzmann, ed., vol. 1768 of Lecture Notes in Computer Science, Dresden, Germany, September 1999, Springer Verlag, pp. 1–17.

157. P. Moulin, *The role of information theory in watermarking and its application to image watermarking*, Signal Processing, 81 (2001), no. 6, pp. 1121–1139.

158. P. Moulin and A. Ivanovic, *The Fisher information game for optimal design of synchronization patterns in blind watermarking*, in Proc. 8th IEEE Int. Conf. on Image Processing, ICIP'01, vol. II, Thessaloniki, Greece, October 2001, pp. 550–553.

159. P. Moulin and J. A. O'Sullivan, *Information-theoretic analysis of information hiding*, in Proc. IEEE Int. Symposium on Information Theory, Sorrento, Italy, June 2000, p. 19.

160. K. T. Mullen, *The contrast sensitivity of human colour vision to red-green and blue-yellow chromatic gratings*, Journal of Physiology, 359 (1985), pp. 381–400.

161. P. Noll, *MPEG digital audio coding*, IEEE Signal Processing Magazine, 14 (1997), no. 5, pp. 59–81.

162. J. J. K. Ó Ruanaidh and T. Pun, *Rotation, scale and translation invariant digital image watermarking*, in Proc. 4th IEEE Int. Conf. on Image Processing, ICIP'97, vol. I, Santa Barbara, CA, USA, October 1997, pp. 536–539.

163. J. A. O'Sullivan, *Some properties of optimal information hiding and information attacks*, in Proc. 39th Allerton Conference on Communications, Control and Computing, Monticello, IL, USA, October 2001.

164. J. A. O'Sullivan, P. Moulin, and J. M. Ettinger, *Information theoretic analysis of steganography*, in Proc. IEEE Int. Symposium on Information Theory, Cambridge, MA, USA, August 1998, p. 297.

165. T. Painter and A. Spanias, *Perceptual coding of digital audio*, Proceedings of the IEEE, 88 (2000), no. 4, pp. 451–513.

166. A. Papoulis, *Probability, Random Variables, and Stochastic Processes*, McGraw Hill, New York, 1991.

167. S. Pereira, S. Voloshynoskiy, and T. Pun, *Optimal transform domain watermark embedding via linear programming*, Signal Processing, 81 (2001), no. 6, pp. 1251–1260.

168. S. Pereira, S. Voloshynovskiy, M. Madueno, S. Marchand-Maillet, and T. Pun, *Second generation benchmarking and application oriented evaluation*, in Proc. 4th Int. Work. on Information Hiding, IH'01, I. S. Moskowitz, ed., vol. 2137 of Lecture Notes in Computer Science, Pittsburgh, PA, USA, April 2001, Springer Verlag, pp. 340–353.

169. F. Perez-Gonzalez, F. Balado, and J. R. Hernandez Martin, *Performance analysis of existing and new methods for data hiding with known-host information in additive channels*, IEEE Trans. on Signal Processing, 51 (2003), no. 4, pp. 960–980.

170. F. Perez-Gonzalez, P. Comesana, and F. Balado, *Dither-modulation data hiding with distortion-compensation: exact performance analysis and an improved detector for JPEG attacks*, in Proc. 10th IEEE Int. Conf. on Image Processing, ICIP'03, vol. II, Barcelona, Spain, September 2003, pp. 503–506.

171. F. Perez-Gonzalez, J. R. Hernandez, and F. Balado, *Approaching the capacity limit in image watermarking: a perspective on coding techniques for data hiding applications*, Signal Processing, 81 (2001), no. 6, pp. 1215–1238.

172. H. A. Peterson, A. J. Ahumada, and A. B. Watson; *An improved detection model for DCT coefficient quantization*, in Human Vision, Visual Processing, and Digital Displays IV, Proc. SPIE vol. 1913, J. P. Allebach and B. E. Rogowitz, eds., San Jose, CA, USA, January 1993, pp. 191–201.

173. F. A. P. Petitcolas, *Watermarking scheme evaluation*, IEEE Signal Processing Magazine, 17 (2000), pp. 58–64.

174. F. A. P. Petitcolas and R. J. Anderson, *Weaknesses of copyright marking systems*, in Multimedia and Security Workshop at ACM Multimedia '98, Bristol, UK, September 1998, pp. 55–62.

175. F. A. P. Petitcolas and R. J. Anderson, *Evaluation of copyright marking systems*, in Proc. IEEE Int. Conf. on Multimedia Computing and Systems, ICMCS '99, vol. I, Florence, Italy, June 1999, pp. 574–579.

176. F. A. P. Petitcolas, R. J. Anderson, and M. G. Kuhn, *Attacks on copyright marking systems*, in Proc. 2nd Int. Work. on Information Hiding, IH'98, D. Aucsmith, ed., vol. 1525 of Lecture Notes in Computer Science, Portland, OR, USA, April 1998, Springer Verlag, pp. 218–238.

177. F. A. P. Petitcolas, R. J. Anderson, and M. G. Kuhn, *Information hiding: a survey*, Proceedings of the IEEE, 87 (1999), no. 7, pp. 1062–1078.

178. B. Pfitzman, *Information hiding terminology*, in Proc. 1st Int. Work. on Information Hiding, IH'96, R. Anderson, ed., vol. 1174 of Lecture Notes in Computer Science, Springer Verlag, Cambridge, UK, May/June 1996, pp. 347–350.

179. I. Pitas and A. N. Venetsanopoulos, *Nonlinear digital filters: principles and applications*, Kluwer Academic Publisher, 1990.

180. A. Piva, R. Caldelli, and A. D. Rosa, *A DWT-based object watermarking system for MPEG-4 video streams*, in Proc. 7th IEEE Int. Conf. on Image Processing, ICIP'00, vol. 3, Vancouver, Canada, September 2000, pp. 5 –8.

181. C. I. Podilchuk and W. Zeng, *Image-adaptive watermarking using visual models*, IEEE Journal of Selected Areas in Communications, 16 (1998), no. 4, pp. 525–539.

182. A. B. Poirson and B. A. Wandell, *Appearance of colored patterns: pattern-color separability*, Journal of Optical Society of America A, 10 (1993), no. 12, pp. 2458–2470.

183. S. S. Pradhan, J. Chou, and K. Ramchandran, *Duality between source coding and channel coding with side information*, IEEE Trans. on Information Theory, 49 (2003), no. 5.

184. S. S. Pradhan and K. Ramchandran, *Distributed source coding using syndromes (DISCUS): design and construction*, IEEE Trans. on Information Theory, 49 (2003), no. 3.

185. J. G. Proakis, *Digital Communications, 2nd Edition*, McGraw-Hill, New York, 1989.

186. L. Qian and K. Nahrstedt, *Watermarking schemes and protocols for protecting rightful ownership and customers rights*, Journal of Visual Commununications and Image Representation, 9 (1998), pp. 194–210.

187. M. Rabbani and R. Joshi, *An overview of the JPEG 2000 still image compression standard*, Signal Processing: Image Communications, 17 (2002), no. 1, pp. 3–48.

188. D. L. Robie and R. M. Merserau, *Video error correction using steganography*, EURASIP Journal on Applied Signal Processing, 2002 (2002), no. 2, pp. 164–173.

189. R. Rovatti, G. Setti, and G. Mazzini, *Chaotic complex spreading sequences for asynchronous DS-CDMA - Part II: some theoretical performance bounds*, IEEE Trans. on Circuits and Systems I, 45 (1998), no. 4, pp. 496–506.

190. D. Sadot, N. S. Kopeika, and S. R. Rotman, *Target acquisition modelling from contrast limited imaging: effects of atmospheric blus and image restoration*, Journal of Optical Society of America A, 12 (1995), no. 11, pp. 2401–2414.

191. J. A. Saghri and P. S. C. A. Habibi, *Image quality measure based on a human visual system model*, Optical Engineering, 28 (1989), no. 7, pp. 813–818.

192. A. Said and W. A. Pearlman, *A new fast and efficient image codec based on set partitioning in hierarchical trees*, IEEE Trans. on Circuits and Systems for Video Technology, 6 (1996), no. 3, pp. 243–250.

193. H. Sari, F. Vanhaverbeke, and M. Moeneclaey, *Extending the capacity of multiple access channel*, IEEE Communications Magazine, 38 (2000), no. 1, pp. 74–82.

194. D. W. Sarwate and M. B. Pursley, *Crosscorrelation properties of pseudorandom and related sequences*, Proceedings of the IEEE, 68 (1980), no. 5, pp. 2399–2419.

195. L. L. Scharf, *Statistical Signal Processing: Detection, Estimation, and Time Series Analysis*, Addison-Wesley, Reading, MA, 1991.

196. L. Schuchman, *Dither signals and their effect on quantization noise*, IEEE Trans. on Communications Technology, 12 (1964), pp. 162–165.

197. C. E. Shannon, *Communication theory of secrecy systems*, Bell Systems Techical Journal, 28 (1949), pp. 656–715.

198. R. K. Sharma and S. Decker, *Practical challenges for digital watermarking applications*, EURASIP Journal on Applied Signal Processing, 2002 (2002), no. 2, pp. 133–139.

199. T. Sikora, *The MPEG-4 video standard verification model*, IEEE Trans. on Circuits and Systems for Video Technology, 7 (1997), no. 1, pp. 19 –31.

200. B. Sklar, *Digital Communications: Fundamentals and Applications, 2nd Edition*, Prentice Hall, PTR, Upper saddle River, NJ, 2001.

201. N. Smart, *Cryptography: an introduction*, McGraw-Hill, 2003.

202. J. Smith and C. Dodge, *Developments in steganography*, in Proc. 3rd Int. Work. on Information Hiding, IH'99, A. Pfitzmann, ed., vol. 1768 of Lecture Notes in Computer Science, Dresden - Germany, September 1999, Springer Verlag, pp. 77–87.

203. J. Song and K. J. R. Liu, *A data embedded video coding scheme for error-prone channels*, IEEE Trans. on Multimedia, 3 (2001), no. 4, pp. 415–423.

204. S. Stankovic, I. Djurovic, and I. Pitas, *Watermarking in the space/spatial-frequency domain using two-dimensional Radon-Wigner distribution*, IEEE Trans. on Image Processing, 10 (2001), no. 4, pp. 650–658.

205. Y. Steinberg and N. Merhav, *Identification in the presence of side information with application to watermarking*, IEEE Trans. on Information Theory, 47 (2001), pp. 1410–1422.

206. M. Steinebach, F. A. P. Petitcolas, F. Raynal, J. Dittmann, C. Fontaine, C. Seibel, N. Fates, and L. C. Ferri, *StirMark benchmark: audio watermarking attacks*, in Proc. Int. Conf. on Information Technology: Coding and Computing, ITCC 2001, Las Vegas, NV, USA, April 2001, pp. 49–54.

207. C. F. Stromeyer III and B. Julesz, *Spatial frequency masking in vision: critical bands and spread masking*, Journal of Optical Society of America, 62 (1972), no. 10, pp. 1221–1232.

208. M. D. Swanson, M. Kobayashi, and A. H. Tewfik, *Multimedia data-embedding and watermarking technologies*, Proceedings of the IEEE, 86 (1998), no. 6, pp. 1064–1087.

209. M. D. Swanson, B. Zhu, and A. H. Tewfik, *Data hiding for video-in-video*, in Proc. 4th IEEE Int. Conf. on Image Processing, ICIP'97, vol. II, Santa Barbara, CA, USA, October 1997, pp. 676–679.

210. M. D. Swanson, B. Zhu, and A. H. Tewfik, *Multiresolution scene-based video watermarking using perceptual models*, IEEE Journal on Selected Areas in Communications, 16 (1998), no. 4, pp. 540–550.

211. M. D. Swanson, B. Zhu, A. H. Tewfik, and L. Boney, *Robust audio watermarking usinf perceptual masking*, Signal Processing, (1998), no. 3, pp. 337–355.

212. E. Switkes, A. Bradley, and K. K. D. Valois, *Contrast dependence and mechanisms of masking interactions among chromatic and luminance gratings*, Journal of Optical Society of America A, 5 (1988), no. 7, pp. 1148–1161.

213. D. Taubman, E. Ordentlich, M. Weinberger, and G. Seroussi, *Embedded block coding in JPEG 2000*, Signal Processing: Image Communications, 17 (2002), no. 1, pp. 49–72.

214. A. M. Tekalp, *Digital video processing*, Prentice Hall, 1995.

215. P. Termont, L. De Strycker, J. Vandewege, M. Op de Beeck, J. Haitsma, T. Kalker, M. Maes, and G. Depovere, *How to achieve robustness against scaling in a real-time digital watermarking system for broadcast monitoring*, in Proc. 7th IEEE Int. Conf. on Image Processing, ICIP'00, vol. I, Vancouver, Canada, September 2000, pp. 407–410.

216. A. H. Tewfik and M. F. Mansour, *Secure watermark detection with non-parametric decision boundaries*, in Proc. IEEE Int. Conf. on Acoustic Speech and Signal Processing, ICASSP'02, vol. II, Orlando, FL, USA, May 2002, pp. 2089–2092.

217. W. Trappe, M. Wu, and K. J. R. Liu, *Joint coding and embedding for collusion-resistant fingerprinting*, in Proc. XI Europ. Signal Processing Conf., EUSIPCO'02, Toulouse, France, September 2002.

218. S. Tsekeridou, N. Nikolaidis, N. Sidiropoulos, and I. Pitas, *Copyright protection of still images using self-similar chaotic watermarks*, in Proc. 7th IEEE Int. Conf. on Image Processing, ICIP'00, vol. I, Vancouver, Canada, September 2000, pp. 411–414.

219. S. Tsekeridou, V. Solachidis, N. Nikolaidis, A. Nikolaidis, A. Tefas, and I. Pitas, *Statistical analysis of a watermarking system based on Bernoulli chaotic sequences*, Signal Processing, 81 (2001), no. 6, pp. 1273–1293.

220. R. G. van Schyndel, A. Z. Tirkel, and I. Svalbe, *Key independent watermark detection*, in Proc. IEEE Int. Conf. on Multimedia Computing and Systems, ICMCS '99, vol. I, Florence, Italy, June 1999, pp. 580–585.

221. H. L. Van Trees, *Detection, Estimation, and Modulation Theory*, Wiley and Sons, New York, 1971.

222. M. Vetterli and J. Kovacevic, *Wavelets and Subband Coding*, Prentice Hall, 1995.

223. A. J. Viterbi and J. K. Omura, *Principles of Digital Communications and Coding*, McGraw-Hill, New York, 1979.

224. S. Voloshynovskiy, F. Deguillaume, S. Pereira, and T. Pun, *Optimal adaptive diversity watermarking with channel state estimation*, in Security and Watermarking of Multimedia Contents III, Proc. SPIE Vol. 4314, P. W. Wong and E. J. Delp, eds., San Jose, CA, USA, January 2001, pp. 673–85.

225. S. Voloshynovskiy, A. Herrigel, N. Baumgaertner, and T. Pun, *A stochastic approach to content adaptive digital image watermarking*, in Proc. 3rd Int. Work. on Information Hiding, IH'99, A. Pfitzmann, ed., vol. 1768 of Lecture Notes in Computer Science, Dresden - Germany, September 1999, Springer Verlag, pp. 211–236.

226. S. Voloshynovskiy, S. Pereira, V.iquise, and T. Pun, *Attack modelling: towards a second generation watermarking benchmark*, Signal Processing, 81 (2001), no. 6, pp. 1177–1214.

227. G. Voyatsis and I. Pitas, *Chaotic watermarks for embedding in the spatial digital image domain*, in Proc. 5th IEEE Int. Conf. on Image Processing, ICIP'98, vol. II, Chicago, IL, USA, October 1998, pp. 432–436.

228. A. B. Watson, *DCT quantization matrices visually optimized for individual images*, in Human Vision, Visual Processing, and Digital Display, Proc. SPIE vol. 1913, J. P. Allebach and B. E. Rogowitz, eds., San Jose, CA, February 1993, pp. 202–216.

229. A. B. Watson, *Perceptual optimization of DCT color quantization*, in Proc. 1st IEEE Int. Conf. on Image Processing, ICIP'94, vol. I, Austin, TX, November 1994, pp. 100–104.

230. E. Weber, *Der tastsinn und das gemeingefuül*, Handwörterbuch der Physiologie, 3 (1846), pp. 481–588.

231. S. Winkler, *A perceptual distortion metric for digital color images,* in Proc. 5th IEEE Int. Conf. on Image Processing, ICIP'98, vol. III, Chicago, IL, USA, October 1998, pp. 399–403.

232. R. B. Wolfgang, C. I. Podilchuk, and E. J. Delp, *The effect of matching watermark and compression transforms in compressed color images,* in Proc. 5th IEEE Int. Conf. on Image Processing, ICIP'98, vol. I, Chicago, IL, October 1998, pp. 440–444.

233. R. B. Wolfgang, C. I. Podilchuk, and E. J. Delp, *Perceptual watermarks for digital images and video,* Proceedings of the IEEE, 87 (1999), no. 7, pp. 1108–1126.

234. M. Wu, S. A. Craver, E. W. Felten, and B. Liu, *Analysis of attacks on SDMI audio watermarks,* in Proc. IEEE Int. Conf. on Acoustic Speech and Signal Processing, ICASSP'01, vol. III, Salt Lake City, UT, USA, May 2001, pp. 1369–1372.

235. G. Wyszecki and W. S. Stiles, *Color Science,* John Wiley & Sons, 2nd ed., 1982.

236. X. Zhang and B. A. Wandell, *A spatial extension of CIELab for digital color image reproduction,* in SID - Information Display Symposium Technical Digest, 1996, pp. 731–734.

237. J. Zhao and E. Koch, *Embedding robust labels into images for copyright protection,* in Proc. of the International Congress on Intellectual Property Rights for Specialized Information, Knowledge and New Technologies, KnowRight '95, Vienna, Austria, August 1995, pp. 242–251.

Index